PRINCIPLES OF ECE THEORY

A NEW PARADIGM OF PHYSICS

Myron W. Evans, Horst Eckardt,
Douglas W. Lindstrom, Stephen J. Crothers

Edited and Collated by Horst Eckardt

September 29, 2016

Published by New Generation Publishing in 2016

Copyright © Myron Evans 2016

First Edition

The author asserts the moral right under the Copyright, Designs and Patents Act 1988 to be identified as the author of this work.

All Rights reserved. No part of this publication may be reproduced, stored in a retrieval system or transmitted, in any form or by any means without the prior consent of the author, nor be otherwise circulated in any form of binding or cover other than that which it is published and without a similar condition being imposed on the subsequent purchaser.

www.newgeneration-publishing.com

 New Generation Publishing

This book is dedicated to
all wholehearted scholars of natural philosophy

Preface

The idea of basing physics or natural philosophy on geometry goes back to ancient times. The Greeks for example regarded geometry as beauty. The amazing intricacy of the Insular style of the Book of Kells was based on the triskelion to a large extent. The idea was used for example by Kepler in the seventeenth century enlightenment which overthrew the Aristotelian, earth centred, philosophy that had held sway since the classical time of Aristotle. Newton's "Principia" is written in terms of geometry. The most famous revival of the idea is Einsteinian general relativity, based on a type of geometry developed by Riemann, Christoffel, Ricci, Levi-Civita, Bianchi and others from the eighteen sixties onwards. Eventually Einstein based his 1915 field equation directly on what was then known as the second Bianchi identity, inferred at the Scuola Normale Superiore in Pisa, around the corner from the University in which Galileo worked.

The ECE theory is named the Einstein Cartan Evans theory to distinguish it from the Einstein Cartan theory, and the first ideas for ECE occurred in early 2003. They emerged from O(3) electrodynamics, whose papers can all be found in the Omnia Opera section of the www.aias.us website. The key idea for O(3) electrodynamics was the B(3) field, which was inferred at Cornell Theory Center in November 1991 after a year in the University of Zuerich and ETH Zuerich in Switzerland. Vigier (who worked with de Broglie) quickly realized that B(3) infers the existence of the Poincaré / de Broglie photon mass, and confirms its existence because B(3) was deduced from experimental data in the inverse Faraday effect. The B(3) field meant that the entire subject of electrodynamics had to be restructured, and this process is recorded in the Omnia Opera of www.aias.us from 1992 to 2003. The restructuring was named "O(3) electrodynamics", a transition theory.

It gradually became apparent that the restructuring meant that a new unified field theory was necessary, one that was based on geometry. This is because previous attempts at a unified field theory, including Einstein's own attempts, were based on a mixture of concepts and many adjustable variables, so many that the old theories became essentially meaningless. The key idea for ECE theory emerged after a reading of a book by Carroll: "Spacetime and Geometry: an Introduction to General Relativity", in particular the end of chapter three, which gives a short synopsis of a geometry due to the mathematician Elie Cartan. This is a more rigorous geometry than the one used by Einstein. The Cartan geometry is based on the definition of torsion and

PREFACE

curvature. The Einstein type of geometry contains only curvature.

In the Spring of 2003 I noticed that the defining equations of torsion and curvature have a similar structure to the defining equations of O(3) electrodynamics, so the first years of ECE theory were dedicated to deducing as much of physics as possible from these geometrical defining equations, with as few hypotheses as possible. It was quickly realized that all the main equations of physics can be derived from Cartan geometry, from its two structure equations, and identities. The ECE theory quickly branched out in many directions and became hugely popular, its readership has always included the best in the world: universities, institutes, government departments, corporations and scholars. The development of ECE theory coincided with the sweeping societal changes brought about by the knowledge revolution. By now ECE is among the most studied theories of physics in history. The readings of items on www.aias.us and www.upitec.org since about 2003 can be estimated in terms of hundreds of millions of printed page equivalents.

Gradually I realized that the Einstein theory of general relativity omits half of geometry: spacetime torsion. Nearly all the textbooks of the Einstein theory assumed zero torsion, most made this assumption axiomatically, some authors were not even aware of torsion. Starting with the classic UFT88, read hundreds of thousands of time, and written in 2007, it has become clear that the neglect of torsion means that curvature also vanishes, and that the Einstein theory collapses completely. It has been replaced in many ways in the 346 UFT items now available on www.aias.us. This is Alwyn van der Merwe's "Post Einsteinian Paradigm Shift" of the avant garde physics of the twenty first century. The tremendous power of website publishing, closely monitored by feedback analysis, and meticulously checked for quality, has meant that new ideas can be brought to any student, however poor, who wishes to study the new ideas. The ideas of ECE and of the obsolete parts of the standard model, now co exist. The main idea of ECE is to improve the old physics, to cut away the deadwood and keep the good parts of the old physics.

ECE and ECE2 have played an important role in applied physics and engineering, notably their ability to explain new and ubiquitous sources of energy. This work culminated in UFT311, which verifies this aspect of ECE and ECE2 theory using a circuit design that is able to trap the unlimited amount of electrical power in spacetime. ECE and ECE2 have also given a plausible explanation of low energy nuclear reactors, now being considered by Congress in Washington D. C. as a source of new energy. The old physics has no plausible explanation for energy from spacetime or LENR.

Acknowledgments to all the AIAS / UPITEC Fellows and those who helped bring about this great paradigm shift of natural philosophy, a new enlightenment. They include all co workers and co authors back to 1971, and those who built the www.aias.us, and www.upitec.org websites, notably Bob Gray, Sean MacLachlan, Gianni Giachetta, Dave Burleigh, Horst Eckardt, Alex Hill (www.et3m.net), Robert Cheshire, Michael Jackson, Simon Clifford and many others. The main co authors of ECE and ECE2 are Horst Eckardt and Douglas Lindstrom, but others such as Stephen Crothers, have also contributed, notably Laurence Felker, who has written a book on ECE read millions of time, literally. Some co authors such as Gareth Evans have worked with me since 1974. Acknowledgments to Kerry Pendergast for writing a biography.

Last but not least, acknowledgments to Queen Elizabeth II, Prime Minister Tony Blair and Parliament for the award of a Civil List Pension in 2005, and to the College of Arms for the award of arms in 2008 in recognition of work on behalf of science and voluntary work for society. There are many others that deserve mention, notably my teachers at Pontardawe Grammar School and University College of Wales Aberystwyth, notably my Ph.D. supervisor Prof. Emeritus Mansel Davies, a humanist and a student of nature.

Craig Cefn Parc, 2016 *Myron W. Evans*

PREFACE

Contents

1. **Basics of Cartan Geometry** — 3
   1.1 Historical Background — 3
   1.2 The Structure Equations of Maurer and Cartan — 8

2. **Electrodynamics and Gravitation** — 19
   2.1 Introduction — 19
   2.2 The Fundamental Hypotheses and Field and Wave Equations — 23
   2.3 The B(3) Field in Cartan Geometry — 27
   2.4 The Field Equations of Electromagnetism — 29
   2.5 The Field Equations of Gravitation — 31

3. **ECE Theory and Beltrami Fields** — 33
   3.1 Introduction — 33
   3.2 Derivation of the Beltrami Equation — 33
   3.3 Elecrostatics, Spin Connection Resonance and Beltrami Structures — 50
   3.4 The Beltrami Equation for Linear Momentum — 59
   3.5 Examples for Beltrami functions — 64
   3.6 Parton Structure of Elementary Particles — 72

4. **Photon Mass and the B(3) Field** — 79
   4.1 Introduction — 79
   4.2 Derivation of the Proca Equations from ECE Theory — 81
   4.3 Link between Photon Mass and B(3) — 84
   4.4 Measurement of Photon Mass by Compton Scattering — 94
   4.5 Photon Mass and Light Deflection due to Gravitation — 100
   4.6 Difficulties with the Einstein Theory of Light Deflection due to Gravitation — 107

5. **The Unification of Quantum Mechanics and General Relativity** — 113
   5.1 Introduction — 113
   5.2 The Fermion Equation — 115
   5.3 Interaction of the ECE Fermion with the Electromagnetic Field — 120
   5.4 New Electron Spin Orbit Effects from the Fermion Equation — 127

    5.5   Refutation of Indeterminacy: Quantum Hamilton and Force Equations 139

6 Antisymmetry   153
    6.1   Introduction 153
    6.2   Application of Antisymmetry to Electrodynamics 157
    6.3   Antisymmetry in ECE Electromagnetism 162
    6.4   Derivation of the Equivalence Principle from Antisymmetry and Other Applications 168

7 Energy from Space Time and Low Energy Nuclear Reactions 173
    7.1   Introduction 173
    7.2   Spin Connection Resonance from the Coulomb Law 175
    7.3   Low Energy Nuclear Reactions (LENR) 180

8 ECE Cosmology   191
    8.1   Introduction 191
    8.2   ECE Theory of Light Deflection due to Gravitation 194
    8.3   The Velocity Curve of a Whirlpool Galaxy 203
    8.4   Description of Orbits with the Minkowski Force Equation . . . 212

9 Relativistic Cosmology and Einstein's "Gravitational Waves" 221
    9.1   Introduction 221
    9.2   Gravitational waves, black holes and big bangs combined . . . 222
    9.3   Gravitational wave propagation speed and the linearisation game 228
    9.4   A black hole is a universe 231
    9.5   Black hole gravity 232
    9.6   The mathematical theory of black holes 234
    9.7   The paradox of black hole mass 261
    9.8   Localisation of gravitational energy and conservation laws . . . 264
    9.9   Numerical relativity and perturbations on black holes 269
    9.10 Big bang cosmology 270
    9.11 Conclusions 275

Chapter 1

Basics of Cartan Geometry

1.1 Historical Background

Geometry was equated with beauty by the ancient Greeks, and was used by them to create art of the highest order. The Parthenon for example was built on principles of geometry, and a deliberate flaw introduced so as not to offend the gods with perfection. A thousand years later the Book of Kells scaled the magnificent peak of insular Celtic art, using the principles of geometry to draw the fine triskeles. Aristotelian thought dominated natural philosophy until Copernicus placed the sun at the centre of the solar system, a challenge to Ecclesia, the dominant European power that had grown out of the beehive cells of remote places such as Skellig Michael. In such places civilization had clung on by its fingernails after the Roman empire was swept away by vigorous peoples of the far north. They had their own type of geometry carved on the prows of their ships, interwoven patterns carved in wood. Copernicus offered a challenge to dogma, always a dangerous thing to do, and human nature never changes. Gradually a new enlightenment began to dawn, with figures such as Galileo and Kepler at its centre. Leonardo da Vinci in the early renaissance had sensed that nature is geometry, and that one cannot do physics without mathematics. Earlier still, the perpendicular and gothic styles of architecture resulted in great European cathedrals built on geometry, for example Cluny, Canterbury and Chartres. Both Leonardo and Descartes thought in terms of swirling whirlpools, reminiscent of van Gogh's starry night. Francis Bacon thought that nature is the measuring stick of all theory, and that dogma is ultimately discarded. This was another challenge to Ecclesia. Galileo boldly asserted that the sun is at the centre of the solar system as we call it today. That offended Ecclesia so he was put under house arrest but survived. It is dangerous to challenge dogma, to challenge the comfortable received wisdom which by passes the need to think. So around 1600, as Bruno was burnt at the stake, Kepler began the laborious task of analyzing the orbit of Mars. Tycho Brahe had finally given him the needed data. This is all described in Koestler's famous book, "The Sleepwalkers". Kepler used the ancient thought in a new way, geometry describes nature, nature is geometry. The orbit of Mars was

1.1. HISTORICAL BACKGROUND

found to be an ellipse, not a circle, with the sun at one of its foci. After an immense amount of work, Kepler discovered three laws of planetary motion. These laws were synthesized by Newton in his theory of universal gravitation, later developed by many mathematicians such as Euler, Bernoulli, Laplace and Hamilton.

Figure 1.1: The Book of Kells: incipit Liber generationis of the Gospel of Matthew, beginning of the Gospel of John.

All of these descriptions of nature rested on three dimensional space and time. The three dimensional space was that of Euclid and time flowed forward on its own. Space and time were different entitles until Michelson and Morley carried out an experiment which overturned this dogma. It seemed that the speed of light c was independent of the direction in which it was measured. It seemed that c was an upper limit, a velocity v could not be added to c. Fitzgerald and Heaviside corresponded about this puzzling result and Heaviside came close to resolving the contradiction. Lorentz swept away the dogma of two thousand years by merging three dimensional space with time to create spacetime in four dimensions, (ct, X, Y, Z). This was the beginning of the theory of special relativity, in transforming quantities from one frame to another, c remained constant but X, Y, and Z varied, so quantities in the new frame are (ct', X', Y', Z'). Lorentz considered the simple case when one frame moves with respect to the other at a constant velocity v but if one frame accelerated with respect to the other the theory became untenable. This is the famous Lorentz transform. The spacetime used by Lorentz is known as flat spacetime, meaning that it is described by a certain limit of a more general geometry. Flat spacetime is described by a simple metric known as $diag(1, -1, -1, -1)$, a four by four matrix with these numbers on its diagonal. Lorentz, Poincaré, Voigt and many others applied the theory of special relativity to electrodynamics and found that the Maxwell Heaviside equations obey the Lorentz transform, and were therefore thought to be equations of special relativity. The Newtonian system of dynamics does not obey the Lorentz transform, there is no limit on the linear velocity in the Newtonian system.

So there developed a schism between dynamics and electrodynamics, they seemed to obey different transformation laws and different geometries. Dynamics had been described for two centuries since Newton by the best minds as existing in Euclidean space and time. Electrodynamics existed in flat spacetime. The underlying geometries of the two subjects seemed to be different. Attempts were made around the turn of the twentieth century to resolve this fundamental challenge to physics. Einstein in 1905 applied the principles of Lorentz to dynamics, using the concepts of four momentum, relativistic momentum and energy. The laws of dynamics were merged with the laws of electrodynamics using c as a universal constant. Einstein also challenged dogma and many scientists of the old school rejected special relativity out of hand. Some dogmatists still reject it. From 1905 onwards physics ceased to be comprehensible without mathematics, which is why so few people understand physics today and are easily deceived by dogmatists. At the end of the nineteenth century several other flaws were found in the older physics, and these were resolved by quantum mechanics, notably by Planck's quantization of energy. Quantum mechanics seemed to give an accurate description of black body radiation, the photoelectric effect and the specific heat of solids, but departed radically from classical physics. Many people today do not understand quantum mechanics or special relativity because they are completely counter intuitive. Planck, Einstein and many others, notably Sommerfeld and his school, developed what is known as the old quantum theory.

Figure 1.2: Gregorio Ricci Cusbastro, Tulio Levi Civita and Elwin Bruno Christoffel.

1.1. HISTORICAL BACKGROUND

The old quantum theory and special relativity had many successes, but existed as separate theories. There was no geometrical framework with which the two types of theory could be unified and special relativity was restricted to one frame moving with respect to another with a constant velocity. The brilliant successes of the classical Newtonian physics were thought of as a limit of special relativity, one in which the velocity v of a particle is much less than c. A new corpuscular theory of light emerged in the old quantum theory, and this corpuscle was named the photon about twenty years later. Initially the photon was thought of as quantized electromagnetic radiation. In about 1905 physics was split three ways, and the work of Rutherford and his school began to show the existence of elementary particles, the electron having been just discovered. Einstein, Langevin and others analyzed the Brownian motion to show the existence of molecules, first inferred by Dalton. The old dogmatists had refused to accept the existence of molecules for over a century. The Rutherford group showed the existence of the alpha particle and inferred the existence of the nucleus and the neutron, later discovered by Chadwick. Rutherford and Soddy demonstrated the existence of isotopes, nuclei with the same number of protons but different number of neutrons. So physics rapidly diverged in all directions, there was no unified theory that could explain all of these tremendous discoveries.

Geometry in the meantime had developed away from Euclidean principles. There were many contributors, the most notable achievement of the mid nineteenth century was that of Riemann, who proposed the concept of metric. Christoffel inferred the geometrical connection shortly thereafter. The metric and the connection describe the difference between Euclidean geometry and a new type of geometry often known as Riemannian geometry. In fact Riemann inferred only the metric. The curvature tensor or Riemann tensor was inferred much later in about 1900 by Ricci and his student Levi-Civita. It took over thirty years to progress from the metric to the curvature tensor. There was no way of knowing the symmetry of the connection. The latter has one upper index and two lower indices, so is a matrix for each upper index. In general a matrix is asymmetric, can have any symmetry, but can always be written as the sum of a symmetric matrix and an antisymmetric matrix. So the connection for each upper index is in general the sum of symmetric and antisymmetric components. Christoffel, Ricci and Levi-Civita assumed without proof that the connection is symmetric in its lower two indices – the symmetric connection. This assumption was used by Bianchi in about 1902 to prove the first Bianchi identity from which the second Bianchi identity follows. Both these identities assume a symmetric connection. The antisymmetric part of the connection was ignored irrationally, or dogmatically. This dogma eventually evolved into general relativity, an incorrect dogma which unfortunately influenced thought in natural philosophy for over a century.

The first physicist to take much notice of these developments in geometry appears to have been Einstein, whose friend Grossmann was a mathematician. Einstein was not fond of the complexity of the Riemannian geometry as it became known, and never developed a mastery of the subject. After several attempts from 1905 to 1915 Einstein used the second Bianchi identity and the covariant Noether Theorem to deduce a field equation of general relativity in late 1915. This field equation was solved by Schwarzschild in December 1915, but Schwarzschild heavily criticised its derivation. It was later criticised by Schröedinger, Bauer, Levi-Civita and others, notably Elie Cartan.

Cartan was among the foremost mathematicians of his era and inferred spinors in 1913. In the early twenties he used the antisymmetric connection to infer the existence of torsion, a quantity that had been thrown away twenty years earlier by Ricci, Levi-Civita and Bianchi, and also by Einstein. The entire theory of general relativity continued to neglect torsion throughout the twentieth century. Cartan and Einstein corresponded but never really understood each other. Cartan realized that there are two fundamental quantities in geometry, torsion and curvature. He expressed this with Maurer in the form of two structure equations and using a differential geometry developed to try to merge the concept of spinors with that of torsion and curvature. The structure equations were still almost unknown to physics before they were implemented in 2003 in the subject of this book, the Einstein- Cartan-Evans unified field theory, known as ECE theory. The ECE theory has swept the world of physics , and has been read an accurately estimated thirty to fifty million times in a decade. This phenomenon is known as the post-Einstein paradigm shift, a phrase coined by Alwyn van der Merwe.

Albert Einstein   **Elie Cartan**   **Myron W. Evans**

Figure 1.3: The eponyms of Einstein-Cartan-Evans theory.

The first and second Maurer-Cartan structure equations can be translated into the Riemannian definitions of respectively torsion and curvature. The concept of commutator of covariant derivatives has been developed to give the torsion and curvature simultaneously with great elegance. The commutator acts on any tensor in any space of any dimension and always isolates the torsion simultaneously with the curvature. The torsion is made up of the difference of two antisymmetric connections, and these connections have the same antisymmetry as the commutator. The connection used in the curvature is also antisymmetric. A symmetric connection means a symmetric commutator. A symmetric commutator always vanishes, and the torsion and curvature vanish if the connection is symmetric. This means that the second Bianchi identity used by Einstein is incorrect and that his field equation is meaningless.

The opening sections of this book develop this basic geometry and use the Cartan identity to produce the geometrically correct field equations of electrodynamics unified with gravitation. The dogmatists have failed to achieve this unification because they used a symmetric connection and because they continued to regard electrodynamics as special relativity.

1.2 The Structure Equations of Maurer and Cartan

These structure equations were developed using the notation of differential geometry and are defined in many papers [1,10] of the UFT series on www.aias.us. The most important discovery made by Elie Cartan in this area of his work was that of spacetime torsion. In order for torsion to exist the geometrical connection must be antisymmetric. In the earlier work of Christoffel, Ricci, Levi-Civita and Bianchi the connection had been assumed to be symmetric. The Einsteinian general relativity continued to repeat this error for over a hundred years, and this incorrect symmetry is the reason why Einstein did not succeed in developing a unified field theory, even though Cartan had informed him of the existence of torsion. The first structure equation defines the torsion in terms of differential geometry. In the simplest or minimalist notation the torsion T is:

$$T = D \wedge q = d \wedge q + \omega \wedge q \qquad (1.1)$$

where $d \wedge$ denotes the wedge derivative of differential geometry, q denotes the Cartan tetrad and ω denotes the spin connection of Cartan. The symbol $D \wedge$ defines the covariant wedge derivative. In this notation the indices of differential geometry are omitted for clarity. The Cartan tetrad was also known initially as the vielbein (many legged) or vierbein (four legged). The wedge derivative is an elegant formulation that can be translated [1, 11] into tensor notation. This is carried out in full detail in the UFT papers, which can be consulted using indices or with google. In this section we concentrate on the essentials without overburdening the text with details. The spin connection is related to the Christoffel connection.

The only textbook to even mention torsion in a clear, understandable way is that of S. M. Caroll [13], accompanied by online notes. The ECE theory uses the same geometry precisely as that described in the first three chapters of Carroll, but ECE has evolved completely away from the interpretation given by. Carroll in his chapter four onwards. Carroll defines torsion but then neglects it without reason, and this is exactly what the twentieth century general relativity proceeded to do. All of Carroll's proofs have been given in all detail in the UFT papers and books [1]- [10] and a considerable amount of new geometry also inferred, notably the Evans identity. In Carroll's notation the first structure equation is:

$$T^a = d \wedge q^a + \omega^a{}_b \wedge q^b \tag{1.2}$$

in which the Latin indices of the tetrad and spin connection have been added. These indices were originally indices of the tangent Minkowski spacetime defined by Cartan at a point P of the general base manifold. The latter is defined with Greek indices. Equation 1.2 when written out more fully becomes:

$$T^a{}_{\mu\nu} = (d \wedge q^a)_{\mu\nu} + \omega^a{}_{\mu b} \wedge q^b{}_\nu \tag{1.3}$$

So the torsion had one upper Latin index and two lower Greek indices. It is a vector-valued two form of differential geometry which is by definition antisymmetric in its Greek indices:

$$T^a{}_{\mu\nu} = -T^a{}_{\nu\mu} \tag{1.4}$$

The torsion is a rank three mixed index tensor.

The tetrad has one upper Latin index a and one lower Greek index μ. It is a vector-valued one form of differential geometry and is a mixed index rank two tensor. The tetrad is defined as a matrix relating a vector V^a and a vector V^μ:

$$V^a = q^a{}_\mu V^\mu \tag{1.5}$$

In his original work Cartan defined V^a as a vector in the tangent spacetime of a base manifold, and defined the vector V^μ in the base manifold. However, during the course of development of ECE theory it was inferred that the tetrad can be used more generally as shown in great detail in the UFT papers to relate a vector V^a defined by a given curvilinear coordinate system to the same vector defined in another curvilinear coordinate system, for example cylindrical polar and Cartesian, or complex circular and Cartesian. The spin connection has one upper and one lower Greek index and one lower Latin index and is related to the Christoffel connection through a fundamental theorem of differential geometry known obscurely as the tetrad postulate. The tetrad postulate is the theorem which states that the complete vector field in any space in any dimension is independent of the way in which that complete vector field is written in terms of components and basis elements. For example in three dimensions the complete vector field is the same in cylindrical polar and Cartesian coordinates or any curvilinear coordinates. The Christoffel connection does not transform as a tensor [1, 10], so the spin connection is not a tensor, but for some purposes may be defined as a one form, with one lower Greek index.

The wedge product of differential geometry is precisely defined in general, and translates Equation 1.3 into tensor notation by acting on the one form $q^a{}_\mu$ and the one form $\omega^a{}_{\mu b}$ to give:

$$T^a{}_{\mu\nu} = \partial_\mu q^a{}_\nu - \partial_\nu q^a{}_\mu + \omega^a{}_{\mu b} q^b{}_\nu - \omega^a{}_{\nu b} q^b{}_\mu \tag{1.6}$$

1.2. THE STRUCTURE EQUATIONS OF MAURER AND CARTAN

which is a tensor equation. It is seen that the entire equation is antisymmetric in the Greek indices μ and ν which means that:

$$T^a{}_{\nu\mu} = \partial_\nu q^a{}_\mu - \partial_\mu q^a{}_\nu + \omega^a{}_{\nu b} q^b{}_\mu - \omega^a{}_{\mu b} q^b{}_\nu. \tag{1.7}$$

This result is important for the ECE antisymmetry laws developed later in this book. In this tensor equation there is summation over repeated indices, so:

$$\omega^a{}_{\mu b} q^b{}_\nu = \omega^a{}_{\mu 1} q^1{}_\nu + \cdots + \omega^a{}_{\mu n} q^n{}_\nu \tag{1.8}$$

in general. It is seen that the torsion has some resemblance to the way in which an electromagnetic field was defined by Lorentz, Poincaré and others in terms of the four potential, a development of the work of Heaviside. This led to the inference of ECE theory in 2003 through a simple postulate described in the next chapter. The difference is that the torsion contains an upper index a and contains an antisymmetric term in the spin connection.

All the equations of Cartan geometry are generally covariant, which means that they transform under the general coordinate transformation, and are equations of general relativity. Therefore the torsion is generally covariant as required by general relativity. The tetrad postulate results in the following relation between the spin connection and the gamma connection:

$$\partial_\mu q^a{}_\nu + \omega^a{}_{\mu b} q^b{}_\nu = \Gamma^\lambda{}_{\mu\nu} q^a{}_\lambda \tag{1.9}$$

and using this equation in Equation 1.6 gives the Riemannian torsion:

$$T^\lambda{}_{\mu\nu} = \Gamma^\lambda{}_{\mu\nu} - \Gamma^\lambda{}_{\nu\mu}. \tag{1.10}$$

In deriving the Riemannian torsion the following equation of Cartan geometry has been used:

$$T^a{}_{\mu\nu} = q^a{}_\lambda T^\lambda{}_{\mu\nu} \tag{1.11}$$

which means that the tetrad plays the role of switching the a index to a λ index. Similarly the equation for torsion can be simplified using:

$$\omega^a{}_{\mu b} q^b{}_\nu = \omega^a{}_{\mu\nu}; \quad \omega^a{}_{\nu b} q^b{}_\mu = \omega^a{}_{\nu\mu} \tag{1.12}$$

to give a simpler expression:

$$T^a{}_{\mu\nu} = \partial_\mu q^a{}_\nu - \partial_\nu q^a{}_\mu + \omega^a{}_{\mu\nu} - \omega^a{}_{\nu\mu}. \tag{1.13}$$

It can be seen that the Riemannian torsion is antisymmetric in μ and ν so $T^\lambda{}_{\mu\nu}$ vanishes if the connection is symmetric, that is if the following were true:

$$\Gamma^\lambda{}_{\mu\nu} =?\Gamma^\lambda{}_{\nu\mu}. \tag{1.14}$$

The Einsteinian general relativity always assumed Equation 1.14 without proof. In fact the commutator method to be described below proves that the connection is antisymmetric. We arrive at the conclusion that Einsteinian general relativity is refuted entirely by its neglect of torsion, and part of the purpose of this book is to forge a new cosmology based on torsion. In order to make the theory of torsion of use to engineers and chemists the tensor notation needs to be translated to vector notation. The precise details of how this is done are given again in the UFT papers and other material on www.aias.us.

In vector notation the torsion splits into orbital torsion and spin torsion. In order to define these precisely the tetrad four vector is defined as the four vector:

$$q^a{}_\mu = (q^a{}_0, -\boldsymbol{q}^a), \tag{1.15}$$
$$q^{a\mu} = (q^{a0}, \boldsymbol{q}^a), \tag{1.16}$$

with a timelike component $q^a{}_0$ and a spacelike component $\boldsymbol{q}^a$. Similarly the spin connection is defined as the four vector:

$$\omega^a{}_{\mu b} = (\omega^a{}_{0b}, -\boldsymbol{\omega}^a{}_b). \tag{1.17}$$

In this notation the orbital torsion is:

$$\boldsymbol{T}^a_{\text{orb}} = -\boldsymbol{\nabla} q^a{}_0 - \frac{1}{c}\frac{\partial \boldsymbol{q}^a}{\partial t} - \omega^a{}_{0b}\boldsymbol{q}^b + \boldsymbol{\omega}^a{}_b q^b{}_0 \tag{1.18}$$

and the spin torsion is:

$$\boldsymbol{T}^a_{\text{spin}} = \boldsymbol{\nabla} \times \boldsymbol{q}^a - \boldsymbol{\omega}^a{}_b \times \boldsymbol{q}^b. \tag{1.19}$$

In ECE electrodynamics the orbital torsion gives the electric field strength and the spin torsion gives the magnetic flux density. In ECE gravitation part of the orbital torsion gives the acceleration due to gravity, and the spin torsion gives the magnetogravitational field. The physical quantities of electrodynamics and gravitation are obtained directly from the torsion and directly from Cartan geometry. For example the fundamental $B(3)$ field of electrodynamics [1, 11] is obtained from the spin torsion of the first structure equation.

In minimal notation the second Cartan Maurer structure equation defines the Cartan curvature:

$$R = D \wedge \omega = d \wedge \omega + \omega \wedge \omega \tag{1.20}$$

so the torsion is the covariant wedge derivative of the tetrad and the curvature is the covariant wedge derivative of the spin connection. Fundamentally therefore these are simple definitions, and that is the elegance of Cartan's geometry. When expanded out into tensor and vector notation they look much more complicated but convey the same information. In the standard notation of differential geometry Equation 1.20 becomes:

$$R^a{}_b = d \wedge \omega^a{}_b + \omega^a{}_c \wedge \omega^c{}_b \tag{1.21}$$

where there is summation over repeated indices. When written out in full Equation 1.21 becomes:

$$R^a{}_{b\mu\nu} = (d \wedge \omega^a{}_b)_{\mu\nu} + \omega^a{}_{\mu c} \wedge \omega^c{}_{\nu b} \tag{1.22}$$

1.2. THE STRUCTURE EQUATIONS OF MAURER AND CARTAN

where the indices of the base manifold have been reinstated. In tensor notation Equation 1.22 becomes:

$$R^a{}_{b\mu\nu} = \partial_\mu \omega^a{}_{\nu b} - \partial_\nu \omega^a{}_{\mu b} + \omega^a{}_{\mu c} \omega^c{}_{\nu b} - \omega^a{}_{\nu c} \omega^c{}_{\mu b} \tag{1.23}$$

which defines the Cartan curvature as a tensor valued two form. It is tensor valued because it has indices a and b, and is a differential two form [1, 11] antisymmetric μ and ν. Using the tetrad postulate Equation 1.9 it can be shown that Equation 1.23 is equivalent to the Riemann curvature tensor:

$$R^\lambda{}_{\rho\mu\nu} = \partial_\mu \Gamma^\lambda{}_{\nu\rho} - \partial_\nu \Gamma^\lambda{}_{\mu\rho} + \Gamma^\lambda{}_{\mu\sigma} \Gamma^\sigma{}_{\nu\rho} - \Gamma^\lambda{}_{\nu\sigma} \Gamma^\sigma{}_{\mu\rho} \tag{1.24}$$

first inferred by Ricci and Levi Civita in about 1900. The proof of this is complicated but is given in full in the UFT papers.

The geometrical connection was inferred by Christoffel in the eighteen sixties in order to define a generally covariant derivative. In four dimensions for example the ordinary derivative ∂_μ does not transform covariantly [1, 11] but by definition the covariant derivative of any tensor has this property. The Christoffel connection is defined by:

$$D_\mu V^\rho = \partial_\mu V^\rho + \Gamma^\rho{}_{\mu\lambda} V^\lambda \tag{1.25}$$

and the spin connection is defined by:

$$D_\mu V^a = \partial_\mu V^a + \omega^a{}_{\mu b} V^b. \tag{1.26}$$

Without additional information there is no way in which to determine the symmetry of the Christoffel and spin connection, and both are asymmetric in general in their lower two indices. The covariant derivative can act on any tensor of any rank in a well defined manner explained in full detail in the UFT papers on www.aias.us. When it acts on the tetrad, a rank two mixed index tensor, it produces the result [1, 11]:

$$D_\mu q^a{}_\nu = \partial_\mu q^a{}_\nu + \omega^a{}_{\mu b} q^b{}_\nu - \Gamma^\lambda{}_{\mu\nu} q^a{}_\lambda. \tag{1.27}$$

The tetrad postulate means that:

$$D_\mu q^a{}_\nu = 0 \tag{1.28}$$

and so the covariant derivative of the tetrad vanishes in order to maintain the invariance of the complete vector field. This has been a fundamental theorem of Cartan geometry for almost a hundred years. The tetrad postulate is the theorem by which Cartan geometry is translated into Riemann geometry. The Riemann torsion and Riemann curvature are defined elegantly by the commutator of covariant derivatives. This is an operator that acts on any tensor in any space of any dimension. When it acts on a vector it is defined for example by:

$$[D_\mu, D_\nu] V^\rho = D_\mu(D_\nu V^\rho) - D_\nu(D_\mu V^\rho). \tag{1.29}$$

CHAPTER 1. BASICS OF CARTAN GEOMETRY

As shown in all detail in UFT 99 Equation 1.29 results in:

$$[D_\mu, D_\nu] V^\rho = R^\rho{}_{\mu\nu\sigma} V^\sigma - T^\lambda{}_{\mu\nu} D_\lambda V^\rho. \tag{1.30}$$

The Riemann curvature and Riemann torsion are always produced simultaneously by the commutator, which therefore produces the first and second Cartan Maurer structure equations when the tetrad postulate is used to translate the Riemann torsion and Riemann curvature to the Cartan torsion and Cartan curvature. The commutator also defines the antisymmetry of the connection and this is of key importance. By definition the commutator is antisymmetric in the indices μ and ν:

$$[D_\mu, D_\nu] V^\rho = -[D_\nu, D_\mu] V^\rho \tag{1.31}$$

and vanishes if these indices are the same, i.e. if the connection is symmetric. From inspection of the equation:

$$[D_\mu, D_\nu] V^\rho = -(\Gamma^\lambda{}_{\mu\nu} - \Gamma^\lambda{}_{\nu\mu}) D_\lambda V^\rho + R^\rho{}_{\mu\nu\sigma} V^\sigma \tag{1.32}$$

the connection has the same symmetry as the commutator, so the connection is anti symmetric:

$$\Gamma^\lambda{}_{\mu\nu} = -\Gamma^\lambda{}_{\nu\mu}, \tag{1.33}$$

a result of key importance. A symmetric connection means a null commutator and this means that the Riemann torsion and Riemann curvature both vanish if the connection is symmetric.

The Einsteinian general relativity used a symmetric connection incorrectly, so the entire twentieth century era is refuted. This is the essence of the post Einsteinian paradigm shift. The correct general relativity is based on field equations obtained from Cartan geometry. These field equations are obtained from identities of Cartan geometry. The first such identity in minimal notation is:

$$D \wedge T = d \wedge T + \omega \wedge T := R \wedge q = q \wedge R \tag{1.34}$$

and this is referred to in this book as the Cartan identity. The covariant derivative of the torsion is the wedge product of the tetrad and curvature. The wedge products in Equation 1.34 are those of a one form and a two form. In the UFT papers it is shown that this produces the following result in tensor notation:

$$\partial_\mu T^a{}_{\nu\rho} + \partial_\rho T^a{}_{\mu\nu} + \partial_\nu T^a{}_{\rho\mu} + \omega^a{}_{\mu b} T^b{}_{\nu\rho} + \omega^a{}_{\rho b} T^b{}_{\mu\nu} + \omega^a{}_{\nu b} T^b{}_{\rho\mu}$$
$$:= R^a{}_{\mu\nu\rho} + R^a{}_{\rho\mu\nu} + R^a{}_{\nu\rho\mu}, \tag{1.35}$$

a sum of three terms. In papers such as UFT 137 this identity is proven in complete detail using the tetrad postulate. The proof is complicated but again shows the great elegance of the Cartan geometry. Using the concept of the Hodge dual [1, 11] the result in Equation 1.35 can be expressed as:

$$\partial_\mu \widetilde{T}^{a\mu\nu} + \omega^a{}_{\mu b} \widetilde{T}^{b\mu\nu} := \widetilde{R}^a{}_\mu{}^{\mu\nu} \tag{1.36}$$

1.2. THE STRUCTURE EQUATIONS OF MAURER AND CARTAN

where the tilde's denote the tensor that is Hodge dual to $T^a{}_{\mu\nu}$. In four dimensions the Hodge dual of an antisymmetric tensor, or two form, is another antisymmetric tensor. From Equation 1.36 the Cartan identity can be expressed as:

$$\partial_\mu \widetilde{T}^{a\mu\nu} = j^{a\nu} = \widetilde{R}^a{}_\mu{}^{\mu\nu} - \omega^a{}_{\mu b}\widetilde{T}^{b\mu\nu}. \tag{1.37}$$

Defining:

$$j^{a\nu} = (j^{a0}, \boldsymbol{j}^a) \tag{1.38}$$

the Cartan identity splits into two vector equations:

$$\boldsymbol{\nabla} \cdot \boldsymbol{T}^a_{spin} = j^{a0} \tag{1.39}$$

and

$$\frac{1}{c}\frac{\partial \boldsymbol{T}^a_{spin}}{\partial t} + \boldsymbol{\nabla} \times \boldsymbol{T}^a_{orb} = \boldsymbol{j}^a. \tag{1.40}$$

These become the basis for the homogeneous equations of electrodynamics in ECE theory, and define the magnetic charge current density in terms of geometry. These equations are given in the Engineering Model of ECE theory on www.aias.us. They also define the homogeneous field equations of gravitation.

The Evans identity of differential geometry was inferred during the course of the development of ECE theory and in minimal notation it is:

$$D \wedge \widetilde{T} = d \wedge \widetilde{T} + \omega \wedge \widetilde{T} := \widetilde{R} \wedge q = q \wedge \widetilde{R}. \tag{1.41}$$

It is valid in four dimensions, because the Hodge dual of a two form in four dimensions is another two form: So the Hodge duals of the torsion and curvature obey the Cartan identity. This result is the Evans identity Equation 1.41. In tensor notation it is:

$$\partial_\mu \widetilde{T}^a{}_{\nu\rho} + \partial_\rho \widetilde{T}^a{}_{\mu\nu} + \partial_\nu \widetilde{T}^a{}_{\rho\mu} + \omega^a{}_{\mu b}\widetilde{T}^b{}_{\nu\rho} + \omega^a{}_{\rho b}\widetilde{T}^b{}_{\mu\nu} + \omega^a{}_{\nu b}\widetilde{T}^b{}_{\rho\mu}$$
$$:= \widetilde{R}^a{}_{\mu\nu\rho} + \widetilde{R}^a{}_{\rho\mu\nu} + \widetilde{R}^a{}_{\nu\rho\mu} \tag{1.42}$$

an equation which is equivalent to:

$$\partial_\mu T^{a\mu\nu} + \omega^a{}_{\mu b}T^{b\mu\nu} := R^a{}_\mu{}^{\mu\nu} \tag{1.43}$$

as shown in full detail in the UFT papers. The tensor equation Equation 1.43 splits into two vector equations:

$$\boldsymbol{\nabla} \cdot \boldsymbol{T}^a_{orb} = J^{a0} = R^a{}_\mu{}^{\mu 0} - \omega^a{}_{\mu b}T^{b\mu 0} \tag{1.44}$$

and

$$\boldsymbol{\nabla} \times \boldsymbol{T}^a_{spin} - \frac{1}{c}\frac{\partial \boldsymbol{T}^a_{orb}}{\partial t} = \boldsymbol{J}^a. \tag{1.45}$$

When translated into electrodynamics these become the inhomogeneous field equations, which define the electric charge density and the electric current density in terms of geometry.

If torsion is neglected or incorrectly assumed to be zero, the Cartan identity reduces to:

$$R \wedge q = 0, \tag{1.46}$$

which is the elegant Cartan notation for the first Bianchi identity:

$$R^\lambda{}_{\mu\nu\rho} + R^\lambda{}_{\rho\mu\nu} + R^\lambda{}_{\nu\rho\mu} = 0. \tag{1.47}$$

The second Bianchi identity can be derived from the first Bianchi identity and is:

$$D_\mu R^\kappa{}_{\lambda\nu\rho} + D_\rho R^\kappa{}_{\lambda\mu\nu} + D_\nu R^\kappa{}_{\lambda\rho\mu} = 0. \tag{1.48}$$

Clearly the two Bianchi identities are true if and only if the torsion is zero. In other words the two identities are true if and only if the Christoffel connection is symmetric. The commutator method shows that the Christoffel connection is antisymmetric so the two Bianchi identities are incorrect. The first Bianchi identity must be replaced by the Cartan identity Equation 1.35 and the second Bianchi identity was replaced in UFT 255 by:

$$D_\mu D_\lambda T^\kappa{}_{\nu\rho} + D_\rho D_\lambda T^\kappa{}_{\mu\nu} + D_\nu D_\lambda T^\kappa{}_{\rho\mu}$$
$$:= D_\mu R^\kappa{}_{\lambda\nu\rho} + D_\rho R^\kappa{}_{\lambda\mu\nu} + D_\nu R^\kappa{}_{\lambda\rho\mu}. \tag{1.49}$$

Therefore Einstein used entirely the wrong identity (Equation 1.48) in his field equation. No experiment can prove incorrect geometry, and indeed the claims of experimentalists to have tested the Einstein field equation with precision have been extensively criticised for many years. The contemporary experimental data themselves may or not be precise, but they do not prove incorrect geometry. Einstein effectively threw away the first Cartan Maurer structure equation, so his geometry contained and still contains only half of the geometrical truth, and geometry is the most self contained of all subjects. The velocity curve of the whirlpool galaxy, discovered in the late fifties, entirely and completely refutes both Einstein and Newton. In several of the UFT papers on www.aias.us, the velocity curve is explained straightforwardly by ECE theory using again the minimum of postulates, for example UFT 238. The dogmatists used and still use ad hoc ideas such as dark matter to cover up the catastrophic failure of the Einstein and Newton theories in whirlpool galaxies. They became idols of the cave, and dreamt up dark matter in it darkest corners. Their claim that the universe is made up mostly of dark matter is an admission of abject failure. To compound this failure they still claim that the Einstein theory is very precise in places such as the solar system. This dogma has reduced natural philosophy to utter nonsense. Either a theory works or it does not work. It cannot be brilliantly successful and fail completely at the same time. ECE and the post Einsteinian paradigm shift uses no dark matter

1.2. THE STRUCTURE EQUATIONS OF MAURER AND CARTAN

and no ideas deliberately cobbled up so they cannot be tested experimentally: "not even wrong" as Pauli wrote.

In some recent work in UFT 254 onwards the Cartan identity has been reduced to a simple and clear vectorial format:

$$\boldsymbol{\nabla} \cdot \boldsymbol{\omega}^a{}_b \times \boldsymbol{T}^b_{\text{spin}} := \boldsymbol{\omega}^a{}_b \cdot \boldsymbol{\nabla} \times \boldsymbol{T}^b_{\text{spin}} - \boldsymbol{T}^b_{\text{spin}} \cdot \boldsymbol{\nabla} \times \boldsymbol{\omega}^a{}_b. \tag{1.50}$$

As always in ECE theory this vector identity is generally covariant. It is very useful when used with the geometrical equations for magnetic and electric charge current densities also developed in UFT 254 onwards. In the following chapter it is shown that combinations of ECE equations such as these produce many new insights.

This introductory survey of Cartan geometry has shown that the ECE theory is based entirely on four equations: the first and second Cartan Maurer structure equations, the Cartan identity, and the tetrad postulate. These equations have been known and taught for almost a century. Using these equations the subject of natural philosophy has been unified on a well known geometrical basis. Electromagnetism has been unified with gravitation and new methods developed to describe the structure of elementary particles. General relativity has been unified with quantum mechanics by developing the tetrad postulate into a generally covariant wave equation:

$$(\Box + \kappa^2) q^a{}_\mu = 0 \tag{1.51}$$

where

$$\kappa^2 = q^\nu{}_a \partial^\mu (\omega^a{}_{\mu\nu} - \Gamma^a{}_{\mu\nu}). \tag{1.52}$$

The wave equation (Equation 1.51) has been reduced to all the main relativistic wave equations such as the Klein Gordon, Proca and Dirac wave equations, and in so doing these wave equations have been derived as equations of general relativity. They are all based on the most fundamental theorem of Cartan geometry, the tetrad postulate. The Dirac equation has been developed into the fermion equation by factorizing the ECE wave equation that reduces in special relativity to the Dirac wave equation. The fermion equation needs only two by two matrices, and does not suffer from negative energy while at the same time producing the positron and other anti particles. So the discoveries of the Rutherford group have also been explained geometrically.

The Heisenberg Uncertainty Principle was replaced and developed in UFT 13, and easily shown to be incorrect in UFT 175. The uncertainty principle should be described more accurately as the indeterminacy principle, which is an admission of failure from the outset. It was rejected by Einstein, de Broglie, Schröedinger and others at the famous 1927 Solvay Conference and split natural philosophy permanently into scientists and dogmatists. The indeterminacy principle has been experimentally proven to be wildly wrong by the Croca group { 12 } using advanced microscopy and other experimental methods. The dogmatists ignore this experimental refutation. The scientists take note of it and adapt their theories accordingly as advocated by Bacon, essentially the founder of the scientific method. Indeterminacy means that quantities

CHAPTER 1. BASICS OF CARTAN GEOMETRY

are absolutely unknowable, and according to the dogmatists of Copenhagen, geometry is unknowable because general relativity is based on geometry. So they never succeeded in unifying general relativity and quantum mechanics. In ECE theory this unification is straightforward as just described, it is based on the tetrad postulate re-expressed as a wave equation. Anything that is claimed dogmatically to emanate from the fervent occult practices of indeterminacy can be obtained rationally and coolly from UFT13 without any fire or brimstone.

So indeterminacy was the first major casualty of ECE theory, other idols began to fall over, and the dogmatists with them. Everything has been thrown out of the window: U(1) gauge invariance, transverse vacuum radiation, the massless photon, the E(2) little group, the Einsteinian general relativity, the U(1) gauge invariance, the GWS electroweak theory, refuted completely in UFT 225, the SU(3) theory of quarks and gluons, quantum electrodynamics with its adjustable parameters such as virtual particles, the hocus pocus of renormalization and regularization, quantum chromodynamics, asymptotic freedom, quark confinement, approximate symmetry, string theory, superstring theory, multiple dimensions, nineteen adjustables, even more adjustables, yet more adjustables, dark matter, dark flow, big bang, black holes, interacting black holes, hundred billion dollar supercolliders, the whole lot, strange dreams leading to the Higgs boson, the murkiest idol of all.

Everything is cool and in the light of reason, everything is geometry.

1.2. THE STRUCTURE EQUATIONS OF MAURER AND CARTAN

Chapter 2

Electrodynamics and Gravitation

Electromagnetic Units in S. I.

Electric field strength $\mathbf{E}$	=	$\mathrm{V\,m^{-1}}$	= $\mathrm{J\,C^{-1}\,m^{-1}}$
Vector Potential $\mathbf{A}$	=	$\mathrm{J\,s\,C^{-1}\,m^{-1}}$	
Scalar Potential ϕ	=	$\mathrm{J\,C^{-1}}$	
Vacuum permittivity ϵ_0	=	$\mathrm{J^{-1}\,C^{-2}\,m^{-1}}$	
Magnetic Flux Density $\mathbf{B}$	=	$\mathrm{J\,s\,C^{-1}\,m^{-2}}$	
Electric charge density ρ	=	$\mathrm{C\,m^{-3}}$	
Electric current density $\mathbf{J}$	=	$\mathrm{C\,m^{-2}\,s^{-1}}$	
Spacetime Torsion T	=	$\mathrm{m^{-1}}$	
Spin connection ω	=	$\mathrm{m^{-1}}$	
Spacetime Curvature R	=	$\mathrm{m^{-2}}$	

2.1 Introduction

The old physics, prior to the post Einsteinian paradigm shift, completely failed to provide a unified logic for electrodynamics and gravitation because the former was developed in flat, or Minkowski, spacetime and the latter in a spacetime which was thought quite wrongly to be described only by curvature. The ECE theory develops both electrodynamics and gravitation directly from Cartan geometry. As shown in the ECE Engineering Model, the field equations of electrodynamics and gravitation in ECE theory have the same format, based directly and with simplicity on the underlying geometry. Therefore the Cartan geometry of Chap. 1 is translated directly into electromagnetism and gravitation using the same type of simple, fundamental hypothesis in each case: the tetrad becomes the 4-potential energy and the torsion becomes the field of force.

In retrospect the method used by Einstein to translate from geometry to gravitation was cumbersome as well as being incorrect. The second Bianchi

2.1. INTRODUCTION

identity was reformulated by Einstein using the Ricci tensor and Ricci scalar into a format where it could be made directly proportional to the covariant Noether Theorem through the Einstein constant k. Both sides of this equation used a covariant derivative, but it was assumed by Einstein without proof that the integration constants were the same on both sides, giving the Einstein field equation:

$$G_{\mu\nu} = kT_{\mu\nu} \tag{2.1}$$

where $G_{\mu\nu}$ is the Einstein tensor, $T_{\mu\nu}$ is the canonical energy momentum density, and k is the Einstein constant. This equation is completely incorrect because it uses a symmetric connection and throws away torsion. If attempts are made to correct this equation for torsion, as in UFT 88 and UFT 255, summarized in Chap. 1, it becomes hopelessly cumbersome; it could still only be used for gravitation and not for a unified field theory of gravitation and electromagnetism. Einstein himself thought that his field equation of 1915 could never be solved, which shows that he was bogged down in complexity. Schwarzschild provided a solution in December 1915 but in his letter declared "friendly war" on Einstein. The meaning of this is not entirely clear but obviously Schwarzschild was not satisfied with the equation. His solution did not contain singularities, and this original solution is on the net, together with a translation by Vankov of the letter to Einstein. This solution of an incorrect field equation is obviously meaningless. The errors were compounded by asserting (after Schwarzschild died in 1916) that the solution contains singularities, so the contemporary Schwarzschild metric is a misattribution and distortion, as well as being completely meaningless. It has been used endlessly by dogmatists to assert the existence of incorrect results such as big bang and black holes. So gravitational science was stagnant from 1915 to 2003. During the course of development of ECE theory it gradually became clear in papers such as UFT 150 that there were many other errors and obscurities in the Einstein theory, notably in the theory of light bending by gravitation, and in the theory of perihelion precession. One of the obvious contradictions in the theory of light deflection by gravitation is that it uses a massless photon that is nevertheless attracted to the sun. The resulting null geodesic method is full of obscurities as shown in UFT 150 (www.aias.us). The Einsteinian general relativity has been comprehensively refuted in reference [2] of Chap. 1. It was completely refuted experimentally in the late fifties by the discovery of the velocity curve of the whirlpool galaxy. At that point it should have been discarded; its apparent successes in the solar system are illusions. Instead, natural philosophy itself was abandoned and dark matter introduced. The Einsteinian theory is still unable to explain the velocity curve of the whirlpool galaxy, it still fails completely, and dark matter does not change this fact. So the Einstein theory cannot be meaningful in the solar system as the result of these experimental observations. ECE theory has revealed the reason why the Einstein theory fails so badly – the neglect of torsion.

Electromagnetism also stagnated throughout the twentieth century and remained the Maxwell Heaviside theory of the nineteenth century. This theory was incorporated unchanged into the attempts of the old physics at unification using U(1) gauge invariance and the massless photon. The idea of the massless photon leads to multiple, well known problems and absurdities, notably the planar E(2) little group of the Poincaré group. Effectively this result means that the free electromagnetic field can have only two states of polarization. The two transverse states labelled (1) and (2). The time like state (0) and the longitudinal state (3) are eliminated in order to save the hypothesis of a massless photon. These problems and obscurities are explained in detail by a standard model textbook such as that of Ryder [24]. The unphysical Gupta Bleuler condition must be used to "eliminate" the (0) and (3) states, leading to multiple unsolved problems in canonical quantization. The use of the Beltrami theory as in UFT 257 onwards produces richly structured longitudinal components of the free electromagnetic field, refuting the U(1) dogma immediately and indicating the existence of photon mass. Beltrami was a contemporary of Heaviside, so the present standard model was effectively refuted as long ago as the late nineteenth century. As soon as the photon becomes identically non zero, however tiny in magnitude, the U(1) theory becomes untenable, because it is no longer gauge invariant [1]- [10], and the Proca equation replaces the d'Alembert equation. The ECE theory leads to the Proca equation and finite photon mass, from the tetrad postulate, using the same basic hypothesis as that which translates geometry into electromagnetism.

Although brilliantly successful in its time, there are many limitations of the Maxwell Heaviside (MH) theory of electromagnetism. In the field of non-linear optics for example its limitations are revealed by the inverse Faraday Effect [1]- [10] (IFE). This phenomenon is the magnetization of material matter by circularly polarized electromagnetic radiation. It was inferred theoretically [7] by Piekara and Kielich, and later by Pershan, and was first observed experimentally in the mid-sixties by van der Ziel at al. in the Bloembergen group at Harvard. It occurs for example in one electron as in UFT 80 to 84 on www.aias.us. The old U(1) gauge invariant theory of electromagnetism becomes untenable immediately when dealing with the inverse Faraday Effect because the latter is caused by the conjugate product of circularly polarized radiation, the cross product of the vector potential with its complex conjugate:

$$\mathbf{A} \times \mathbf{A}^* = \mathbf{A}^{(1)} \times \mathbf{A}^{(2)}. \tag{2.2}$$

Indices (1) and (2) are used to define the complex circular basis [1]- [10], whose unit vectors are:

$$\mathbf{e}^{(1)} = \frac{1}{\sqrt{2}} (\mathbf{i} - i\mathbf{j}) \tag{2.3}$$

$$\mathbf{e}^{(2)} = \frac{1}{\sqrt{2}} (\mathbf{i} + i\mathbf{j}) \tag{2.4}$$

$$\mathbf{e}^{(3)} = \mathbf{k} \tag{2.5}$$

obeying the cyclical, O(3) symmetry, relation:

$$\mathbf{e}^{(1)} \times \mathbf{e}^{(2)} = i\mathbf{e}^{(3)*} \tag{2.6}$$

2.1. INTRODUCTION

$$\mathbf{e}^{(3)} \times \mathbf{e}^{(1)} = i\mathbf{e}^{(2)*} \tag{2.7}$$

$$\mathbf{e}^{(2)} \times \mathbf{e}^{(3)} = i\mathbf{e}^{(1)*} \tag{2.8}$$

in three dimensional space. The unit vectors $\mathbf{e}^{(1)}$ and $\mathbf{e}^{(2)}$ are complex conjugates. The gauge principle of the MH theory can be expressed as follows:

$$\mathbf{A} \rightarrow \mathbf{A} + \nabla \chi \tag{2.9}$$

so the conjugate product becomes:

$$\mathbf{A} \times \mathbf{A}^* = (\mathbf{A} + \nabla \chi) \times (\mathbf{A} + \nabla \chi)^* \tag{2.10}$$

and is not U(1) gauge invariant, so the resulting longitudinal magnetization of the inverse Faraday effect is not gauge invariant, Q.E.D. Many other phenomena in non-linear optics [7] are not U(1) gauge invariant and they all refute the standard model and such artifacts as the "Higgs boson". The absurdity of the old physics becomes glaringly evident in that it asserts that the conjugate product exists in isolation of the longitudinal and time like components of spacetime, (0) and (3). So in the old physics the cross product (2.2) cannot produce a longitudinal component. This is absurd because space has three components (1), (2) and (3). The resolution of this fundamental paradox was discovered in Nov. 1991 with the inference of the B(3) field, the appellation given to the longitudinal magnetic component of the free electromagnetic field, defined by CH01:BIB01- [10]:

$$\mathbf{B}^{(3)*} = -ig\mathbf{A}^{(1)} \times \mathbf{A}^{(2)} \tag{2.11}$$

where g is a parameter.

The B(3) field is the key to the geometrical unification of gravitation and electromagnetism and also infers the existence of photon mass, experimentally, because it is longitudinal and observable experimentally in the inverse Faraday effect. The zero photon mass theory is absurd because it asserts that B(3) cannot exist, that the third component of space itself cannot exist, and that the inverse Faraday effect does not exist. The equation that defines the B(3) field is not U(1) gauge invariant because the B(3) field is changed by the gauge transform (2.10). The equation is not therefore one of U(1) electrodynamics, and was used in the nineties to develop a higher topology electrodynamics known as O(3) electrodynamics CH01:BIB01- [10]. These papers are recorded in the Omnia Opera section of www.aias.us. Almost simultaneously, several other theories of higher topology electrodynamics were developed [25], notably theories by Horwitz et al., Lehnert and Roy, Barrett, and Harmuth et al., and by Evans and Crowell [8] These are described in several volumes of the "Contemporary Chemical Physics" series edited by M. W. Evans [25]. These higher topology electrodynamical theories also occur in Beltrami theories as reviewed for example by Reed [7], [27]. In 2003 these higher topology theories evolved into ECE theory.

2.2 The Fundamental Hypotheses and Field and Wave Equations

The first hypothesis of Einstein Cartan Evans (ECE) unified field theory is that the electromagnetic potential ($A^a{}_\mu$) is the Cartan tetrad within a scaling factor. Therefore the electromagnetic potential is defined by:

$$A^a{}_\mu = A^{(0)} q^a{}_\mu \tag{2.12}$$

and has one upper index a, indicating the state of polarization, and one lower index to indicate that it is a vector valued differential 1-form of Cartan's geometry. The gravitational potential is defined by:

$$\Phi^a{}_\mu = \Phi^{(0)} q^a{}_\mu \tag{2.13}$$

where $\Phi^{(0)}$ is a scaling factor. Therefore the first ECE hypothesis means that electromagnetism is Cartan's geometry within a scalar, $A^{(0)}$. Physics is geometry. Ubi materia, ibi geometria (Johannes Kepler). This is a much simpler hypothesis than that of Einstein, and much more powerful. It is a hypothesis that extends general relativity to electromagnetism. The mathematical correctness of the theory is guaranteed by the mathematical correctness and economy of thought of Cartan's geometry as described in chapter one.

The second ECE hypothesis is that the electromagnetic field ($F^a{}_{\mu\nu}$) is the Cartan torsion within the same scaling factor as the potential. The second hypothesis follows from the first hypothesis by the first Cartan Maurer structure equation. Therefore in minimal notation:

$$F = D \wedge A = d \wedge A + \omega \wedge A \tag{2.14}$$

which is an elegant relation between field and potential, the simplest possible relation in a geometry with both torsion and curvature. The field is the covariant wedge derivative of the potential, both for electromagnetism and gravitation. It follows that the entire geometrical development of chapter one can be applied directly to electromagnetism and gravitation. In the standard notation of differential geometry used by S. M. Carroll [13] the electromagnetic field is defined in ECE theory by:

$$F^a = d \wedge A^a + \omega^a{}_b \wedge A^b. \tag{2.15}$$

In MH theory the same relation is [24]:

$$F = d \wedge A. \tag{2.16}$$

The MH theory does not have a spin connection and does not have polarization indices. The ECE theory is general relativity based directly on Cartan geometry, the MH theory is special relativity and is not based on geometry. The presence of the spin connection in Eq. (2.15) means that the field is the frame of reference itself, a dynamic frame that translates and rotates. In MH theory the field is an entity different in concept from the frame of reference, the Minkowski frame of flat spacetime.

2.2. THE FUNDAMENTAL HYPOTHESES AND FIELD AND WAVE...

In tensor notation the electromagnetic field is:

$$F^a{}_{\mu\nu} = \partial_\mu A^a{}_\nu - \partial_\nu A^a{}_\mu + \omega^a{}_{\mu b} A^b{}_\nu - \omega^a{}_{\nu b} A^b{}_\mu \tag{2.17}$$

and can be expressed more simply as:

$$F^a{}_{\mu\nu} = \partial_\mu A^a{}_\nu - \partial_\nu A^a{}_\mu + A^{(0)} \left(\omega^a{}_{\mu\nu} - \omega^a{}_{\nu\mu} \right). \tag{2.18}$$

In the MH theory the electromagnetic field is

$$F_{\mu\nu} = \partial_\mu A_\nu - \partial_\nu A_\mu \tag{2.19}$$

and has no polarization index or spin connection. The electromagnetic potential is the 4-vector:

$$A^a{}_\mu = (A^a{}_0, -\mathbf{A}^a) = \left(\frac{\phi^a}{c}, -\mathbf{A}^a \right) \tag{2.20}$$

in covariant definition, or:

$$A^{a\mu} = \left(A^{a0}, \mathbf{A}^a \right) = \left(\frac{\phi^a}{c}, \mathbf{A}^a \right) \tag{2.21}$$

in contravariant definition. The upper index a denotes the state of polarization. For example in the complex circular basis it has four indices:

$$a = (0), (1), (2), (3) \tag{2.22}$$

one timelike (0) and three spacelike (1), (2), (3). The (1) and (2) indices are transverse and the (3) index is longitudinal. The spacelike part of the potential 4-vector is the vector $\mathbf{A}^a$, and so this can only have space indices, (1), (2) and (3). It cannot have a timelike index (0) by definition. The 4-potential can be written for each of the four indices (0), (1), (2) and (3) as:

$$A_\mu = (A_0, -\mathbf{A}). \tag{2.23}$$

When the a index is (0) the 4-potential reduces to the scalar potential:

$$A^{(0)}{}_\mu = \left(A^{(0)}{}_0, -\mathbf{0} \right). \tag{2.24}$$

When the a index is (1), (2) or (3) the 4-potential is interpreted as:

$$A^{(i)}{}_\mu = \left(A^{(i)}{}_0, -\mathbf{A}^i \right), \quad i = 1, 2, 3 \tag{2.25}$$

so $A^{(i)}{}_0$ for example is the scalar part of the 4-potential $A^a{}_\mu$, associated with index (1). As described by S. M. Carroll, the tetrad is a 1-form for each index a. This means that the 4-potential $A^a{}_\mu$ is a 4-potential for each index a:

$$A^{(0)}{}_\mu = (A_\mu)^{(0)} \tag{2.26}$$

$$A^{(i)}{}_\mu = (A_\mu)^{(i)}, \quad i = 1, 2, 3 \tag{2.27}$$

CHAPTER 2. ELECTRODYNAMICS AND GRAVITATION

and this is a basic property of Cartan geometry.

In order to translate the tensor notation of Eq. (2.17) to vector notation, it is necessary to define the torsion as a 4 x 4 antisymmetric matrix. The choice of matrix is guided by experiment, so that the ECE theory reduces to laws that are able to describe electromagnetic phenomena by direct use of Cartan geometry. As described in Chap. 1 there exist orbital and spin torsion defined by equations which are similar in structure to electromagnetic laws which have been tested with great precision, notably the Gauss law of magnetism, the Faraday law of induction, the Coulomb law and the Ampère Maxwell law. These laws must be recovered in a well-defined limit of ECE theory. Newtonian gravitation must be recovered in another limit of ECE theory.

The torsion matrix for each a is chosen by hypothesis to be:

$$T_{\rho\sigma} = \begin{bmatrix} 0 & T_1(\text{orb}) & T_2(\text{orb}) & T_3(\text{orb}) \\ -T_1(\text{orb}) & 0 & -T_3(\text{spin}) & T_2(\text{spin}) \\ -T_2(\text{orb}) & T_3(\text{spin}) & 0 & -T_1(\text{spin}) \\ -T_3(\text{orb}) & -T_2(\text{spin}) & T_1(\text{spin}) & 0 \end{bmatrix} \quad (2.28)$$

This equation may be looked upon as the third ECE hypothesis. The Hodge dual [1]- [11], [24] of this matrix is:

$$\widetilde{T}^{\mu\nu} = \begin{bmatrix} 0 & -T^1(\text{spin}) & -T^2(\text{spin}) & -T^3(\text{spin}) \\ T^1(\text{spin}) & 0 & T^3(\text{orb}) & -T^2(\text{orb}) \\ T^2(\text{spin}) & -T^3(\text{orb}) & 0 & T^1(\text{orb}) \\ T^3(\text{spin}) & T^2(\text{orb}) & -T^1(\text{orb}) & 0 \end{bmatrix} \quad (2.29)$$

Indices are raised and lowered by the metric tensor in any space [13]:

$$\widetilde{T}^{\mu\nu} = g^{\mu\alpha} g^{\nu\beta} \widetilde{T}_{\alpha\beta}. \quad (2.30)$$

Alternatively the antisymmetric torsion matrix may be defined as:

$$T^{\mu\nu} = \begin{bmatrix} 0 & -T^1(\text{orb}) & -T^2(\text{orb}) & -T^3(\text{orb}) \\ T^1(\text{orb}) & 0 & -T^3(\text{spin}) & T^2(\text{spin}) \\ T^2(\text{orb}) & T^3(\text{spin}) & 0 & -T^1(\text{spin}) \\ T^3(\text{orb}) & -T^2(\text{spin}) & T^1(\text{spin}) & 0 \end{bmatrix} \quad (2.31)$$

with raised indices. From this definition the spin torsion vector in 3D is:

$$\mathbf{T}(\text{spin}) = T_X(\text{spin})\mathbf{i} + T_Y(\text{spin})\mathbf{j} + T_Z(\text{spin})\mathbf{k} \quad (2.32)$$

in which:

$$T_X(\text{spin}) = T^1(\text{spin}) = \widetilde{T}^{10} = -\widetilde{T}^{01} \quad (2.33)$$

$$T_Y(\text{spin}) = T^2(\text{spin}) = \widetilde{T}^{20} = -\widetilde{T}^{02} \quad (2.34)$$

$$T_Z(\text{spin}) = T^3(\text{spin}) = \widetilde{T}^{30} = -\widetilde{T}^{03} \quad (2.35)$$

Similarly the orbital torsion vector in 3D is defined by:

$$\mathbf{T}(\text{orb}) = T_X(\text{orb})\mathbf{i} + T_Y(\text{orb})\mathbf{j} + T_Z(\text{orb})\mathbf{k} \quad (2.36)$$

2.2. THE FUNDAMENTAL HYPOTHESES AND FIELD AND WAVE...

where the vector components are related to the matrix components as follows:

$$T_X(\text{orb}) = T^1(\text{orb}) = T_{10} = -T_{01} \tag{2.37}$$

$$T_Y(\text{orb}) = T^2(\text{orb}) = T_{20} = -T_{02} \tag{2.38}$$

$$T_Z(\text{orb}) = T^3(\text{orb}) = T_{30} = -T_{03} \tag{2.39}$$

With these definitions the electric field strength $\mathbf{E}^a$ and the magnetic flux density $\mathbf{B}^a$ are defined by:

$$\mathbf{E}^a = cA^{(0)}\mathbf{T}^a(\text{orb}) \tag{2.40}$$

and

$$\mathbf{B}^a = A^{(0)}\mathbf{T}^a(\text{spin}). \tag{2.41}$$

For each index a the field tensor with raised μ and ν is defined to be:

$$F^{\mu\nu} = \begin{bmatrix} 0 & -E_X & -E_Y & -E_Z \\ E_X & 0 & -cB_Z & cB_Y \\ E_Y & cB_Z & 0 & -cB_X \\ E_Z & -cB_Y & cB_X & 0 \end{bmatrix}. \tag{2.42}$$

With these fundamental definitions the tensor notation (2.17) can be translated to vector notation. The latter is used by engineers and is more transparent than tensor notation. The 4-derivative appearing in the tensor equation (2.17) is defined to be:

$$\partial_\mu = \left(\frac{1}{c}\frac{\partial}{\partial t}, \boldsymbol{\nabla}\right). \tag{2.43}$$

Consider the indices of the orbital torsion:

$$T^a{}_{0i} = \partial_0 q^a{}_i - \partial_i q^a{}_0 + \omega^a{}_{0b} q^b{}_i - \omega^a{}_{ib} q^b{}_0$$
$$i = 1, 2, 3. \tag{2.44}$$

These translate into the indices of the field tensor as follows:

$$F^a{}_{0i} = \partial_0 A^a{}_i - \partial_i A^a{}_0 + \omega^a{}_{0b} A^b{}_i - \omega^a{}_{ib} A^b{}_0$$
$$i = 1, 2, 3 \tag{2.45}$$

from which it follows that the electric field strength is:

$$\mathbf{E}^a = -c\boldsymbol{\nabla} A^a{}_0 - \frac{\partial \mathbf{A}^a}{\partial t} - c\omega^a{}_{0b}\mathbf{A}^b + cA^b{}_0\boldsymbol{\omega}^a{}_b \tag{2.46}$$

where the spin connection 4-vector is expressed as:

$$\omega^a{}_{\mu b} = (\omega^a{}_{0b}, -\boldsymbol{\omega}^a{}_b) \tag{2.47}$$

using the above definitions.

CHAPTER 2. ELECTRODYNAMICS AND GRAVITATION

The indices of the spin torsion tensor are:

$$T^a{}_{12} = \partial_1 q^a{}_2 - \partial_2 q^a{}_1 + \omega^a{}_{1b} q^b{}_2 - \omega^a{}_{2b} q^b{}_1$$
$$T^a{}_{13} = \partial_1 q^a{}_3 - \partial_3 q^a{}_1 + \omega^a{}_{1b} q^b{}_3 - \omega^a{}_{3b} q^b{}_1$$
$$T^a{}_{23} = \partial_2 q^a{}_3 - \partial_3 q^a{}_2 + \omega^a{}_{2b} q^b{}_3 - \omega^a{}_{3b} q^b{}_2 \qquad (2.48)$$

and they translate into the spin components of the field tensor:

$$F^a{}_{12} = \partial_1 A^a{}_2 - \partial_2 A^a{}_1 + \omega^a{}_{1b} A^b{}_2 - \omega^a{}_{2b} A^b{}_1$$
$$F^a{}_{13} = \partial_1 A^a{}_3 - \partial_3 A^a{}_1 + \omega^a{}_{1b} A^b{}_3 - \omega^a{}_{3b} A^b{}_1$$
$$F^a{}_{23} = \partial_2 A^a{}_3 - \partial_3 A^a{}_2 + \omega^a{}_{2b} A^b{}_3 - \omega^a{}_{2b} A^b{}_2. \qquad (2.49)$$

With the above definitions these equations can be expressed as the magnetic flux density:

$$\mathbf{B}^a = \boldsymbol{\nabla} \times \mathbf{A}^a - \boldsymbol{\omega}^a{}_b \times \mathbf{A}^b. \qquad (2.50)$$

In the MH theory the corresponding equations are:

$$\mathbf{E} = -\boldsymbol{\nabla}\phi - \frac{\partial \mathbf{A}}{\partial t} \qquad (2.51)$$

and

$$\mathbf{B} = \boldsymbol{\nabla} \times \mathbf{A} \qquad (2.52)$$

without polarization indices and without spin connection.

2.3 The B(3) Field in Cartan Geometry

The B(3) field is a consequence of the general expression for magnetic flux density in ECE theory:

$$\mathbf{B}^a = \boldsymbol{\nabla} \times \mathbf{A}^a - \boldsymbol{\omega}^a{}_b \times \mathbf{A}^b. \qquad (2.53)$$

In general, summation over repeated indices means that:

$$\mathbf{B}^a = \boldsymbol{\nabla} \times \mathbf{A}^a - \boldsymbol{\omega}^a{}_{(1)} \times \mathbf{A}^{(1)} - \boldsymbol{\omega}^a{}_{(2)} \times \mathbf{A}^{(2)} - \boldsymbol{\omega}^a{}_{(3)} \times \mathbf{A}^{(3)} \qquad (2.54)$$

but this general expression can be simplified as discussed later in this book using the assumption:

$$\omega^a{}_b = \epsilon^a{}_{bc} \omega^c \qquad (2.55)$$

which is the expression for the duality of a tensor and vector. It can be shown using the vector form of the Cartan identity that the B(3) field is given by:

$$\mathbf{B}^{(3)} = \boldsymbol{\nabla} \times \mathbf{A}^{(3)} - i\frac{\kappa}{A^{(0)}} \mathbf{A}^{(1)} \times \mathbf{A}^{(2)} \qquad (2.56)$$

2.3. THE B(3) FIELD IN CARTAN GEOMETRY

where the potentials are related by the cyclic theorem:

$$\mathbf{A}^{(1)} \times \mathbf{A}^{(2)} = iA^0 \mathbf{A}^{(3)*} \text{ et cyclicum.} \tag{2.57}$$

For plane wave the potentials are as follows:

$$\mathbf{A}^{(1)} \times \mathbf{A}^{(2)*} = \frac{A^0}{\sqrt{2}} (\mathbf{i} - i\mathbf{j}) e^{i(\omega t - \kappa Z)}, \tag{2.58}$$

$$\mathbf{A}^{(3)} = A^0 \mathbf{k}, \tag{2.59}$$

so the B(3) field is defined by:

$$\mathbf{B}^{(3)} = -i\frac{\kappa}{A^{(0)}} \mathbf{A}^{(1)} \times \mathbf{A}^{(2)}. \tag{2.60}$$

Therefore B(3) is the result of general relativity, and does not exist in the Maxwell Heaviside field theory because the MH theory is a theory of special relativity without a geometrical connection. The B(3) field is a radiated longitudinal field that propagates in the (3) or Z axis. When it was inferred in Nov. 1991 it was a completely new concept, and it was gradually realized that it led to a higher topology electrodynamics which was identified with Cartan geometry in 2003. "Higher topology" in this sense means that a different differential geometry is needed to define electrodynamics. This can be seen through the fact that the field in U(1) gauge invariant electrodynamics is:

$$F = d \wedge A \tag{2.61}$$

but in ECE theory it is:

$$F^a = d \wedge A^a + \omega^a{}_b \wedge A^b \tag{2.62}$$

with the presence of indices and spin connection. A choice of internal indices leads to O(3) electrodynamics as outlined above and explained in more detail later.

It was gradually realized that O(3) electrodynamics and ECE electrodynamics accurately reduce to the MH theory in certain limits, but also give much more information, an example being the inverse Faraday effect. The B(3) field led for the first time to an electrodynamics that is based on general covariance, and not Lorentz covariance, so it became easily possible to unify electromagnetism with gravitation.

It is important to realize that B(3) is not a static magnetic field, it interacts with material matter through the conjugate product $\mathbf{A}^{(1)} \times \mathbf{A}^{(2)}$ by which it is defined. So B(3) is intrinsically nonlinear in nature while a static magnetic field is not related to the conjugate product of nonlinear optics. The B(3) field needs for its definition a geometrical connection, and a different set of field equations from those that govern a static magnetic field. The latter is governed by the Gauss law of magnetism and the Ampère law. The static magnetic field does not propagate at c in the vacuum, but B(3) propagates in the vacuum along with A(1) and A(2) and when B(3) interacts with matter it produces a magnetization through a well-defined hyperpolarizability in the inverse Faraday effect. The field equations needed to define B(3) must be obtained from Cartan geometry, and are not equations of Minkowski spacetime.

2.4 The Field Equations of Electromagnetism

These are based directly on the Cartan and Evans identities using the hypotheses (2.40) and (2.41) and give a richly structured theory summarized in the ECE Engineering Model on www.aias.us. Before proceeding to a description of the field equations a summary is given of the Cartan identity in vector notation. In a similar manner to the torsion, the second Cartan Maurer structure equation gives an orbital curvature and a spin curvature:

$$\mathbf{R}^a{}_b(\text{spin}) = \boldsymbol{\nabla} \times \boldsymbol{\omega}^a{}_b - \boldsymbol{\omega}^a{}_c \times \boldsymbol{\omega}^c{}_b. \tag{2.63}$$

As in UFT 254 consider now the Cartan identity:

$$d \wedge T^a + \omega^a{}_b \wedge T^b := R^a{}_b \wedge q^b. \tag{2.64}$$

The space part of this identity can be written as:

$$\boldsymbol{\nabla} \cdot \mathbf{T}^a + \boldsymbol{\omega}^a{}_b \cdot \mathbf{T}^b = \mathbf{q}^b \cdot (\boldsymbol{\nabla} \times \boldsymbol{\omega}^a{}_b - \boldsymbol{\omega}^a{}_c \times \boldsymbol{\omega}^c{}_b). \tag{2.65}$$

Rearranging and using:

$$\mathbf{q}^b \cdot \boldsymbol{\omega}^a{}_c \times \boldsymbol{\omega}^c{}_b = \boldsymbol{\omega}^a{}_b \cdot \boldsymbol{\omega}^a{}_c \times \mathbf{q}^c \tag{2.66}$$

and

$$\boldsymbol{\nabla} \cdot \boldsymbol{\nabla} \times \mathbf{q}^a = 0 \tag{2.67}$$

gives:

$$\boldsymbol{\nabla} \cdot \boldsymbol{\omega}^a{}_b \times \mathbf{q}^b = \boldsymbol{\omega}^a{}_b \cdot \boldsymbol{\nabla} \times \mathbf{q}^b - \mathbf{q}^b \cdot \boldsymbol{\nabla} \times \boldsymbol{\omega}^a{}_b \tag{2.68}$$

i.e. gives the Cartan identity in vector notation, a very useful result that will be used later in this chapter and book. The self-consistency and correctness of the result (2.68) is shown by the fact that it is an example of the well-known vector identity:

$$\boldsymbol{\nabla} \cdot \mathbf{F} \times \mathbf{G} = \mathbf{G} \cdot \boldsymbol{\nabla} \times \mathbf{F} - \mathbf{F} \cdot \boldsymbol{\nabla} \times \mathbf{G}. \tag{2.69}$$

So it can be seen clearly that Cartan geometry generalizes well known geometry and vector identities.

For ECE electrodynamics Eq. (2.68) becomes:

$$\boldsymbol{\nabla} \cdot \boldsymbol{\omega}^a{}_b \times \mathbf{A}^b = \boldsymbol{\omega}^a{}_b \cdot \boldsymbol{\nabla} \times \mathbf{A}^b - \mathbf{A}^b \cdot \boldsymbol{\nabla} \times \boldsymbol{\omega}^a{}_b. \tag{2.70}$$

The magnetic flux density is defined in ECE theory as:

$$\mathbf{B}^a = \boldsymbol{\nabla} \times \mathbf{A}^a - \boldsymbol{\omega}^a{}_b \times \mathbf{A}^b \tag{2.71}$$

so:

$$\boldsymbol{\nabla} \cdot \mathbf{B}^a = -\boldsymbol{\nabla} \cdot \boldsymbol{\omega}^a{}_b \times \mathbf{A}^b \tag{2.72}$$

2.4. THE FIELD EQUATIONS OF ELECTROMAGNETISM

giving the Gauss law of magnetism in general relativity and ECE unified field theory.

As in UFT 256 the Cartan identity and the fundamental ECE hypotheses give the homogeneous field equations of electromagnetism in ECE theory:

$$\nabla \cdot \mathbf{B}^a = \frac{\rho^m}{\epsilon_0 c} = \boldsymbol{\omega}^a{}_b \cdot \mathbf{B}^b - \mathbf{A}^b \cdot \mathbf{R}^a{}_b(\text{spin}) \tag{2.73}$$

and

$$\frac{\partial \mathbf{B}^a}{\partial t} + \nabla \times \mathbf{E}^a = \mathbf{J}^m/\epsilon_0$$
$$= \boldsymbol{\omega}^a{}_b \times \mathbf{E}^b - c\omega_0 \mathbf{B}^a - c\left(\mathbf{A}^b \times \mathbf{R}^a{}_b(\text{orb}) - \mathbf{A}^b{}_0 \mathbf{R}^a{}_b(\text{spin})\right) \tag{2.74}$$

in which the spin curvature is defined by Eq. (2.63) and the orbital curvature by:

$$\mathbf{R}^a{}_b(\text{orb}) = -\nabla \omega^a{}_{0b} - \frac{1}{c}\frac{\partial \boldsymbol{\omega}^a{}_b}{\partial t} - \omega^a{}_{0c} \boldsymbol{\omega}^c{}_b + \omega^c{}_{0b} \boldsymbol{\omega}^a{}_c. \tag{2.75}$$

The right hand sides of these equations give respectively the magnetic charge density and the magnetic current density. The controversy over the existence of the magnetic charge current density has been going on for over a century, and the consensus seems to be that they do not exist. (If they are proven to be reproducible and repeatable the ECE theory can account for them as in the above equations.) If the magnetic charge current density vanishes then:

$$\boldsymbol{\omega}^a{}_b \cdot \mathbf{B}^b = \mathbf{A}^b \cdot \mathbf{R}^a{}_b(\text{spin}) \tag{2.76}$$

and

$$\boldsymbol{\omega}^a{}_b \times \mathbf{E}^b - c\omega_0 \mathbf{B}^a = c\left(\mathbf{A}^b \times \mathbf{R}^a{}_b(\text{orb}) - \mathbf{A}^b{}_0 \mathbf{R}^a{}_b(\text{spin})\right) \tag{2.77}$$

and they imply the Gauss law of magnetism in ECE theory:

$$\nabla \cdot \mathbf{B}^a = 0 \tag{2.78}$$

and the Faraday law of induction:

$$\frac{\partial \mathbf{B}^a}{\partial t} + \nabla \times \mathbf{E}^a = \mathbf{0}. \tag{2.79}$$

The Evans identity gives

$$\nabla \cdot \mathbf{E}^a = \frac{\rho^a}{\epsilon_0} = \boldsymbol{\omega}^a{}_b \cdot \mathbf{E}^b - c\mathbf{A}^b \cdot \mathbf{R}^a{}_b(\text{orb}) \tag{2.80}$$

and:

$$\nabla \times \mathbf{B}^a - \frac{1}{c^2}\frac{\partial \mathbf{E}^a}{\partial t} = \mu_0 \mathbf{J}^a$$
$$= \boldsymbol{\omega}^a{}_b \times \mathbf{B}^b + \frac{\omega_0}{c}\mathbf{E}^b - A^b{}_0 \mathbf{R}^a{}_b(\text{orb}) - \mathbf{A}^b \times \mathbf{R}^a{}_b(\text{spin}). \tag{2.81}$$

CHAPTER 2. ELECTRODYNAMICS AND GRAVITATION

Eq. (2.80) defines the electric charge density:

$$\rho^a = \epsilon_0 \left(\boldsymbol{\omega}^a{}_b \cdot \mathbf{E}^b - c\mathbf{A}^b \cdot \mathbf{R}^a{}_b(\text{orb}) \right) \tag{2.82}$$

and Eq. (2.81) defines the electric current density:

$$\mathbf{J}^a = \frac{1}{\mu_0} \left(\boldsymbol{\omega}^a{}_b \times \mathbf{B}^b + \frac{\omega_0}{c}\mathbf{E}^b - \left(\mathbf{A}^b \times \mathbf{R}^a{}_b(\text{spin}) + A^b{}_0 \mathbf{R}^a{}_b(\text{orb}) \right) \right). \tag{2.83}$$

With these definitions the inhomogeneous field equations become the Coulomb law:

$$\nabla \cdot \mathbf{E}^a = \rho^a / \epsilon_0 \tag{2.84}$$

and the Ampère Maxwell law:

$$\nabla \times \mathbf{B}^a - \frac{1}{c^2} \frac{\partial \mathbf{E}^a}{\partial t} = \mu_0 \mathbf{J}^a. \tag{2.85}$$

2.5 The Field Equations of Gravitation

As shown in the Engineering Model the field equations of gravitation are the two homogeneous field equations:

$$\nabla \cdot \mathbf{h} = 4\pi G \rho_{gm} \tag{2.86}$$

and

$$\nabla \times \mathbf{g} + \frac{1}{c}\frac{\partial \mathbf{h}}{\partial t} = \frac{4\pi G}{c} \mathbf{j}_{gm} \tag{2.87}$$

and the two inhomogeneous equations:

$$\nabla \cdot \mathbf{g} = 4\pi G \rho_m \tag{2.88}$$

and

$$\nabla \times \mathbf{h} - \frac{1}{c}\frac{\partial \mathbf{g}}{\partial t} = \frac{4\pi G}{c} \mathbf{J}_m. \tag{2.89}$$

Here $\mathbf{g}$ is the acceleration due to gravity, and $\mathbf{h}$ is the gravitomagnetic field, defined by the Cartan Maurer structure equations as:

$$\mathbf{g} = -\frac{\partial \mathbf{Q}}{\partial t} - \boldsymbol{\nabla}\Phi - \omega_0 \mathbf{Q} + \Phi\boldsymbol{\omega} \tag{2.90}$$

and

$$\boldsymbol{\Omega} = \boldsymbol{\nabla} \times \mathbf{Q} - \boldsymbol{\omega} \times \mathbf{Q}. \tag{2.91}$$

In the Newtonian physics only Eq. (2.88) exists, where G is the Newton constant and where ρ_m is the mass density. ρ_{gm} and $\mathbf{J}_{gm}$ are the (hypothetical) gravitomagnetic mass density and current. In the ECE equations there is a

2.5. THE FIELD EQUATIONS OF GRAVITATION

gravitomagnetic field **h** (developed in UFT 117 and UFT 118) and a Faraday law of gravitational induction, Eq. (2.87), developed in UFT 75. The latter paper describes the experimental evidence for the gravitational law of induction and UFT 117 and UFT 118 use the gravitomagnetic field to explain precession not explicable in the Newtonian theory.

It is likely that all the fields predicted by the ECE theory of gravitation will eventually be discovered because they are based on geometry as advocated by Kepler. During the course of the development of ECE there have been many advances in electromagnetism and gravitation. There has been space here for a short overview summary.

Chapter 3

ECE Theory and Beltrami Fields

3.1 Introduction

Towards the end of the nineteenth century the Italian mathematician Eugenio Beltrami developed a system of equations for the description of hydrodynamic flow in which the curl of a vector is proportional to the vector itself [26]. An example is the use of the velocity vector. For a long time this solution was not used outside the field of hydrodynamics, but in the fifties it started to be used by workers such as Alfven and Chandrasekhar in the area of cosmology, notably whirlpool galaxies. The Beltrami field as it came to be known has been observed in plasma vortices and as argued by Reed [27] is indicative of a type of electrodynamics such as ECE. Therefore this chapter is concerned with the ways in which ECE electrodynamics reduce to Beltrami electrodynamics, and with other applications of the Beltrami electrodynamics such as a new theory of the parton structure of elementary particles. The ECE theory is based on geometry and is ubiquitous throughout nature on all scales, and so is the Beltrami theory, which can be looked upon as a sub theory of ECE theory.

3.2 Derivation of the Beltrami Equation

Consider the Cartan identity in vector notation, derived in Chapter 2:

$$\boldsymbol{\nabla} \cdot \boldsymbol{\omega}^a{}_b \times \mathbf{q}^b = \mathbf{q}^b \cdot \boldsymbol{\nabla} \times \boldsymbol{\omega}^a{}_c - \boldsymbol{\omega}^a{}_b \cdot \boldsymbol{\nabla} \times \mathbf{q}^b. \tag{3.1}$$

In the absence of a magnetic monopole:

$$\boldsymbol{\nabla} \cdot \boldsymbol{\omega}^a{}_b \times \mathbf{q}^b = 0 \tag{3.2}$$

so:

$$\mathbf{q}^b \cdot \boldsymbol{\nabla} \times \boldsymbol{\omega}^a{}_b = \boldsymbol{\omega}^a{}_b \cdot \boldsymbol{\nabla} \times \mathbf{q}^b. \tag{3.3}$$

3.2. DERIVATION OF THE BELTRAMI EQUATION

Assume that the spin connection is an axial vector dual in its index space to an antisymmetric tensor:

$$\omega^a{}_b = \epsilon^a{}_{bc}\omega^c \tag{3.4}$$

where $\epsilon^a{}_{bc}$ is the totally antisymmetric unit tensor in three dimensions. Then Eq. (3.3) reduces to:

$$\mathbf{q}^b \cdot \boldsymbol{\nabla} \times \boldsymbol{\omega}^c = \boldsymbol{\omega}^c \cdot \boldsymbol{\nabla} \times \mathbf{q}^b. \tag{3.5}$$

An example of this in electromagnetism is:

$$\mathbf{A}^{(2)} \cdot \boldsymbol{\nabla} \times \boldsymbol{\omega}^{(1)} = \boldsymbol{\omega}^{(1)} \cdot \boldsymbol{\nabla} \times \mathbf{A}^{(2)} \tag{3.6}$$

in the complex circular basis ((1), (2), (3)). The vector potential is defined by the ECE hypothesis:

$$\mathbf{A}^a = A^{(0)}\mathbf{q}^a. \tag{3.7}$$

From Chap. 2, Eq, (2.76) the geometrical condition for the absence of a magnetic monopole is:

$$\boldsymbol{\omega}^a{}_b \cdot \mathbf{B}^b = \mathbf{A}^b \cdot \mathbf{R}^a{}_b(\text{spin}) \tag{3.8}$$

where the spin curvature Eq. (2.63) is defined by:

$$\mathbf{R}^a{}_b(\text{spin}) = \boldsymbol{\nabla} \times \boldsymbol{\omega}^a{}_b - \boldsymbol{\omega}^a{}_c \times \boldsymbol{\omega}^c{}_b \tag{3.9}$$

and where $\mathbf{B}^a$ is the magnetic flux density vector. Using Eq. (3.4):

$$\mathbf{R}^c(\text{spin}) = \boldsymbol{\nabla} \times \boldsymbol{\omega}^c - \boldsymbol{\omega}^b \times \boldsymbol{\omega}^a. \tag{3.10}$$

In the complex circular basis defined by Eq. (3.6) the spin curvatures are:

$$\begin{aligned}\mathbf{R}^{(1)}(\text{spin}) &= \boldsymbol{\nabla} \times \boldsymbol{\omega}^{(1)} + i\,\boldsymbol{\omega}^{(3)} \times \boldsymbol{\omega}^{(1)} \\ \mathbf{R}^{(2)}(\text{spin}) &= \boldsymbol{\nabla} \times \boldsymbol{\omega}^{(2)} + i\,\boldsymbol{\omega}^{(2)} \times \boldsymbol{\omega}^{(3)} \\ \mathbf{R}^{(3)}(\text{spin}) &= \boldsymbol{\nabla} \times \boldsymbol{\omega}^{(3)} + i\,\boldsymbol{\omega}^{(1)} \times \boldsymbol{\omega}^{(2)}\end{aligned} \tag{3.11}$$

and the magnetic flux density vectors are:

$$\begin{aligned}\mathbf{B}^{(1)} &= \boldsymbol{\nabla} \times \mathbf{A}^{(1)} + i\,\boldsymbol{\omega}^{(3)} \times \mathbf{A}^{(1)} \\ \mathbf{B}^{(2)} &= \boldsymbol{\nabla} \times \mathbf{A}^{(2)} + i\,\boldsymbol{\omega}^{(2)} \times \mathbf{A}^{(2)} \\ \mathbf{B}^{(3)} &= \boldsymbol{\nabla} \times \mathbf{A}^{(3)} + i\,\boldsymbol{\omega}^{(1)} \times \mathbf{A}^{(3)}.\end{aligned} \tag{3.12}$$

Eq. (8) may be exemplified by:

$$\boldsymbol{\omega}^{(1)} \cdot \mathbf{B}^{(2)} = \mathbf{A}^{(1)} \cdot \mathbf{R}^{(2)}(\text{spin}) \tag{3.13}$$

which may be developed as:

$$\begin{aligned}&\boldsymbol{\omega}^{(1)} \cdot \left(\boldsymbol{\nabla} \times \mathbf{A}^{(2)} + i\,\boldsymbol{\omega}^{(2)} \times \mathbf{A}^{(3)}\right) \\ &= \mathbf{A}^{(1)} \cdot \left(\boldsymbol{\nabla} \times \boldsymbol{\omega}^{(2)} + i\,\boldsymbol{\omega}^{(2)} \times \boldsymbol{\omega}^{(3)}\right).\end{aligned} \tag{3.14}$$

CHAPTER 3. ECE THEORY AND BELTRAMI FIELDS

Possible solutions are

$$\boldsymbol{\omega}^{(i)} = \pm \frac{\kappa}{A^{(0)}} \mathbf{A}^{(i)}, \; i = 1, 2, 3 \tag{3.15}$$

and in order to be consistent with the original [1-10] solution of B(3) the negative sign is developed:

$$\mathbf{B}^{(3)} = \nabla \times \mathbf{A}^{(3)} - i \frac{\kappa}{A^{(0)}} \mathbf{A}^{(1)} \times \mathbf{A}^{(2)} \text{ et cyclicum.} \tag{3.16}$$

From Eq. (3.2):

$$\nabla \cdot \boldsymbol{\omega}^{(3)} \times \mathbf{A}^{(1)} = 0 \tag{3.17}$$

and the following is an identity of vector analysis:

$$\nabla \cdot \nabla \times \mathbf{A}^{(1)} = 0. \tag{3.18}$$

A possible solution of Eq. (3.17) is:

$$\nabla \times \mathbf{A}^{(1)} = i \, \boldsymbol{\omega}^{(3)} \times \mathbf{A}^{(1)} = -i \frac{\kappa}{A^{(0)}} \mathbf{A}^{(3)} \times \mathbf{A}^{(1)}. \tag{3.19}$$

Similarly:

$$\nabla \times \mathbf{A}^{(2)} = i \, \boldsymbol{\omega}^{(2)} \times \mathbf{A}^{(3)} = -i \frac{\kappa}{A^{(0)}} \mathbf{A}^{(2)} \times \mathbf{A}^{(3)}. \tag{3.20}$$

Now multiply both sides of the basis equations (3.6) to (3.8) of Chap. 2 by

$$A^{(0)2} e^{i\phi} e^{-i\phi} \tag{3.21}$$

where the electromagnetic phase is:

$$\phi = \omega t - \kappa Z \tag{3.22}$$

to find the cyclic equation:

$$\mathbf{A}^{(1)} \times \mathbf{A}^{(2)} = i \, A^{(0)} \mathbf{A}^{(3)*} \text{ et cyclicum} \tag{3.23}$$

where:

$$\mathbf{A}^{(1)} = \mathbf{A}^{(2)*} = A^{(0)} \mathbf{e}^{(1)} e^{i\phi} = \frac{A^{(0)}}{\sqrt{2}} (\mathbf{i} - i\mathbf{j}) e^{i\phi}, \tag{3.24}$$

$$\mathbf{A}^{(3)} = A^{(0)} \mathbf{e}^{(3)} = A^{(0)} \mathbf{k}. \tag{3.25}$$

From Eqs. (3.23-3.25):

$$\nabla \times \mathbf{A}^{(1)} = \kappa \mathbf{A}^{(1)} \tag{3.26}$$
$$\nabla \times \mathbf{A}^{(2)} = \kappa \mathbf{A}^{(2)} \tag{3.27}$$
$$\nabla \times \mathbf{A}^{(3)} = 0 \mathbf{A}^{(3)} \tag{3.28}$$

3.2. DERIVATION OF THE BELTRAMI EQUATION

which are Beltrami equations [26], [27].

The foregoing analysis may be simplified by considering only one component out of the two conjugate components labelled (1) and (2). This procedure, however, loses information in general. By considering one component, Eq. (3.1) is simplified to:

$$\nabla \cdot \boldsymbol{\omega} \times \mathbf{q} = \mathbf{q} \cdot \nabla \times \boldsymbol{\omega} - \boldsymbol{\omega} \cdot \nabla \times \mathbf{q} \tag{3.29}$$

and the assumption of zero magnetic monopole leads to:

$$\nabla \cdot \boldsymbol{\omega} \times \mathbf{q} = 0 \tag{3.30}$$

which implies

$$\boldsymbol{\omega} \cdot \nabla \times \mathbf{q} = \mathbf{q} \cdot \nabla \times \boldsymbol{\omega}. \tag{3.31}$$

Proceeding as in note 257(7) in the UFT section of www.aias.us leads to:

$$\boldsymbol{\omega} \cdot \mathbf{B} = \mathbf{A} \cdot \nabla \times \boldsymbol{\omega} \tag{3.32}$$

where:

$$\mathbf{R}(\text{spin}) = \nabla \times \boldsymbol{\omega} \tag{3.33}$$

is the simplified format of the spin curvature. From Eqs. (3.31) and (3.32):

$$\boldsymbol{\omega} \cdot \mathbf{B} = \mathbf{A} \cdot \nabla \times \boldsymbol{\omega} = \boldsymbol{\omega} \cdot \nabla \times \mathbf{A} \tag{3.34}$$

so:

$$\mathbf{B} = \mathbf{A} \cdot \nabla \times \mathbf{A}. \tag{3.35}$$

However, in ECE theory:

$$\mathbf{B} = \nabla \times \mathbf{A} - \boldsymbol{\omega} \times \mathbf{A} \tag{3.36}$$

so Eqs. (3.35) and (3.36) imply:

$$\boldsymbol{\omega} \times \mathbf{A} = \mathbf{0}. \tag{3.37}$$

Therefore in this simplified model the spin connection vector is parallel to the vector potential. These results are consistent with [1-10]:

$$p^\mu = eA^\mu = \hbar\kappa^\mu = \hbar\omega^\mu \tag{3.38}$$

from the minimal prescription. So in this simplified model:

$$\omega^\mu = (\omega_0, \boldsymbol{\omega}) = \frac{e}{\hbar} A^\mu = \frac{e}{\hbar}(A_0, \mathbf{A}). \tag{3.39}$$

The electric field strength is defined in the simplified model by:

$$\mathbf{E} = -\nabla \phi - \frac{\partial \mathbf{A}}{\partial t} - c\omega_0 \mathbf{A} + \phi \boldsymbol{\omega} \tag{3.40}$$

CHAPTER 3. ECE THEORY AND BELTRAMI FIELDS

where the scalar potential is

$$\phi = cA_0. \tag{3.41}$$

From Eqs. (3.39) and (3.40):

$$\mathbf{E} = -\nabla\phi - \frac{\partial \mathbf{A}}{\partial t}, \tag{3.42}$$

$$\mathbf{B} = \nabla \times \mathbf{A}, \tag{3.43}$$

which is the same as the structure given by Heaviside, but these equations have been derived from general relativity and Cartan geometry, whereas the Heaviside structure is empirical. The equations (3.29) to (3.43) are oversimplified however because they are derived by consideration of only one out of two conjugate conjugates (1) and (2). Therefore they are derived using real algebra instead of complex algebra. They lose the B(3) field and also spin connection resonance, developed later in this book.

In the case of field matter interaction the electric field strength, $\mathbf{E}$ is replaced by the electric displacement, $\mathbf{D}$, and the magnetic flux density, $\mathbf{B}$ by the magnetic field strength, $\mathbf{H}$:

$$\mathbf{D} = \epsilon_0 \mathbf{E} + \mathbf{P}, \tag{3.44}$$

$$\mathbf{H} = \frac{1}{\mu_0}(\mathbf{B} - \mathbf{M}), \tag{3.45}$$

where $\mathbf{P}$ is the polarization, $\mathbf{M}$ is the magnetization, ϵ_0 is the vacuum permittivity and μ_0 is the vacuum permeability. The four equations of electrodynamics for each index (1) or (2) are:

$$\nabla \cdot \mathbf{B} = 0 \tag{3.46}$$

$$\nabla \times \mathbf{E} + \frac{\partial \mathbf{B}}{\partial t} = \mathbf{0} \tag{3.47}$$

$$\nabla \cdot \mathbf{D} = \rho \tag{3.48}$$

$$\nabla \times \mathbf{H} = \mathbf{J} + \frac{\partial \mathbf{D}}{\partial t} \tag{3.49}$$

where ρ is the charge density and $\mathbf{J}$ is the current density.

The Gauss law of magnetism:

$$\nabla \cdot \mathbf{B} = 0 \tag{3.50}$$

implies the magnetic Beltrami equation [27]:

$$\nabla \times \mathbf{B} = \kappa \mathbf{B} \tag{3.51}$$

because:

$$\frac{1}{\kappa}\nabla \cdot \nabla \times \mathbf{B} = 0. \tag{3.52}$$

3.2. DERIVATION OF THE BELTRAMI EQUATION

So the magnetic Beltrami equation is a consequence of the absence of a magnetic monopole and the Beltrami solution is always a valid solution. From Eqs. (3.49) and (3.51)

$$\nabla \times \mathbf{B} = \kappa \mathbf{B} = \mu_0 \mathbf{J} + \frac{1}{c^2}\frac{\partial \mathbf{E}}{\partial t} \qquad (3.53)$$

and for magnetostatics or if the Maxwell displacement current is small:

$$\mathbf{B} = \frac{\mu_0}{\kappa}\mathbf{J}. \qquad (3.54)$$

In this case the magnetic flux density is proportional to the current density. From Eq. 3.51:

$$\nabla \times \mathbf{B} = \frac{\mu_0}{\kappa}\nabla \times \mathbf{J} = \kappa \mathbf{B} \qquad (3.55)$$

so

$$\mathbf{B} = \frac{\mu_0}{\kappa^2}\nabla \times \mathbf{J}. \qquad (3.56)$$

Eqs. (3.54) and (3.56) imply that the current density must have the structure:

$$\nabla \times \mathbf{J} = \kappa \mathbf{J} \qquad (3.57)$$

in order to produce a Beltrami equation (3.51) in magnetostatics. Eq. (3.54) suggests that the jet observed from the plane of a whirlpool galaxy is a longitudinal solution of the Beltrami equation, a J(3) current associated with a B(3) field.

In field matter interaction the electric Beltrami equation:

$$\nabla \times \mathbf{E} = \kappa \mathbf{E} \qquad (3.58)$$

is not valid because it is not consistent with the Coulomb law:

$$\nabla \cdot \mathbf{E} = \frac{\rho}{\epsilon_0}. \qquad (3.59)$$

From Eqs. (3.58) and (3.59):

$$\nabla \cdot \nabla \times \mathbf{E} = \frac{\rho}{\epsilon_0}\kappa \qquad (3.60)$$

which violates the vector identity:

$$\nabla \cdot \nabla \times \mathbf{E} = 0. \qquad (3.61)$$

The electric Beltrami equation:

$$\nabla \times \mathbf{E} = \kappa \mathbf{E} \qquad (3.62)$$

is valid for the free electromagnetic field.

CHAPTER 3. ECE THEORY AND BELTRAMI FIELDS

Consider the four equations of the free electromagnetic field:

$$\nabla \cdot \mathbf{B} = 0 \tag{3.63}$$

$$\nabla \times \mathbf{E} + \frac{\partial \mathbf{B}}{\partial t} = \mathbf{0} \tag{3.64}$$

$$\nabla \cdot \mathbf{E} = 0 \tag{3.65}$$

$$\nabla \times \mathbf{B} - \frac{1}{c^2}\frac{\partial \mathbf{E}}{\partial t} = \mathbf{0} \tag{3.66}$$

for each index of the complex circular basis. It follows from Eqs. (3.64) and (3.66) that:

$$\nabla \times (\nabla \times \mathbf{B}) = \frac{1}{c^2}\frac{\partial}{\partial t}\nabla \times \mathbf{E} \tag{3.67}$$

and:

$$\nabla \times (\nabla \times \mathbf{E}) = -\frac{\partial}{\partial t}\nabla \times \mathbf{B}. \tag{3.68}$$

The transverse plane wave solutions are:

$$\mathbf{E} = \frac{E^{(0)}}{\sqrt{2}}(\mathbf{i} - i\mathbf{j})e^{i\phi} \tag{3.69}$$

and

$$\mathbf{B} = \frac{B^{(0)}}{\sqrt{2}}(i\mathbf{i} + \mathbf{j})e^{i\phi} \tag{3.70}$$

where:

$$\phi = \omega t - \kappa Z \tag{3.71}$$

and where ω is the angular velocity at instant t and κ is the magnitude of the wave vector at Z.

From vector analysis:

$$\nabla \times (\nabla \times \mathbf{B}) = \nabla(\nabla \cdot \mathbf{B}) - \nabla^2 \mathbf{B} \tag{3.72}$$

$$\nabla \times (\nabla \times \mathbf{E}) = \nabla(\nabla \cdot \mathbf{E}) - \nabla^2 \mathbf{E} \tag{3.73}$$

and for the free field the divergences vanish, so we obtain the Helmholtz wave equations:

$$(\nabla^2 + \kappa^2)\mathbf{B} = 0 \tag{3.74}$$

and

$$(\nabla^2 + \kappa^2)\mathbf{E} = 0. \tag{3.75}$$

These are the Trkalian equations:

$$\nabla \times (\nabla \times \mathbf{B}) = \kappa \nabla \times \mathbf{B} = \kappa^2 \mathbf{B} \tag{3.76}$$

3.2. DERIVATION OF THE BELTRAMI EQUATION

and

$$\nabla \times (\nabla \times \mathbf{E}) = \kappa \nabla \times \mathbf{E} = \kappa^2 \mathbf{E}. \tag{3.77}$$

So solutions of the Beltrami equations are also solutions of the Helmholtz wave equations. From Eqs. (3.64), (3.67) and (3.76):

$$-\nabla^2 \mathbf{B} - \frac{\kappa}{c^2} \frac{\partial \mathbf{E}}{\partial t} = \left(-\nabla^2 + \frac{1}{c^2} \frac{\partial^2}{\partial t^2}\right) \mathbf{B} = 0 \tag{3.78}$$

which is the d'Alembert equation:

$$\Box \mathbf{B} = \mathbf{0}. \tag{3.79}$$

For finite photon mass, implied by the longitudinal solutions of the free electromagnetic field:

$$\hbar^2 \omega^2 = c^2 \hbar^2 \kappa^2 + m_0^2 c^4 \tag{3.80}$$

in which case:

$$\left(\Box + \left(\frac{m_0 c}{\hbar}\right)^2\right) \mathbf{B} = \mathbf{0} \tag{3.81}$$

which is the Proca equation. This was first derived in ECE theory from the tetrad postulate of Cartan geometry and is discussed later in this book. From Eqs. (3.67) and (3.68):

$$\frac{\partial^2}{\partial t^2} \nabla \times \mathbf{B} = -\omega^2 \nabla \times \mathbf{B} \tag{3.82}$$

and:

$$\frac{\partial^2}{\partial t^2} \nabla \times \mathbf{E} = -\omega^2 \nabla \times \mathbf{E}. \tag{3.83}$$

In general:

$$\frac{\partial^2}{\partial t^2} e^{i\phi} = -\omega^2 e^{i\phi} \tag{3.84}$$

and

$$e^{i\phi} = e^{i\omega t} e^{-i\kappa Z} \tag{3.85}$$

so the general solution of the Beltrami equation

$$\nabla \times \mathbf{B} = \kappa \mathbf{B} \tag{3.86}$$

will also be a general solution of the equations (3.63) to (3.66) multiplied by the phase factor $\exp(i\omega t)$.

ECE theory can be used to show that the magnetic flux density, vector potential and spin connection vector are always Beltrami vectors with intricate structures in general, solutions of the Beltrami equation. The Beltrami

CHAPTER 3. ECE THEORY AND BELTRAMI FIELDS

structure of the vector potential is proven in ECE physics from the Beltrami structure of the magnetic flux density **B**. The space part of the Cartan identity also has a Beltrami structure. If real algebra is used, the Beltrami structure of **B** immediately refutes U(1) gauge invariance because **B** becomes directly proportional to **A**. It follows that the photon mass is identically non-zero, however tiny in magnitude. Therefore there is no Higgs boson in nature because the latter is the result of U(1) gauge invariance. The Beltrami structure of **B** is the direct result of the Gauss law of magnetism and the absence of a magnetic monopole. It is difficult to conceive why U(1) gauge invariance should ever have been adopted as a theory, because its refutation is trivial. Once U(1) gauge invariance is discarded a rich panoply of new ideas and results emerge.

The Beltrami equation for magnetic flux density in ECE physics is:

$$\nabla \times \mathbf{B}^a = \kappa \mathbf{B}^a. \tag{3.87}$$

In the simplest case κ is a wave-vector but it can become very intricate. Combining Eq. (3.87) with the Ampere Maxwell law of ECE physics:

$$\nabla \times \mathbf{B}^a = \mu_0 \mathbf{J}^a + \frac{1}{c^2} \frac{\partial \mathbf{E}^a}{\partial t} \tag{3.88}$$

the magnetic flux density is given directly by:

$$\mathbf{B}^a = \frac{1}{\kappa} \left(\frac{1}{c^2} \frac{\partial \mathbf{E}^a}{\partial t} + \mu_0 \mathbf{J}^a \right). \tag{3.89}$$

Using the Coulomb law of ECE physics:

$$\nabla \cdot \mathbf{E}^a = \frac{\rho^a}{\epsilon_0} \tag{3.90}$$

it is found that:

$$\nabla \cdot \mathbf{B}^a = \frac{\mu_0}{\kappa} \left(\frac{\partial \rho^a}{\partial t} + \nabla \cdot \mathbf{J}^a \right) = 0, \tag{3.91}$$

a result which follows from:

$$\epsilon_0 \mu_0 = \frac{1}{c^2} \tag{3.92}$$

where c is the universal constant known as the vacuum speed of light. The conservation of charge current density in ECE physics is:

$$\frac{\partial \rho^a}{\partial t} + \nabla \cdot \mathbf{J}^a = 0 \tag{3.93}$$

so $\mathbf{B}^a$ is always a Beltrami vector.

In the simplified physics with real algebra:

$$\mathbf{B} = \nabla \times \mathbf{A}, \tag{3.94}$$
$$\nabla \times \mathbf{B} = \kappa \nabla \times \mathbf{A}, \tag{3.95}$$

3.2. DERIVATION OF THE BELTRAMI EQUATION

where $\mathbf{A}$ is the vector potential. Eqs. (3.94) and (3.95) show immediately that in U(1) physics the vector potential also obeys a Beltrami equation:

$$\nabla \times \mathbf{A} = \kappa \mathbf{A}, \qquad (3.96)$$

$$\mathbf{B} = \kappa \mathbf{A} \qquad (3.97)$$

so in this simplified theory the magnetic flux density is directly proportional to the vector potential $\mathbf{A}$. It follows immediately that $\mathbf{A}$ cannot be U(1) gauge invariant because U(1) gauge invariance means:

$$\mathbf{A} \to \mathbf{A} + \nabla \psi \qquad (3.98)$$

and if $\mathbf{A}$ is changed, $\mathbf{B}$ is changed. The obsolete dogma of U(1) physics asserted that Eq. (3.98) does not change any physical quantity. This dogma is obviously incorrect because $\mathbf{B}$ is a physical quantity and Eq. (3.97) changes it. Therefore there is finite photon mass and no Higgs boson. Finite photon mass and the Proca equation are developed later in this book, and the theory is summarized here for ease of reference. The Proca equation [1-10] can be developed as:

$$\nabla \cdot \mathbf{B}^a = 0 \qquad (3.99)$$

$$\nabla \times \mathbf{E}^a + \frac{\partial \mathbf{B}^a}{\partial t} = 0 \qquad (3.100)$$

$$\nabla \cdot \mathbf{E}^a = \frac{\rho^a}{\epsilon_0} \qquad (3.101)$$

$$\nabla \times \mathbf{B}^a - \frac{1}{c^2} \frac{\partial \mathbf{E}^a}{\partial t} = \mu_0 \mathbf{J}^a \qquad (3.102)$$

where the 4-current density is:

$$J^{a\mu} = (c\rho^a, \mathbf{J}^a) \qquad (3.103)$$

and where the 4-potential is:

$$A^{a\mu} = \left(\frac{\phi^a}{c}, \mathbf{A}^a\right). \qquad (3.104)$$

Proca theory asserts that:

$$J^{a\mu} = -\epsilon_0 \left(\frac{mc}{\hbar}\right)^2 A^{a\mu} \qquad (3.105)$$

where m is the finite photon mass and $\hbar$ is the reduced Planck constant. Therefore:

$$\rho^a = -\epsilon_0 c^2 \left(\frac{mc}{\hbar}\right)^2 \phi^a, \qquad (3.106)$$

$$\mathbf{J}^a = -\epsilon_0 c^2 \left(\frac{mc}{\hbar}\right)^2 \mathbf{A}^a. \qquad (3.107)$$

The Proca equation was inferred in the mid-thirties but is almost entirely absent from the textbooks. This is an unfortunate result of incorrect dogma, that the photon mass, is zero despite being postulated by Einstein in about 1905 to be a particle or corpuscle, as did Newton before him. The U(1) Proca theory in S. I. Units is:

$$\partial_\mu F^{\mu\nu} = \frac{J^\nu}{\epsilon_0} = -\left(\frac{mc}{\hbar}\right)^2 A^\nu. \tag{3.108}$$

It follows immediately that:

$$\partial_\nu \partial_\mu F^{\mu\nu} = \frac{1}{\epsilon_0}\partial_\nu J^\nu = -\left(\frac{mc}{\hbar}\right)^2 \partial_\nu A^\nu = 0. \tag{3.109}$$

and that:

$$\partial_\mu J^\mu = \partial_\mu A^\mu = 0. \tag{3.110}$$

Eq. (3.109) is conservation of charge current density and Eq. (3.110) is the Lorenz condition. In the Proca equation the Lorenz condition has nothing to do with gauge invariance. The U(1) gauge invariance means that:

$$A^\mu \to A^\mu + \partial^\mu \chi \tag{3.111}$$

and from Eq. (3.108) it is trivially apparent that the Proca field and charge current density change under transformation (3.111), so are not gauge invariant, QED. The entire edifice of U(1) electrodynamics collapses as soon as photon mass is considered.

In vector notation Eq. (3.109) is:

$$\frac{1}{c}\frac{\partial}{\partial t}\nabla \cdot \mathbf{E} = \frac{1}{c\epsilon_0}\frac{\partial \rho}{\partial t} = 0 \tag{3.112}$$

and

$$\nabla \cdot \nabla \times \mathbf{B} - \frac{1}{c^2}\frac{\partial}{\partial t}\nabla \cdot \mathbf{E} = \mu_0 \nabla \cdot \mathbf{J} = 0. \tag{3.113}$$

Now use:

$$\nabla \cdot \nabla \times \mathbf{B} = 0 \tag{3.114}$$

and the Coulomb law of this simplified theory (without index a):

$$\nabla \cdot \mathbf{E} = \frac{\rho}{\epsilon_0} \tag{3.115}$$

to find that:

$$-\frac{1}{c^2\epsilon_0}\frac{\partial \rho}{\partial t} = \mu_0 \nabla \cdot \mathbf{J} \tag{3.116}$$

which is the equation of charge current conservation:

$$\frac{\partial \rho}{\partial t} + \nabla \cdot \mathbf{J} = 0. \tag{3.117}$$

3.2. DERIVATION OF THE BELTRAMI EQUATION

In the Proca theory, Eq. (3.110) implies the Lorenz gauge as it is known in standard physics:

$$\partial_\mu A^\mu = \frac{1}{c^2}\frac{\partial \phi}{\partial t} + \mathbf{\nabla} \cdot \mathbf{A} = 0. \tag{3.118}$$

The Proca wave equation in the usual development [31], [32] is obtained from the U(1) definition of the field tensor:

$$F^{\mu\nu} = \partial^\mu A^\nu - \partial^\nu A^\mu \tag{3.119}$$

so

$$\partial_\mu \left(\partial^\mu A^\nu - \partial^\nu A^\mu \right) = \Box A^\nu - \partial^\nu \partial_\mu A^\mu = -\left(\frac{mc}{\hbar}\right)^2 A^\nu \tag{3.120}$$

in which

$$\partial_\mu A^\mu = 0. \tag{3.121}$$

Eq. (3.121) follows from Eq. (3.108) in Proca physics, but in standard U(1) physics with identically zero photon mass the Lorenz gauge has to be assumed, and is arbitrary. So the Proca wave equation in the usual development [31], [32] is:

$$\left(\Box + \left(\frac{mc}{\hbar}\right)^2\right) A^\nu = 0. \tag{3.122}$$

In ECE physics [1-10] Eq. (3.122) is derived from the tetrad postulate of Cartan geometry and becomes:

$$\left(\Box + \left(\frac{mc}{\hbar}\right)^2\right) A^a_\mu = 0. \tag{3.123}$$

In ECE physics the conservation of charge current density is:

$$\partial_\mu J^{a\mu} = 0 \tag{3.124}$$

and is consistent with Eqs. (3.48) and (3.49).

In ECE physics the electric charge density is geometrical in origin and is:

$$\rho^a = \epsilon_0 \left(\boldsymbol{\omega}^a{}_b \cdot \mathbf{E}^b - c\mathbf{A}^b \cdot \mathbf{R}^a{}_b(\text{orb}) \right) \tag{3.125}$$

and the electric current density is:

$$\mathbf{J}^a = \frac{1}{\mu_0} \left(\boldsymbol{\omega}^a{}_b \times \mathbf{B}^b + \frac{\omega_0}{c}\mathbf{E}^b - \mathbf{A}^b \times \mathbf{R}^a{}_b(\text{spin}) - A^b{}_0 \mathbf{R}^a{}_b(\text{orb}) \right). \tag{3.126}$$

Here $\mathbf{R}^a{}_b(\text{spin})$ and $\mathbf{R}^a{}_b(\text{orb})$ are the spin and orbital components of the curvature tensor [1-10]. So Eqs. (3.93), (3.125) and (3.126) give many new equations of physics which can be developed systematically in future work. In magnetostatics for example the relevant equations are:

$$\mathbf{\nabla} \cdot \mathbf{B}^a = 0, \tag{3.127}$$
$$\mathbf{\nabla} \times \mathbf{B}^a = \mu_0 \mathbf{J}^a, \tag{3.128}$$

CHAPTER 3. ECE THEORY AND BELTRAMI FIELDS

and

$$\nabla \cdot \mathbf{J}^a = \nabla \cdot \nabla \times \mathbf{B}^a = 0 \qquad (3.129)$$

so it follows from charge current conservation that:

$$\frac{\partial \rho^a}{\partial t} = 0. \qquad (3.130)$$

If it is assumed that the scalar potential is zero in magnetostatics, the usual assumption, then:

$$\mathbf{J}^a = \frac{1}{\mu_0} \left(\omega^a{}_b \times \mathbf{B}^b - \mathbf{A}^b \times \mathbf{R}^a{}_b(\text{spin}) \right) \qquad (3.131)$$

because there is no electric field present. It follows from Eqs. (3.129) and (3.131) that

$$\nabla \cdot \omega^a{}_b \times \mathbf{B}^b = \nabla \cdot \mathbf{A}^b \times \mathbf{R}^a{}_b(\text{spin}) \qquad (3.132)$$

in ECE magnetostatics.

In UFT258 and immediately preceding papers of this series it has been shown that in the absence of a magnetic monopole:

$$\omega^a{}_b \cdot \mathbf{B}^b = \mathbf{A}^b \cdot \mathbf{R}^a{}_b(\text{spin}) \qquad (3.133)$$

and that the space part of the Cartan identity in the absence of a magnetic monopole gives the two equations:

$$\nabla \cdot \omega^a{}_b \times \mathbf{A}^b = 0 \qquad (3.134)$$

and

$$\omega^a{}_b \cdot \nabla \times \mathbf{A}^b = \mathbf{A}^b \cdot \nabla \times \omega^a{}_b. \qquad (3.135)$$

In ECE physics the magnetic flux density is:

$$\mathbf{B}^a = \nabla \times \mathbf{A}^a - \omega^a{}_b \times \mathbf{A}^b \qquad (3.136)$$

so the Beltrami equation gives:

$$\nabla \times \mathbf{B}^a = \kappa \mathbf{B}^a = \kappa \left(\nabla \times \mathbf{A}^a - \omega^a{}_b \times \mathbf{A}^b \right). \qquad (3.137)$$

Eq. (3.134) from the space part of the Cartan identity is also a Beltrami equation, as is any non-divergent equation:

$$\nabla \times \left(\omega^a{}_b \times \mathbf{A}^b \right) = \kappa \omega^a{}_b \times \mathbf{A}^b. \qquad (3.138)$$

From Eq. (3.137):

$$\nabla \times (\nabla \times \mathbf{A}^a) - \nabla \times \left(\omega^a{}_b \times \mathbf{A}^b \right) = \kappa \left(\nabla \times \mathbf{A}^a - \omega^a{}_b \times \mathbf{A}^b \right). \qquad (3.139)$$

3.2. DERIVATION OF THE BELTRAMI EQUATION

Using Eq. (3.138):

$$\nabla \times (\nabla \times \mathbf{A}^a) = \kappa \nabla \times \mathbf{A}^a \tag{3.140}$$

which implies that the vector potential is also defined in general by a Beltrami equation:

$$\nabla \times \mathbf{A}^a = \kappa \mathbf{A}^a \tag{3.141}$$

QED. This is a generally valid result of ECE physics which implies that:

$$\nabla \cdot \mathbf{A}^a = 0. \tag{3.142}$$

From Eq. (3.110) it follows that:

$$\frac{\partial \rho^a}{\partial t} = 0 \tag{3.143}$$

is a general result of ECE physics. From Eqs. (3.135) and (3.141):

$$\nabla \times \omega^a{}_b = \kappa \omega^a{}_b \tag{3.144}$$

so the spin connection vector of ECE physics is also defined in general by a Beltrami equation. This important result can be cross checked for internal consistency using note 258(4) on www.aias.us, starting from Eq. (3.50) of this paper. Considering the X component for example:

$$\omega^a{}_{Xb}(\nabla \times \mathbf{A}^a)_X = A^b{}_X (\nabla \times \omega^a{}_b)_X \tag{3.145}$$

and it follows that:

$$\frac{1}{A^{(1)}_X}\left(\nabla \times \mathbf{A}^{(1)}\right)_X = \frac{1}{\omega^{(a)}_{X(1)}}\left(\nabla \times \omega^a{}_{(1)}\right)_X \tag{3.146}$$

and similarly for the Y and Z components. In order for this to be a Beltrami equation, Eqs. (3.141) and (3.144) must be true, QED.

In magnetostatics there are additional results which emerge as follows. From vector analysis:

$$\nabla \cdot \omega^a{}_b \times \mathbf{B}^b = \mathbf{B}^b \cdot \nabla \times \omega^a{}_b - \omega^a{}_b \cdot \nabla \times \mathbf{B}^b \tag{3.147}$$

and

$$\nabla \cdot \mathbf{A}^b \times \mathbf{R}^a{}_b(\text{spin}) = \mathbf{R}^a{}_b(\text{spin}) \cdot \nabla \times \mathbf{A}^a - \mathbf{A}^b \cdot \nabla \times \mathbf{R}^a{}_b(\text{spin}). \tag{3.148}$$

It is immediately clear that Eqs. (3.87) and (3.144) give Eq. (3.147) self consistently, QED. Eq. (3.148) gives

$$\nabla \cdot \omega^a{}_b \times \mathbf{B}^b = \nabla \cdot \mathbf{A}^b \times \mathbf{R}^a{}_b(\text{spin}) = 0 \tag{3.149}$$

and using Eq. (3.148):

$$\nabla \times \mathbf{R}^a{}_b(\text{spin}) = \kappa \mathbf{R}^a{}_b(\text{spin}) \tag{3.150}$$

CHAPTER 3. ECE THEORY AND BELTRAMI FIELDS

so the spin curvature is defined by a Beltrami equation in magnetostatics. Also in magnetostatics:

$$\boldsymbol{\nabla} \times \mathbf{B}^a = \kappa \mathbf{B}^a = \mu_0 \mathbf{J}^a \qquad (3.151)$$

so it follows that the current density of magnetostatics is also defined by a Beltrami equation:

$$\boldsymbol{\nabla} \times \mathbf{J}^a = \kappa \, \mathbf{J}^a. \qquad (3.152)$$

All these Beltrami equations in general have intricate flow structures graphed following sections of this chapter and animated on www.aias.us. As discussed in Eqs. (3.31) to (3.35) of Note 258(5) on www.aias.us, plane wave structures and O(3) electrodynamics [1-10] are also defined by Beltrami equations. The latter give simple solutions for vacuum plane waves. In other cases the solutions become intricate. The B(3) field is defined by the simplest type of Beltrami equation

$$\boldsymbol{\nabla} \times \mathbf{B}^{(3)} = 0 \, \mathbf{B}^{(3)}. \qquad (3.153)$$

In photon mass theory therefore:

$$\boldsymbol{\nabla} \times \mathbf{A}^a = \kappa \, \mathbf{A}^a, \qquad (3.154)$$

$$\left(\Box + \left(\frac{mc}{\hbar} \right)^2 \right) \mathbf{A}^a = \mathbf{0}. \qquad (3.155)$$

It follows from Eq. (3.154) that:

$$\boldsymbol{\nabla} \cdot \mathbf{A}^a = 0 \qquad (3.156)$$

so:

$$\boldsymbol{\nabla} \times (\boldsymbol{\nabla} \times \mathbf{A}^a) = \kappa \, \boldsymbol{\nabla} \times \mathbf{A}^a = \kappa^2 \, \mathbf{A}^a \qquad (3.157)$$

produces the Helmholtz wave equation:

$$\left(\boldsymbol{\nabla}^2 + \kappa^2 \right) \mathbf{A}^a = 0. \qquad (3.158)$$

Eq. (3.155) is

$$\left(\frac{1}{c^2} \frac{\partial^2}{\partial t^2} - \boldsymbol{\nabla}^2 + \left(\frac{mc}{\hbar} \right)^2 \right) \mathbf{A}^a = \mathbf{0} \qquad (3.159)$$

so:

$$\left(\frac{1}{c^2} \frac{\partial^2}{\partial t^2} + \kappa^2 + \left(\frac{mc}{\hbar} \right)^2 \right) \mathbf{A}^a = \mathbf{0}. \qquad (3.160)$$

Now use:

$$\mathbf{p} = \hbar \boldsymbol{\kappa} \qquad (3.161)$$

3.2. DERIVATION OF THE BELTRAMI EQUATION

and:

$$\frac{\partial^2}{\partial t^2} = -\frac{E^2}{\hbar^2} \tag{3.162}$$

to find that Eq. (3.160) is the Einstein energy equation for the photon of mass m, so the analysis is rigorously self-consistent, QED.

In ECE physics the Lorenz gauge is:

$$\partial_\mu A^{a\mu} = 0 \tag{3.163}$$

i.e.

$$\frac{1}{c^2}\frac{\partial \phi^a}{\partial t} + \boldsymbol{\nabla} \cdot \mathbf{A}^a = 0 \tag{3.164}$$

with the solution:

$$\frac{\partial \phi^a}{\partial t} = \boldsymbol{\nabla} \cdot \mathbf{A}^a = 0. \tag{3.165}$$

This is again a general result of ECE physics applicable under any circumstances. Also in ECE physics in general the spin connection vector has no divergence:

$$\boldsymbol{\nabla} \cdot \boldsymbol{\omega}^a{}_b = 0 \tag{3.166}$$

because:

$$\boldsymbol{\nabla} \times \boldsymbol{\omega}^a{}_b = \kappa\, \boldsymbol{\omega}^a{}_b. \tag{3.167}$$

Another rigorous test for self-consistency is given by the definition of the magnetic field in ECE physics:

$$\mathbf{B}^a = \boldsymbol{\nabla} \times \mathbf{A}^a - \boldsymbol{\omega}^a{}_b \times \mathbf{A}^b \tag{3.168}$$

so:

$$\boldsymbol{\nabla} \cdot \mathbf{B}^a = -\boldsymbol{\nabla} \cdot \boldsymbol{\omega}^a{}_b \times \mathbf{A}^b = 0 \tag{3.169}$$

By vector analysis:

$$\boldsymbol{\nabla} \cdot \boldsymbol{\omega}^a{}_b \times \mathbf{A}^b = \mathbf{A}^b \cdot \boldsymbol{\nabla} \times \boldsymbol{\omega}^a{}_b - \boldsymbol{\omega}^a{}_b \cdot \boldsymbol{\nabla} \times \mathbf{A}^b = 0 \tag{3.170}$$

because

$$\boldsymbol{\nabla} \times \boldsymbol{\omega}^a{}_b = \kappa\, \boldsymbol{\omega}^a{}_b, \tag{3.171}$$
$$\boldsymbol{\nabla} \times \mathbf{A}^b = \kappa\, \mathbf{A}^b, \tag{3.172}$$

and:

$$\boldsymbol{\nabla} \cdot \mathbf{A}^b = 0, \tag{3.173}$$
$$\boldsymbol{\nabla} \cdot \boldsymbol{\omega}^a{}_b = 0. \tag{3.174}$$

CHAPTER 3. ECE THEORY AND BELTRAMI FIELDS

In the absence of a magnetic monopole Eq. (3.84) also follows from the space part of the Cartan identity. So the entire analysis is rigorously self-consistent. The cross consistency of the Beltrami and ECE equations can be checked using:

$$\mathbf{B}^b = \kappa \, \mathbf{A}^b - \boldsymbol{\omega}^b{}_c \times \mathbf{A}^c \tag{3.175}$$

as in note 258(1) on www.aias.us. Eq. (3.175) follows from Eqs. (3.168) and (3.172). Multiply Eq. (3.175) by $\boldsymbol{\omega}^a{}_b$ and use Eq. (3.133) to find:

$$\kappa \, \boldsymbol{\omega}^a{}_b \cdot \mathbf{A}^b - \boldsymbol{\omega}^a{}_b \cdot \boldsymbol{\omega}^b{}_c \times \mathbf{A}^c = \mathbf{A}^b \cdot \mathbf{R}^a{}_b(\text{spin}). \tag{3.176}$$

Now use:

$$\boldsymbol{\omega}^a{}_b \cdot \boldsymbol{\omega}^b{}_c \times \mathbf{A}^c = \mathbf{A}^c \cdot (\boldsymbol{\omega}^a{}_b \times \boldsymbol{\omega}^b{}_c) \tag{3.177}$$

and relabel summation indices to find:

$$\kappa \, \boldsymbol{\omega}^a{}_b \cdot \mathbf{A}^b - \mathbf{A}^b \cdot (\boldsymbol{\omega}^a{}_c \times \boldsymbol{\omega}^c{}_b) = \mathbf{A}^b \cdot \mathbf{R}^a{}_b(\text{spin}). \tag{3.178}$$

It follows that:

$$\mathbf{R}^a{}_b(\text{spin}) = \kappa \, \boldsymbol{\omega}^a{}_b - \boldsymbol{\omega}^a{}_c \times \boldsymbol{\omega}^c{}_b = \nabla \times \boldsymbol{\omega}^a{}_b - \boldsymbol{\omega}^a{}_c \times \boldsymbol{\omega}^c{}_b \tag{3.179}$$

QED. The analysis correctly and self consistently produces the correct definition of the spin curvature.

Finally, on the U(1) level for the sake of illustration, consider the Beltrami equations of note 258(3) on www.aias.us:

$$\nabla \times \mathbf{A} = \kappa \, \mathbf{A} \tag{3.180}$$

and

$$\nabla \times \mathbf{B} = \kappa \, \mathbf{B} \tag{3.181}$$

in the Ampere Maxwell law

$$\nabla \times \mathbf{B} - \frac{1}{c^2} \frac{\partial \mathbf{E}}{\partial t} = \mu_0 \mathbf{J}. \tag{3.182}$$

It follows that:

$$\kappa^2 \, \mathbf{A} = \mathbf{J} + \frac{1}{c^2} \frac{\partial \mathbf{E}}{\partial t} \tag{3.183}$$

where:

$$\mathbf{E} = -\nabla \phi - \frac{\partial \mathbf{A}}{\partial t}. \tag{3.184}$$

Therefore

$$\kappa^2 \, \mathbf{A} = \mu_0 \mathbf{J} + \frac{1}{c^2} \frac{\partial}{\partial t} \left(-\nabla \phi - \frac{\partial \mathbf{A}}{\partial t} \right) \tag{3.185}$$

and using the Lorenz condition:
$$\nabla \cdot \mathbf{A} + \frac{1}{c^2} \frac{\partial \phi}{\partial t} = 0 \tag{3.186}$$
it follows that:
$$\frac{\partial \phi}{\partial t} = 0. \tag{3.187}$$
Using
$$\Box = \frac{1}{c^2} \frac{\partial^2}{\partial t^2} - \nabla^2 \tag{3.188}$$
Eq. (3.185) becomes the d'Alembert equation in the presence of current density:
$$\Box \mathbf{A} = \mu_0 \mathbf{J}. \tag{3.189}$$
The solutions of the d'Alembert equation (3.189) may be found from:
$$\mathbf{B} = \kappa \mathbf{A} \tag{3.190}$$
showing in another way that as soon as the Beltrami equation (3.87) is used, U(1) gauge invariance is refuted.

3.3 Elecrostatics, Spin Connection Resonance and Beltrami Structures

As argued already the first Cartan structure equation defines the electric field strength as:
$$\mathbf{E}^a = -c\nabla A^a{}_0 - \frac{\partial \mathbf{A}^a}{\partial t} - c\omega^a{}_{0b} \mathbf{A}^b + cA^b{}_0 \boldsymbol{\omega}^a{}_b \tag{3.191}$$
where the four potential of ECE electrodynamics is defined by:
$$A^a{}_\mu = (A^a{}_0, -\mathbf{A}^a) = \left(\frac{\phi^a}{c}, -\mathbf{A}^a\right). \tag{3.192}$$
Here ϕ^a is the scalar potential. If it is assumed that the subject of electrostatics is defined by:
$$\mathbf{B}^a = 0, \quad \mathbf{A}^a = 0, \quad \mathbf{J}^a = 0 \tag{3.193}$$
then the Coulomb law in ECE theory is given by:
$$\nabla \cdot \mathbf{E}^a = \boldsymbol{\omega}^a{}_b \cdot \mathbf{E}^b. \tag{3.194}$$
The electric current in ECE theory is defined by:
$$\mathbf{J}^a = \epsilon_0 \, c \left(\omega^a{}_{0b} \mathbf{E}^b - cA^b{}_0 \mathbf{R}^a{}_b(\text{orb}) + c\boldsymbol{\omega}^a{}_b \times \mathbf{B}^b - c\mathbf{A}^b \times \mathbf{R}^a{}_b(\text{spin}) \right) \tag{3.195}$$

CHAPTER 3. ECE THEORY AND BELTRAMI FIELDS

where $\mathbf{R}^a{}_b(\text{spin})$ is the spin part of the curvature vector and where $\mathbf{B}^b$ is the magnetic flux density. From Eqs. (3.193) and (3.195):

$$\mathbf{J}^a = 0 = \epsilon_0 \, c \left(\omega^a{}_{0b} \mathbf{E}^b - c A^b{}_0 \mathbf{R}^a{}_b(\text{orb}) \right) \tag{3.196}$$

so in ECE electrostatics:

$$\omega^a{}_{0b} \mathbf{E}^b = c A^b{}_0 \mathbf{R}^a{}_b(\text{orb}) \tag{3.197}$$

and

$$\mathbf{E}^a = -c \nabla A^a{}_0 + c A^b{}_0 \boldsymbol{\omega}^a{}_b \tag{3.198}$$

with

$$\nabla \times \mathbf{E}^a = 0. \tag{3.199}$$

From Eqs. (3.198) and (3.199)

$$\nabla \times \mathbf{E}^a = c \nabla \times (A^b{}_0 \boldsymbol{\omega}^a{}_b) \tag{3.200}$$

so we obtain the constraint:

$$\nabla \times (A^b{}_0 \boldsymbol{\omega}^a{}_b) = 0. \tag{3.201}$$

The magnetic charge density in ECE theory is given by:

$$\rho^a_{\text{magn}} = \epsilon_0 \, c \left(\boldsymbol{\omega}^a{}_b \cdot \mathbf{B}^b - \mathbf{A}^b \cdot \mathbf{R}^a{}_b(\text{spin}) \right) \tag{3.202}$$

and the magnetic current density by:

$$\mathbf{J}^a_{\text{magn}} = \epsilon_0 \left(\boldsymbol{\omega}^a{}_b \times \mathbf{E}^b - c \omega^a{}_{0b} \mathbf{B}^b - c \left(\mathbf{A}^b \times \mathbf{R}^a{}_b(\text{orb}) - A^b{}_0 \mathbf{R}^a{}_b(\text{spin}) \right) \right). \tag{3.203}$$

These are thought to vanish experimentally in electromagnetism, so:

$$\boldsymbol{\omega}^a{}_b \cdot \mathbf{B}^b = \mathbf{A}^b \cdot \mathbf{R}^a{}_b(\text{spin}) \tag{3.204}$$

and

$$\boldsymbol{\omega}^a{}_b \times \mathbf{E}^a - c \omega^a{}_{0b} \mathbf{B}^b - c \mathbf{A}^b \times \mathbf{R}^a{}_b(\text{orb}) + c A^b{}_0 \mathbf{R}^a{}_b(\text{spin}) = 0. \tag{3.205}$$

In ECE electrostatics Eq. (3.204) is true automatically because:

$$\mathbf{B}^b = 0, \quad \mathbf{A}^b = 0 \tag{3.206}$$

and Eq. (3.203) becomes:

$$\boldsymbol{\omega}^a{}_b \times \mathbf{E}^b + c A^b{}_0 \mathbf{R}^a{}_b(\text{spin}) = 0. \tag{3.207}$$

3.3. ELECROSTATICS, SPIN CONNECTION RESONANCE AND...

So the equations of ECE electrostatics are:

$$\boldsymbol{\nabla} \cdot \mathbf{E}^a = \omega^a{}_b \cdot \mathbf{E}^b \tag{3.208}$$

$$\omega^a{}_{0b} \mathbf{E}^b = \phi^b \mathbf{R}^a{}_b(\text{orb}) \tag{3.209}$$

$$\boldsymbol{\omega}^a{}_b \times \mathbf{E}^b + \phi^b \mathbf{R}^a{}_b(\text{spin}) = 0 \tag{3.210}$$

$$\mathbf{E}^a = -\boldsymbol{\nabla}\phi^a + \phi^b \boldsymbol{\omega}^a{}_b \tag{3.211}$$

Later on in this chapter it is shown that these equations lead to a solution in terms of Bessel functions, but not to Euler Bernoulli resonance.

In order to obtain spin connection resonance Eq. (3.208) must be extended to:

$$\boldsymbol{\nabla} \cdot \mathbf{E}^a = \omega^a{}_b \cdot \mathbf{E}^b - c\mathbf{A}^b(\text{vac}) \cdot \mathbf{R}^a{}_b(\text{orb}) \tag{3.212}$$

where $\mathbf{A}^b(\text{vac})$ is the Eckardt Lindstrom vacuum potential [1-10]. The static electric field is defined by:

$$\mathbf{E}^a = -\boldsymbol{\nabla}\phi^a + \phi^b \boldsymbol{\omega}^a{}_b \tag{3.213}$$

so from Eqs. (3.212) and (3.213):

$$\nabla^2 \phi^a + \left(\boldsymbol{\omega}^a{}_b \cdot \boldsymbol{\omega}^b{}_c\right)\phi^c = \boldsymbol{\nabla} \cdot \left(\phi^b \boldsymbol{\omega}^a{}_b\right) + \boldsymbol{\omega}^a{}_b \cdot \boldsymbol{\nabla}\phi^b + c\mathbf{A}^b(\text{vac}) \cdot \mathbf{R}^a{}_b(\text{orb}). \tag{3.214}$$

By the ECE antisymmetry law:

$$-\boldsymbol{\nabla}\phi^a = \phi^b \boldsymbol{\omega}^a{}_b \tag{3.215}$$

leading to the Euler Bernoulli resonance equation:

$$\nabla^2 \phi^a + \left(\boldsymbol{\omega}^a{}_b \cdot \boldsymbol{\omega}^b{}_c\right)\phi^c = \frac{1}{2} c\mathbf{A}^b(\text{vac}) \cdot \mathbf{R}^a{}_b(\text{orb}) \tag{3.216}$$

and spin connection resonance [1-10]. The left hand side contains the Hooke law term and the right hand side the driving term originating in the vacuum potential. Denote:

$$\rho^a(\text{vac}) = \frac{\epsilon_0 c}{2} \mathbf{A}^b(\text{vac}) \cdot \mathbf{R}^a{}_b(\text{orb}) \tag{3.217}$$

then the equation becomes:

$$\nabla^2 \phi^a + \left(\boldsymbol{\omega}^a{}_b \cdot \boldsymbol{\omega}^b{}_c\right)\phi^c = \frac{\rho^a(\text{vac})}{\epsilon_0}. \tag{3.218}$$

The left hand side of Eq. (3.218) is a field property and the right hand side a property of the ECE vacuum. In the simplest case:

$$\nabla^2 \phi + (\omega_0)^2 \phi = \frac{\rho(\text{vac})}{\epsilon_0} \tag{3.219}$$

and produces undamped resonance if:

$$\rho(\text{vac}) = \epsilon_0 A \, \cos \omega Z \tag{3.220}$$

where A is a constant. The particular integral of Eq. (3.219) is:

$$\phi = \frac{A \cos \omega Z}{(\omega_0)^2 - \omega^2} \tag{3.221}$$

and spin connection resonance occurs at:

$$\omega = \omega_0 \tag{3.222}$$

when:

$$\phi \to \infty \tag{3.223}$$

and there is a resonance peak of electric field strength from the vacuum.

Later in this chapter solutions of Eq. (3.218) are given in terms of a combination of Bessel functions, and also an analysis using the Eckardt Lindstrom vacuum potential as a driving term.

In the absence of a magnetic monopole the Cartan identity is, as argued already:

$$\boldsymbol{\nabla} \cdot \boldsymbol{\omega}^a{}_b \times \mathbf{A}^b = 0 \tag{3.224}$$

which implies:

$$\boldsymbol{\omega}^a{}_b \cdot \boldsymbol{\nabla} \times \mathbf{A}^b = \mathbf{A}^b \cdot \boldsymbol{\nabla} \times \boldsymbol{\omega}^a{}_b. \tag{3.225}$$

A possible solution of this equation is:

$$\boldsymbol{\omega}^a{}_b = \epsilon^a{}_{bc} \boldsymbol{\omega}^c \tag{3.226}$$

leading as argued already to a rigorous justification for O(3) electrodynamics. The Cartan identity (3.224) is itself a Beltrami equation:

$$\boldsymbol{\nabla} \times \left(\boldsymbol{\omega}^a{}_b \times \mathbf{A}^b \right) = \kappa \, \boldsymbol{\omega}^a{}_b \times \mathbf{A}^b. \tag{3.227}$$

From Eqs. (3.226) and (3.227):

$$\boldsymbol{\nabla} \times \left(\mathbf{A}^c \times \mathbf{A}^b \right) = \kappa \, \mathbf{A}^c \times \mathbf{A}^b. \tag{3.228}$$

In the complex circular basis:

$$\mathbf{A}^{(1)} \times \mathbf{A}^{(2)} = i \, A^{(0)} \mathbf{A}^{(3)*} \text{ et cyclicum} \tag{3.229}$$

so from Eqs. (3.228) and (3.229):

$$\boldsymbol{\nabla} \times \mathbf{A}^{(i)} = \kappa \, \mathbf{A}^{(i)}, \quad i = 1, 2, 3 \tag{3.230}$$

which are Beltrami equations as argued earlier in this chapter.

This result can be obtained self consistently using the Gauss law:

$$\boldsymbol{\nabla} \cdot \mathbf{B}^a = 0 \tag{3.231}$$

3.3. ELECROSTATICS, SPIN CONNECTION RESONANCE AND...

which as argued already implies the Beltrami equation:

$$\nabla \times \mathbf{B}^a = \kappa\, \mathbf{B}^a. \tag{3.232}$$

From Eqs. (3.168) and (3.232):

$$\nabla \times \mathbf{B}^a = \kappa\, \mathbf{B}^a = \kappa \left(\nabla \times \mathbf{A}^a - \omega^a{}_b \times \mathbf{A}^b \right) \tag{3.233}$$

so:

$$\nabla \times (\nabla \times \mathbf{A}^a) - \nabla \times (\omega^a{}_b \times \mathbf{A}^b) = \kappa \left(\nabla \times \mathbf{A}^a - \omega^a{}_b \times \mathbf{A}^b \right) \tag{3.234}$$

Using Eq. (3.227) gives:

$$\nabla \times (\nabla \times \mathbf{A}^a) = \kappa\, \nabla \times \mathbf{A}^a \tag{3.235}$$

which implies Eqs. (3.228) to (3.230) QED. As shown earlier in this chapter the Beltrami structure also governs the spin connection vector:

$$\nabla \times \omega^a{}_b = \kappa\, \omega^a{}_b. \tag{3.236}$$

It follows that the equations:

$$\omega^{(3)} = \frac{1}{2} \frac{\kappa}{A^{(0)}} \mathbf{A}^{(3)} \tag{3.237}$$

and:

$$\omega^{(2)} = \frac{1}{2} \frac{\kappa}{A^{(0)}} \mathbf{A}^{(2)} \tag{3.238}$$

produce O(3) electrodynamics [1-10]:

$$\mathbf{B}^{(1)*} = \nabla \times \mathbf{A}^{(1)*} - i \frac{\kappa}{A^{(0)}} \mathbf{A}^{(2)} \times \mathbf{A}^{(3)} \quad \text{et cyclicum.} \tag{3.239}$$

As shown in Note 259(3) on www.aias.us there are many inter-related equations of O(3) electrodynamics which all originate in geometry.

Later in this chapter it is argued a consequence of these conclusions is that the spin connection and orbital curvature vectors also obey a Beltrami structure.

The fact that ECE is a unified field theory also allows the development and interrelation of several basic equations, including the definition of B(3):

$$\mathbf{B}^{(3)} = \nabla \times \mathbf{A}^{(3)} - i \frac{\kappa}{A^{(0)}} \mathbf{A}^{(1)} \times \mathbf{A}^{(2)}. \tag{3.240}$$

It can be written as:

$$\mathbf{B} = -i \frac{e}{\hbar} \mathbf{A} \times \mathbf{A}^* = B^{(0)} \mathbf{k} = B_Z \mathbf{k}. \tag{3.241}$$

Although B(3) is a radiated and propagating field as is well-known [1-10] Eq. (3.241) can be used as a general definition of the magnetic flux density for a

CHAPTER 3. ECE THEORY AND BELTRAMI FIELDS

choice of potentials. This is important for the subject of magnetostatics and the development [1-10] of the fermion equation with:

$$\mathbf{A} = \frac{1}{2}\mathbf{B} \times \mathbf{r}. \tag{3.242}$$

Eq. (3.241) gives the transition from classical to quantum mechanics. In ECE electrodynamics $\mathbf{A}$ must always be a Beltrami field and this is the result of the Cartan identity as already argued. So it is necessary to solve the following equations simultaneously:

$$\mathbf{B} = -i\frac{e}{\hbar}\mathbf{A} \times \mathbf{A}^*, \quad \mathbf{A} = \frac{1}{2}\mathbf{B} \times \mathbf{r}, \quad \boldsymbol{\nabla} \times \mathbf{A} = \kappa\, \mathbf{A}. \tag{3.243}$$

This can be done using the principles of general relativity, so that the electromagnetic field is a rotating and translating frame of reference. The position vector is therefore:

$$\mathbf{r} = \mathbf{r}^* = \frac{r^{(0)}}{\sqrt{2}}\left(\mathbf{i} - i\mathbf{j}\right)e^{i\phi} \tag{3.244}$$

where:

$$\mathbf{r} = \mathbf{r}^{(1)}, \quad \mathbf{r}^* = \mathbf{r}^{(2)}, \quad \phi = \omega t - \kappa Z \tag{3.245}$$

so:

$$\mathbf{r}^{(1)} \times \mathbf{r}^{(2)} = ir^{(0)}\mathbf{r}^{(3)*} \quad \text{et cyclicum.} \tag{3.246}$$

It follows that:

$$\boldsymbol{\nabla} \times \mathbf{r}^{(1)} = \kappa\, \mathbf{r}^{(1)} \tag{3.247}$$
$$\boldsymbol{\nabla} \times \mathbf{r}^{(2)} = \kappa\, \mathbf{r}^{(2)} \tag{3.248}$$
$$\boldsymbol{\nabla} \times \mathbf{r}^{(3)} = 0\, \mathbf{r}^{(3)} \tag{3.249}$$

The results (3.246) for plane waves can be generalized to any Beltrami solutions, so it follows that spacetime itself has a Beltrami structure. From Eqs. (3.242) and (3.244):

$$\mathbf{A} = \mathbf{A}^{(1)} = \frac{B^{(0)}r^{(0)}}{2\sqrt{2}}\left(i\mathbf{i} + \mathbf{j}\right)e^{i\phi} = \frac{A^{(0)}}{\sqrt{2}}\left(i\mathbf{i} + \mathbf{j}\right)e^{i\phi} \tag{3.250}$$

where:

$$A^{(0)} = \frac{1}{2}B^{(0)}r^{(0)} \tag{3.251}$$

and from Eq. (3.250):

$$\boldsymbol{\nabla} \times \mathbf{A} = \kappa\, \mathbf{A} \tag{3.252}$$

3.3. ELECROSTATICS, SPIN CONNECTION RESONANCE AND ...

QED. Therefore it is always possible to write the vector potential in the form (3.242) provided that spacetime itself has a Beltrami structure. This conclusion ties together several branches of physics because Eq. (3.242) is used to produce the Landé factor, ESR, NMR and so on from the Dirac equation, which becomes the fermion equation [1-10] in ECE physics.

As argued already the tetrad postulate and ECE postulate give:

$$\left(\Box + \kappa_0^2\right) \mathbf{A} = \mathbf{0} \tag{3.253}$$

and the fermion or chiral Dirac equation is a factorization of Eq. (3.253). As shown in Chapter 1:

$$\kappa_0^2 = q^\nu{}_a \partial^\mu \left(\omega^a{}_{\mu\nu} - \Gamma^a{}_{\mu\nu}\right) \tag{3.254}$$

where $q^\nu{}_a$ is the inverse tetrad, defined by:

$$q^a{}_\nu q^\nu{}_a = 1. \tag{3.255}$$

In generally covariant format Eq. (3.253) is:

$$\left(\Box + \kappa_0^2\right) A^a{}_\mu = 0 \tag{3.256}$$

and with:

$$A^a{}_\mu = (A^a{}_0, -\mathbf{A}^a) \tag{3.257}$$

it follows that:

$$\left(\Box + \kappa_0^2\right) A_0 = 0, \tag{3.258}$$
$$\left(\Box + \kappa_0^2\right) \mathbf{A} = \mathbf{0}, \tag{3.259}$$

which gives Eq. (3.254) QED. The d'Alembertian is defined by:

$$\Box = \frac{1}{c^2} \frac{\partial^2}{\partial t^2} - \nabla^2. \tag{3.260}$$

The Beltrami condition:

$$\nabla \mathbf{A} = \kappa \, \mathbf{A} \tag{3.261}$$

gives the Helmholtz wave equation:

$$\left(\nabla^2 + \kappa^2\right) \mathbf{A} = \mathbf{0} \tag{3.262}$$

if:

$$\nabla \cdot \mathbf{A} = 0. \tag{3.263}$$

From Eq. (3.259):

$$\left(\frac{1}{c^2} \frac{\partial^2}{\partial t^2} - \nabla^2 + \kappa_0^2\right) \mathbf{A} = \mathbf{0} \tag{3.264}$$

CHAPTER 3. ECE THEORY AND BELTRAMI FIELDS

so:

$$\frac{1}{c^2}\frac{\partial^2 \mathbf{A}}{\partial t^2} + \left(\kappa_0^2 + \kappa^2\right)\mathbf{A} = 0 \tag{3.265}$$

which is the equation for the time dependence of $\mathbf{A}$. The Helmholtz and Beltrami equations are for the space dependence of $\mathbf{A}$. Eq. (3.267) is satisfied by:

$$\mathbf{A} = \mathbf{A}_0 \exp(i\omega t) \tag{3.266}$$

where:

$$\frac{\omega^2}{c^2} = \kappa^2 + \kappa_0^2. \tag{3.267}$$

Eq. (3.267) is a generalization of the Einstein energy equation for a free particle:

$$E^2 = c^2 p^2 + m^2 c^4 \tag{3.268}$$

where:

$$E = \hbar\omega, \quad \mathbf{p} = \hbar\boldsymbol{\kappa} \tag{3.269}$$

using:

$$\kappa_0^2 = \left(\frac{mc}{\hbar}\right)^2 = q^\nu{}_a \partial^\mu \left(\omega^a{}_{\mu\nu} - \Gamma^a{}_{\mu\nu}\right). \tag{3.270}$$

So mass in ECE theory is defined by geometry.

The general solution of Eq (3.256) is therefore:

$$A^a{}_\mu = A^a{}_\mu(0) \exp\left(i(\omega t - \kappa Z)\right) \tag{3.271}$$

where:

$$\omega^2 = c^2 \left(\kappa^2 + \kappa_0^2\right). \tag{3.272}$$

It follows that there exist the equations:

$$\left(\Box + \kappa_0^2\right)\phi^a = 0 \tag{3.273}$$

and

$$\left(\boldsymbol{\nabla}^2 + \kappa^2\right)\phi^a = 0 \tag{3.274}$$

where ϕ^a is the scalar potential in ECE physics. For each a:

$$\left(\boldsymbol{\nabla}^2 + \kappa^2\right)\phi = 0. \tag{3.275}$$

Now write:

$$\kappa_0 = \frac{mc}{\hbar} \tag{3.276}$$

3.3. ELECROSTATICS, SPIN CONNECTION RESONANCE AND...

where m is mass. The relativistic wave equation for each a is:

$$\left(\Box + \kappa_0^2\right)\phi = 0 \tag{3.277}$$

which is the quantized format of:

$$E^2 = c^2 p^2 + m^2 c^4 = c^2 p^2 + \hbar^2 \kappa_0^2 c^2. \tag{3.278}$$

Eq. (3.278) is:

$$E = \gamma m c^2 \tag{3.279}$$

where the Lorentz factor is:

$$\gamma = \left(1 - \frac{v^2}{c^2}\right)^{-1/2} \tag{3.280}$$

and where the relativistic momentum is:

$$\mathbf{p} = \gamma m \mathbf{v}. \tag{3.281}$$

Define the relativistic energy as:

$$T = E - mc^2 \tag{3.282}$$

and it follows that:

$$T = (\gamma - 1)mc^2 \xrightarrow[v \ll c]{} \frac{1}{2}mv^2 \tag{3.283}$$

which is the non-relativistic limit of the kinetic energy, i.e.:

$$T = \frac{p^2}{2m}. \tag{3.284}$$

Using:

$$T = i\hbar \frac{\partial}{\partial t}, \quad \mathbf{p} = -i\hbar \boldsymbol{\nabla} \tag{3.285}$$

Eq. (3.284) quantizes to the free particle Schroedinger equation:

$$-\frac{\hbar^2}{2m}\boldsymbol{\nabla}^2 \phi = T\phi \tag{3.286}$$

which is the Helmholtz equation:

$$\left(\boldsymbol{\nabla}^2 + \frac{2mT}{\hbar^2}\right)\phi = 0. \tag{3.287}$$

It follows that the free particle Schroedinger equation is a Beltrami equation but with the vector potential replaced by the scalar potential ϕ, which plays the role of the wavefunction. It also follows in the non-relativistic limit that:

$$\left(\boldsymbol{\nabla}^2 + \frac{2mT}{\hbar^2}\right)\mathbf{A} = \mathbf{0}, \tag{3.288}$$

CHAPTER 3. ECE THEORY AND BELTRAMI FIELDS

so:

$$\kappa^2 = \frac{2mT}{\hbar^2}. \tag{3.289}$$

The Helmholtz equation (3.287) can be written as:

$$\left(\boldsymbol{\nabla}^2 + \kappa^2\right)\phi = 0 \tag{3.290}$$

which is an Euler Bernoulli equation without a driving term on the right hand side. In the presence of potential energy V Eq. (3.286) becomes:

$$H\phi = \left(-\frac{\hbar^2}{2m}\boldsymbol{\nabla}^2 + V\right)\phi = E\phi \tag{3.291}$$

where H is the Hamiltonian and E the total energy:

$$E = T + V \tag{3.292}$$

Eq. (3.291) is:

$$\left(\boldsymbol{\nabla}^2 + \kappa^2\right)\phi = \frac{2mV}{\hbar^2}\phi \tag{3.293}$$

which is an inhomogeneous Helmholtz equation similar to an Euler Bernoulli resonance equation with a driving term on the right hand side. However Eq. (3.293) is an eigenequation rather than an Euler Bernoulli equation as conventionally defined, but Eq. (3.293) has very well-known resonance solutions in quantum mechanics. Eq. (3.293) may be written as:

$$\left(\boldsymbol{\nabla}^2 + \kappa_1^2\right)\phi = 0 \tag{3.294}$$

where:

$$\kappa_1^2 = \frac{2m}{\hbar^2}(E - V) \tag{3.295}$$

and in UFT226 ff. on www.aias.us was used in the theory of low energy nuclear reactions (LENR). Eq. (3.294) is well known to be a linear oscillator equation which can be used to define the structure of the atom and nucleus. It can be transformed into an Euler Bernoulli equation as follows:

$$\left(\boldsymbol{\nabla}^2 + \kappa_1^2\right)\phi = A\cos(\kappa_2 Z) \tag{3.296}$$

where the right hand side represents a vacuum potential. It is exactly the structure obtained from the ECE Coulomb law as argued already.

3.4 The Beltrami Equation for Linear Momentum

The free particle Schroedinger equation can be obtained from the Beltrami equation for momentum:

$$\boldsymbol{\nabla} \times \mathbf{p} = \kappa \mathbf{p} \tag{3.297}$$

3.4. THE BELTRAMI EQUATION FOR LINEAR MOMENTUM

which can be developed into the Helmholtz equation:

$$\left(\nabla^2 + \kappa^2\right)\mathbf{p} = \mathbf{0} \tag{3.298}$$

if it is assumed that:

$$\nabla \cdot \mathbf{p} = 0. \tag{3.299}$$

If $\mathbf{p}$ is a linear momentum in the classical straight line then:

$$\kappa = 0. \tag{3.300}$$

In general however $\mathbf{p}$ has intricate Beltrami solutions, some of which are animated in UFT258 on www.aias.us and its animation section.

Now quantize Eq. (3.298):

$$\mathbf{p}\psi = -i\hbar\nabla\psi \tag{3.301}$$

so:

$$\left(\nabla^2 + \kappa^2\right)\nabla\psi = \mathbf{0}. \tag{3.302}$$

Use:

$$\nabla^2 \nabla \psi = \nabla \nabla^2 \psi \tag{3.303}$$

and:

$$\nabla(\kappa^2 \psi) = \kappa^2 \nabla \psi \tag{3.304}$$

assuming that:

$$\nabla \kappa = \mathbf{0} \tag{3.305}$$

to arrive at:

$$\nabla\left(\left(\nabla^2 + \kappa^2\right)\psi\right) = \mathbf{0}. \tag{3.306}$$

A possible solution is:

$$\left(\nabla^2 + \kappa^2\right)\psi = \mathbf{0} \tag{3.307}$$

which is the Helmholtz equation for the scalar ψ, the wave function of quantum mechanics. The Schroedinger equation for a free particle is obtained by applying Eq. (3.301) to:

$$E = \frac{p^2}{2m} \tag{3.308}$$

so:

$$-\frac{\hbar^2}{2m}\nabla^2\psi = E\psi \tag{3.309}$$

and:
$$\left(\nabla^2 + \frac{2Em}{\hbar^2}\right)\psi = 0. \tag{3.310}$$

Eqs. (3.307) and (3.310) are the same if:
$$\kappa^2 = \frac{2Em}{\hbar^2} \tag{3.311}$$

QED. Using the de Broglie relation:
$$\mathbf{p} = \hbar\boldsymbol{\kappa} \tag{3.312}$$

then:
$$p^2 = 2Em \tag{3.313}$$

which is Eq. (3.308), QED. Therefore the free particle Schroedinger equation is the Beltrami equation:
$$\nabla \times \mathbf{p} = \left(\frac{2Em}{\hbar^2}\right)^{1/2} \mathbf{p} \tag{3.314}$$

with:
$$\mathbf{p}\psi = -i\hbar\nabla\psi. \tag{3.315}$$

The free particle Schroedinger equation originates in the Beltrami equation.

This method can be extended to the general Schroedinger equation in which the potential energy V is present. Consider the momentum Beltrami equation (3.297) in the general case where κ depends on coordinates. Taking the curl of both sides of Eq. (3.297):
$$\nabla \times (\nabla \times \mathbf{p}) = \nabla \times (\kappa \, \mathbf{p}). \tag{3.316}$$

By vector analysis Eq. (3.316) can be developed as:
$$\nabla (\nabla \cdot \mathbf{p}) - \nabla^2 \mathbf{p} = \kappa^2 \mathbf{p} + \nabla\kappa \times \mathbf{p} \tag{3.317}$$

so:
$$\left(\nabla^2 + \kappa^2\right)\mathbf{p} = \nabla (\nabla \cdot \mathbf{p}) - \nabla\kappa \times \mathbf{p}. \tag{3.318}$$

One possible solution is:
$$\left(\nabla^2 + \kappa^2\right)\mathbf{p} = \mathbf{0} \tag{3.319}$$

and
$$\nabla (\nabla \cdot \mathbf{p}) = \nabla\kappa \times \mathbf{p}. \tag{3.320}$$

3.4. THE BELTRAMI EQUATION FOR LINEAR MOMENTUM

Eq. (3.320) implies

$$\mathbf{p} \cdot \boldsymbol{\nabla}(\boldsymbol{\nabla} \cdot \mathbf{p}) = \mathbf{p} \cdot \boldsymbol{\nabla}\kappa \times \mathbf{p} = 0. \tag{3.321}$$

Two possible solutions of Eq. (3.321) are:

$$\boldsymbol{\nabla} \cdot \mathbf{p} = 0 \tag{3.322}$$

and

$$\boldsymbol{\nabla}(\boldsymbol{\nabla} \cdot \mathbf{p}) = \mathbf{0}. \tag{3.323}$$

Using the quantum postulate (3.301) in Eq. (3.319) gives:

$$\left(\boldsymbol{\nabla}^2 + \kappa^2\right) \boldsymbol{\nabla}\psi = \mathbf{0} \tag{3.324}$$

and the Schroedinger equation [1-10]:

$$\left(\boldsymbol{\nabla}^2 + \kappa^2\right) \psi = 0. \tag{3.325}$$

From Eq. (3.325)

$$\boldsymbol{\nabla}\left(\left(\boldsymbol{\nabla}^2 + \kappa^2\right) \psi\right) = \mathbf{0} \tag{3.326}$$

i. e.

$$\left(\boldsymbol{\nabla}^2 + \kappa^2\right) \boldsymbol{\nabla}\psi + \left(\boldsymbol{\nabla}\left(\boldsymbol{\nabla}^2 + \kappa^2\right)\right) \psi = \mathbf{0}, \tag{3.327}$$

a possible solution of which is:

$$\left(\boldsymbol{\nabla}^2 + \kappa^2\right) \boldsymbol{\nabla}\psi = \mathbf{0} \tag{3.328}$$

and

$$\left(\boldsymbol{\nabla}\left(\boldsymbol{\nabla}^2 + \kappa^2\right)\right) \psi = \mathbf{0}. \tag{3.329}$$

Eq. (3.329) is Eq. (3.324), QED. Eq. (3.329) can be written as:

$$\boldsymbol{\nabla}\boldsymbol{\nabla}^2\psi + \boldsymbol{\nabla}\kappa^2\psi = \mathbf{0} \tag{3.330}$$

i. e.

$$\boldsymbol{\nabla}\left(\boldsymbol{\nabla}^2\psi + \kappa^2\psi\right) = \mathbf{0}. \tag{3.331}$$

A possible solution of Eq. (3.331) is the Schroedinger equation:

$$\left(\boldsymbol{\nabla}^2 + \kappa^2\right) \psi = 0. \tag{3.332}$$

So the Schroedinger equation is compatible with Eq. (3.324).
Eq. (3.322) gives:

$$\boldsymbol{\nabla}^2\psi = 0 \tag{3.333}$$

which is consistent with Eq. (3.332) only if:

$$\kappa^2 = 0. \tag{3.334}$$

Eq. (3.323) gives:

$$\nabla\left(\nabla^2\psi\right) = \mathbf{0} \tag{3.335}$$

where:

$$\nabla^2\psi = -\kappa^2\psi. \tag{3.336}$$

Therefore:

$$\nabla\left(\kappa^2\psi\right) = \left(\nabla\kappa^2\right)\psi + \kappa^2\nabla\psi \tag{3.337}$$

and:

$$\nabla\psi = -\left(\frac{\nabla\kappa^2}{\kappa^2}\right)\psi. \tag{3.338}$$

Therefore:

$$\nabla \cdot \nabla\psi = \nabla^2\psi = -\nabla \cdot \left(\frac{\nabla\kappa^2}{\kappa^2}\psi\right) \tag{3.339}$$

$$= -\left(\nabla \cdot \left(\frac{\nabla\kappa^2}{\kappa^2}\right)\right)\psi - \left(\frac{\nabla\kappa^2}{\kappa^2}\right)\nabla\psi.$$

From a comparison of Eqs. (3.332) and (3.339) we obtain the subsidiary condition:

$$\nabla^2\kappa^2 = \kappa^4 \tag{3.340}$$

where:

$$\kappa^2 = \frac{2m}{\hbar^2}\left(V - E\right). \tag{3.341}$$

Therefore:

$$\nabla\kappa^2 = \frac{2m}{\hbar^2}\nabla V \tag{3.342}$$

and

$$\nabla^2\kappa^2 = \frac{2m}{\hbar^2}\nabla^2 V \tag{3.343}$$

giving a quadratic constraint in $V - E$:

$$\nabla^2(V - E) = \frac{2m}{\hbar^2}(V - E)^2. \tag{3.344}$$

This can be written as a quadratic equation in E, which is a constant. E is expressed in terms of V, ∇V, and $\nabla^2 V$. Using:

$$\nabla E = \mathbf{0} \tag{3.345}$$

3.5. EXAMPLES FOR BELTRAMI FUNCTIONS

gives a differential equation in V which can be solved numerically, giving an expression for V. Finally this expression for V is used in the Schroedinger equation:

$$\left(-\frac{\hbar^2}{2m}\nabla^2 + V\right)\psi = E\psi \tag{3.346}$$

to find the energy levels of E and the wavefunctions ψ. These are energy levels and wavefunctions of the interior parton structure of an elementary particle such as an electron, proton or neutron. The well-developed methods of computational quantum mechanics can be used to find the expectation values of any property and can be applied to scattering theory, notably deep inelastic electron-electron, electron-proton and electron-neutron scattering. The data are claimed conventionally to provide evidence for quark structure, but the quark model depends on the validity of the U(1) and electroweak sectors of the standard model. In this book these sector theories are refuted in many ways.

3.5 Examples for Beltrami functions

In this section we give some examples of Beltrami fields with the corresponding graphs. We start the demonstration with a general consideration. Marsh [28] defines a general Beltrami field with cylindrical geometry by

$$\mathbf{B} = \begin{bmatrix} 0 \\ B_\theta(r) \\ B_Z(r) \end{bmatrix} \tag{3.347}$$

with cylindrical coordinates r, θ, Z. There is only an r dependence of the field components. For this to be a Beltrami field, the Beltrami condition in cylindrical coordinates

$$\nabla \times \mathbf{B} = \begin{bmatrix} \frac{1}{r}\frac{\partial B_Z}{\partial \theta} - \frac{\partial B_\theta}{\partial Z} \\ \frac{\partial B_r}{\partial Z} - \frac{\partial B_Z}{\partial r} \\ \frac{1}{r}\left(\frac{\partial(r B_\theta)}{\partial r} - \frac{\partial B_r}{\partial \theta}\right) \end{bmatrix} = \kappa \mathbf{B} \tag{3.348}$$

must hold. The divergence in cylindridal coordinates is

$$\nabla \cdot \mathbf{B} = \frac{1}{r}\frac{\partial(r B_r)}{\partial r} + \frac{1}{r}\frac{\partial B_\theta}{\partial \theta} + \frac{\partial B_Z}{\partial Z}. \tag{3.349}$$

Obviously the field (3.347) is divergence-free, a prerequisite to be a Beltrami field. Eq.(3.348) simplifies to

$$\nabla \times \mathbf{B} = \begin{bmatrix} 0 \\ -\frac{\partial B_Z}{\partial r} \\ \frac{\partial B_\theta}{\partial r} + \frac{1}{r}B_\theta \end{bmatrix} = \kappa \begin{bmatrix} 0 \\ B_\theta \\ B_Z \end{bmatrix}. \tag{3.350}$$

CHAPTER 3. ECE THEORY AND BELTRAMI FIELDS

κ can be a function in general. Here we consider the case of constant κ. From the second component of Eq.(3.350) follows

$$-\frac{\partial}{\partial r} B_Z = \kappa B_\theta \tag{3.351}$$

and from the third component

$$r \frac{\partial}{\partial r} B_\theta + B_\theta = \kappa r B_Z. \tag{3.352}$$

Integrating Eq.(3.351), inserting the result for B_Z into (3.352) gives

$$\frac{\partial}{\partial r} B_\theta + \frac{B_\theta}{r} = -\kappa^2 \int B_\theta \, dr, \tag{3.353}$$

and differentiating this equation leads to the second order differential equation

$$r^2 \frac{\partial^2}{\partial r^2} B_\theta + r \frac{\partial}{\partial r} B_\theta + \kappa^2 r^2 B_\theta - B_\theta = 0. \tag{3.354}$$

Finally we change the variable r to κr which leads to Bessel's differential equation

$$r^2 \frac{d^2}{d r^2} B_\theta(\kappa r) + r \frac{d}{d r} B_\theta(\kappa r) + (\kappa^2 r^2 - 1) B_\theta(\kappa r) = 0. \tag{3.355}$$

The solution is the Bessel function

$$B_\theta(r) = B_0 J_1(\kappa r) \tag{3.356}$$

(with a constant B_0) and from (3.351) follows

$$B_Z(r) = B_0 J_0(\kappa r). \tag{3.357}$$

This is the known solution of Reed/Marsh, scaled by the wave number κ, with longitudinal components. This solution is graphed in Fig. 3.1. The stream lines are shown in Fig. 3.2. It has to be taken in mind that stream lines show how a test particle moves in the vector field which is considered a velocity field:

$$\mathbf{x} + \Delta \mathbf{x} = \mathbf{x} + \mathbf{v}(\mathbf{x}) \, \Delta t. \tag{3.358}$$

All streamline examples are started with 9 points in parallel on the X axis so all animations should be comparable.

The general Beltrami field can be written as

$$\mathbf{v} = \kappa \, \boldsymbol{\nabla} \times (\psi \mathbf{a}) + \boldsymbol{\nabla} \times \boldsymbol{\nabla} \times (\psi \mathbf{a}) \tag{3.359}$$

where ψ is an arbitrary function, κ is a constant and $\mathbf{a}$ is a constant vector. In Fig. 3.3 we show an example with

$$\psi = \frac{1}{L^3} XYZ, \tag{3.360}$$

$$\mathbf{a} = [0, 0, 1]. \tag{3.361}$$

3.5. EXAMPLES FOR BELTRAMI FUNCTIONS

The field is coplanar to the XY plane and gives planar streamlines of hyperbolic form.

Another known solution based on Bessel functions is the Lundquist solution

$$\mathbf{v} = \begin{bmatrix} J_1(\kappa r)\lambda e^{-\lambda Z} \\ J_1(\kappa r)\alpha e^{-\lambda Z} \\ J_1(\kappa r) e^{-\lambda Z} \end{bmatrix} \tag{3.362}$$

with

$$\kappa = \sqrt{\alpha^2 + \lambda^2} \tag{3.363}$$

and constants α and λ. The Lundquist function (for $Z > 0$) is graphed in Fig. 3.4 and initially behaves similar to the Bessel case discussed above. However the field shrinks with Z due to the exponential factor. Fig. 3.5 shows a projection into the XY plane. The vectors are always rotated by 45° against the radial direction. Longitudinal parts are not visible here as discussed for the Rodriguez-Vaz case. Outer streamlines (Fig. 3.6) go down to the region $Z < 0$, and here the exponential factor $\exp(-\lambda Z)$ gives an exponential growth, this is well recognizable in the second version of this animation on www.aias.us. λ can be assumed complex-valued as discussed by Reed, leading to oscillatory solutions, but then problems can arise in other parts of the field definition.

Finally we give some graphic examples for plane waves. Although these are well known, it is useful to recall certain features that not always are considered where plane waves are used. In ECE theory their most prominent appearance is in the vector potential of the free electromagnetic field, in cyclic cartesian coordinates:

$$\mathbf{A}_1 = \frac{A_0}{\sqrt{2}} \begin{bmatrix} e^{i(\omega t - \kappa Z)} \\ -i\, e^{i(\omega t - \kappa Z)} \\ 0 \end{bmatrix}, \quad \mathbf{A}_2 = \frac{A_0}{\sqrt{2}} \begin{bmatrix} e^{i(\omega t - \kappa Z)} \\ i\, e^{i(\omega t - \kappa Z)} \\ 0 \end{bmatrix}, \quad \mathbf{A}_3 = 0. \tag{3.364}$$

Their divergence is zero and the eigenvalue of the curl operator is κ or $-\kappa$, respectively. The plane wave can also be defined as real valued:

$$\mathbf{A}_1 = \frac{A_0}{\sqrt{2}} \begin{bmatrix} \cos(\omega t - \kappa Z) \\ -\sin(\omega t - \kappa Z) \\ 0 \end{bmatrix}, \quad \mathbf{A}_2 = \frac{A_0}{\sqrt{2}} \begin{bmatrix} \sin(\omega t - \kappa Z) \\ \cos(\omega t - \kappa Z) \\ 0 \end{bmatrix}, \quad \mathbf{A}_3 = 0 \tag{3.365}$$

and are Beltrami fields also, however with positive eigenvalues for $\mathbf{A}_1$ and $\mathbf{A}_2$. The real-valued plane waves are graphed as vector fields in Fig. 3.7 for a fixed instant of time $t = 0$. $\mathbf{A}_1$ and $\mathbf{A}_2$ are perpendicular to one another and define a rotating frame in Z direction. The streamlines in one plane are all parallel straight lines. To show a variation, they have been graphed in Fig. 3.8 for different starting points on the Z axis. Here the rotation of frames can be seen again.

Streamlines of plane waves are not very instructive concerning the physical meaning of these waves. It is more illustrative to show their time behaviour.

We started with streamlines in the XY plane and computed their time evolution. The streamlines would remain in that plane so we added a Z component $v\,t$ to simulate a propagation in that direction as is the case for electromagnetic waves with $v = c$. Thus in Fig. 3.9 the trace of circularly polarized waves is obtained. Interestingly the waves are phase-shifted, although all starting points are at $Y = 0$.

In this paper we are considering plane wave in the context of Beltrami fields. As worked out the fields E, B and A are parallel. Therefore the components $\mathbf{A}_1$ and $\mathbf{A}_2$ do not demonstrate the behaviour of electric and magnetic fields of ordinary transversal electromagnetic fields which are phase-shifted by 90°. Reed [27] gives a very good explanation of this extraordinary case:

Every plane wave solution corresponds to two circularly polarized waves propagating oppositely to each other and combining to form a standing wave. This standing wave does not possess the standard power flow feature of linearly- or circularly-polarized waves with $E \perp B$, since the combined Poynting vectors of the circularly-polarized waves cancel each other similar to the situation we met earlier in connection with Beltrami plasma vortex filaments. Essentially, the combination of these two waves produces a standing wave propagating non-zero magnetic helicity. In the book by Marsh [28] the relationship is shown between the helicity and energy densities for this wave as well, as the very interesting fact that any magnetostatic solution to the FFMF equations can be used to construct a solution to Maxwell's equations with $E \| B$.

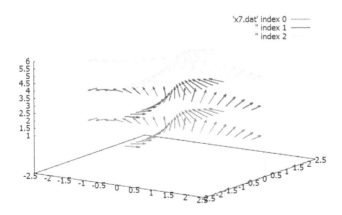

Figure 3.1: Bessel function solution.

3.5. EXAMPLES FOR BELTRAMI FUNCTIONS

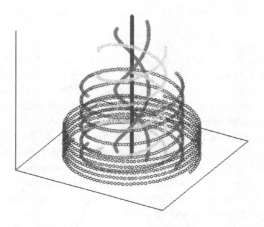

Figure 3.2: Streamlines of the Bessel function solution.

Figure 3.3: General solution with $\psi = \frac{1}{L^3}XYZ$.

CHAPTER 3. ECE THEORY AND BELTRAMI FIELDS

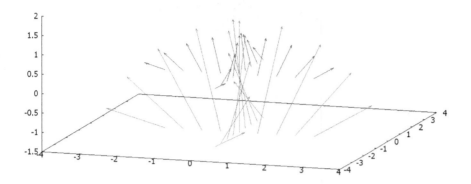

Figure 3.4: Lundquist solution.

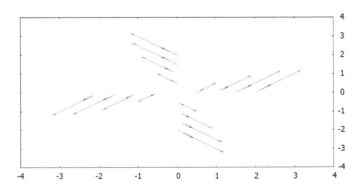

Figure 3.5: Lundquist solution, projected to XY plane.

3.5. EXAMPLES FOR BELTRAMI FUNCTIONS

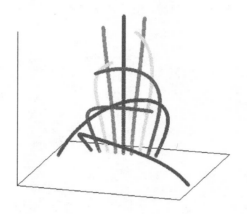

Figure 3.6: Streamlines of Lundquist solution.

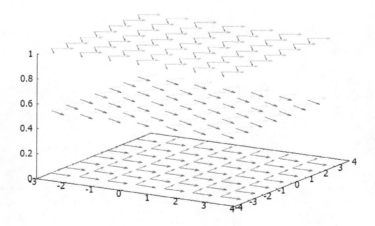

Figure 3.7: Plane wave field, $\mathbf{A}_1$ and $\mathbf{A}_2$.

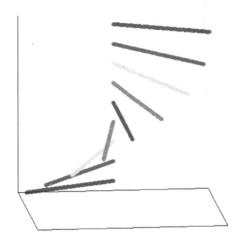

Figure 3.8: Streamlines of plane waves.

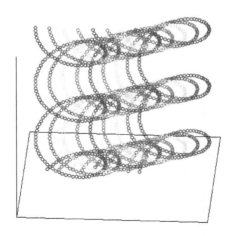

Figure 3.9: Time evolution of points transported by plane waves.

3.6 Parton Structure of Elementary Particles

We develop a solution of the constraint Schroedinger equation (3.346) on basis of the Beltrami equations developed in this chapter. The solution is applied to elementary particles and reveals their so-called Parton structure.

Solution of the constraint equation (3.340)

Before solving the Schroedinger equation (3.346), the potential is derived from the constraint equation (3.340) or (3.344), respectively. We choose the form (3.340) for κ^2 which holds for all energies E so a solution of (3.340) is universal in E. For the electron it is known that there is no angular dependence of the particle charge density. For the proton there is only a weak angular dependence. Therefore we restrict the ∇^2 operator in (340) to the radial part, giving

$$\frac{d^2}{dr^2}\kappa^2(r) + \frac{2}{r}\frac{d}{dr}\kappa^2(r) = \kappa^4(r) \tag{3.366}$$

with

$$\kappa^2 = \frac{2m(V-E)}{\hbar^2} \tag{3.367}$$

as before. When κ^2 is known, the potential is obtainable by

$$V = E + \frac{\hbar^2 \kappa^2}{2m}. \tag{3.368}$$

In order to simplify Eq.(3.366) we substitute κ by a new function λ:

$$\lambda^2(r) := r\,\kappa^2(r). \tag{3.369}$$

This is the same procedure as getting rid of the first derivative in the standard solution procedure for the radial Schroedinger equation. Eq.(3.366) then reads:

$$\frac{d^2}{dr^2}\lambda^2(r) = \frac{\lambda^4(r)}{r}. \tag{3.370}$$

The initial conditions have to be chosen as follows. Because the radial coordinate in (3.369) starts at $r=0$, we have to use $\lambda^2(0) = 0$ to be consistent. For the derivative of λ^2 follows from (3.369):

$$\frac{d\lambda^2}{dr} = \kappa^2 + 2r\frac{d\kappa}{dr}. \tag{3.371}$$

Only the first term contributes for $r=0$ so that the initial value of κ^2 determines the derivative of λ^2 at this point. In total:

$$\lambda^2(0) = 0, \tag{3.372}$$

$$\frac{d\lambda^2}{dr}(0) = \kappa^2(0). \tag{3.373}$$

If $\kappa^2(0)$ is positive, we obtain only functions with positive curvature for λ^2 and κ^2, see Fig. 3.10. The potential function is always positive and greater than zero, allowing no bound states. Both functions diverge for large r. Therefore we have to start with a negative value of $\kappa^2(0)$. Then we obtain a negative region of the potential function, beginning with a horizontal tangent. This is the same as in the Woods Saxon potential, a model potential for of atomic nuclei. There is no singularity at the origin because there is no point charge.

Numerical studies give the result that the solutions λ^2 and κ^2 are always of the type shown in Fig. 3.11. The radial scale is determined by the depth of the inital value $\kappa^2(0)$. We have chosen this value so large that the radial scale (in atomic units) is in the range of the radii of elementary particles, see Table 3.1. As an artifact, the diverging behaviour for $r \to \infty$ found previously remains for negative initial values of the potential function. Obviously κ^2 crosses zero when the derivative of λ^2 has a horizontal tangent (Fig. 3.11). It would be convenient to cut the potential at this radius.

Solution of the radial Schroedinger equation

After having dertermined the potential function κ^2 which internally depends on E, we can solve the radial Schroedinger equation derived from (3.346):

$$-\frac{\hbar^2}{2m}\frac{d^2}{dr^2}R(r) - \frac{\hbar^2}{mr}\frac{d}{dr}R(r) + V(r)R(r) = E R(r) \tag{3.374}$$

with R being the radial part of the wave function. We substitute R as usual:

$$P(r) := r R(r) \tag{3.375}$$

to obtain the simplified equation

$$\frac{d^2}{dr^2}P(r) = \frac{2m}{\hbar^2}(V(r) - E) P(r). \tag{3.376}$$

$V - E$ can be replaced by κ^2 which is already known from the constraint equation, so we have

$$\frac{d^2}{dr^2}P(r) = \frac{\lambda^2 P(r)}{r} = \kappa^2 P(r). \tag{3.377}$$

Obviously the energy parameter E is subsumed by κ. The computed κ function is valid for an arbitrary E. Since the left hand side of (3.377) is a replacement of the ∇^2 operator, the Schroedinger equation has been transformed into a Beltrami equation with variable scalar function κ^2 (assuming no divergence of P). There is no energy dependence left and the equation can be solved as an ordinary differential equation. This is a linear equation in P so that the result can be normalized arbitrarily and so can the final result R. This is the same again as for the solution procedure of the Schroedinger equation. Regarding the initial conditions, P starts at zero as discussed above and its derivative can be chosen arbitrarily, for example:

$$P(0) = 0, \tag{3.378}$$

$$\frac{dP}{dr}(0) = 1. \tag{3.379}$$

3.6. PARTON STRUCTURE OF ELEMENTARY PARTICLES

The results for R, R^2 and $R^2 r^2$ are graphed in Fig. 3.12. Again the functions have to be cut at the cut-off radius of about $2 \cdot 10^{-5}$ a.u.

Comparison with experiments

Experimental values of particle radii are listed in Table 3.1. The classical electron radius is calculated from equating the mass energy with the electrostostatic energy in a sphere and turns out to be simply

$$r_e = \alpha^2 a_0 \tag{3.380}$$

with α being the fine structure constant and a_0 the Bohr radius. This radius value is however larger than the proton radius. Therefore a more realistic calculational procedure seems to be scaling the proton radius with the mass ratio compared to the electron (second row in Table 3.1). The experimental limits are even smaller so that the accepted opinion is that the electron is a point particle which it certainly cannot be in a mathematical sense since there are no singularities in nature.

The charge density characteristics of proton and neutron are exponentially decreasing functions. This is not totally identical to the properties obtained for R^2 from our calculation (Fig. 3.13) which more looks like a Gaussian function. However, Gaussians have been observed for atomic nuclei containing more than one proton and neutron.

There is a diagram in the literature showing the charge densities for the proton and neutron [30] (replicated in Fig. 3.14). The charge densities start with zero values therefore they seem to describe the effective charge in a sphere of radius r which has to be compared with

$$\rho_e = R^2 \cdot r^2 \tag{3.381}$$

of our calculation. This function (with negative sign) has been graphed in Fig. 3.13 in the range below the cut-off radius. Since our function is not normalized the vertical scales differ. The proton has a shoulder in the charge density which is not reproduced by our calculation. The neutron is known not to be charge-neutral over the radius but to have a positive core and a negative outer region. The negative region which is called "shell" even pertains to the centre in Fig. 3.14. The shape of the shell is quite conforming to our calculation in Fig. 3.13. Some other experimental charge densities of the proton have been derived by Venkat et al. [29] and Sardin [30], see Fig. 4 therein. They compare quite well with our results for $R^2 r^2$, Fig. 3.13 of this paper.

As already stated, our calculation does not contain an explicit energy parameter, therefore we do not obtain a mass spectrum of elementary particles or partons. The diameter of effective charge is defined by the initial value of κ^2. For the results shown we had to choose $\kappa^2 = -5 \cdot 10^{10}$ a.u. which is quite a lot. The rest energy of the proton is 938 MeV or $3.5 \cdot 10^7$ a.u. which is three orders of magnitude less. Obviously the potential has to be much deeper than the (negative) rest energy.

In conclusion, the Beltrami approach of ECE theory leads to a qualitatively correct desription of the internal structure of elementary particles, in particular

CHAPTER 3. ECE THEORY AND BELTRAMI FIELDS

Particle	charge density characteristic	radius [m]	radius [a.u.]
electron (classical)	delta function	$2.82 \cdot 10^{-15}$	$5.33 \cdot 10^{-5}$
electron (derived)[a]	delta function	$9.1 \cdot 10^{-17}$	$1.72 \cdot 10^{-6}$
proton (measured)	neg. exponential function	$1.11 \cdot 10^{-15}$	$2.10 \cdot 10^{-5}$
proton (charge radius)	neg. exponential function	$8.8 \cdot 10^{-16}$	$1.66 \cdot 10^{-5}$
neutron (measuerd)	neg. exponential function	$1.7 \cdot 10^{-15}$	$3.21 \cdot 10^{-5}$
atomic nuclei	Gaussian or Fermi function	$2 - 8 \cdot 10^{-15}$	$4 - 15 \cdot 10^{-5}$

[a]Electron radius from volume comparison with $(m_{proton}/m_{electron})^{1/3}$

Table 3.1: Experimental data of elementary particles [29], [30].

the neutron. The binding energy cannot be determined since it cancels out from the calculation. It seems that the Beltrami structure is not valid in the boundary region of elementary particles or partons since the charge density does not go asymptotically to zero. This can be remedied by defining a cut-off radius where the radial function has a zero crossing. This was a first approach to compute the interior of elementary particles (the so-called parton structure) by ECE theory. For future developments more sophisticated approaches have to be found.

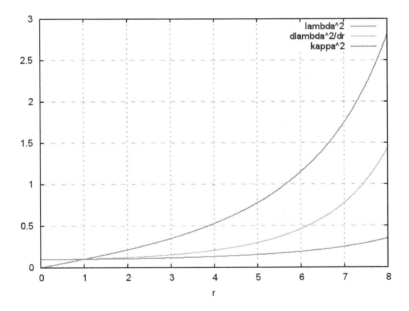

Figure 3.10: Solution functions of constraint equation (38) for $\kappa^2(0) > 0$.

3.6. PARTON STRUCTURE OF ELEMENTARY PARTICLES

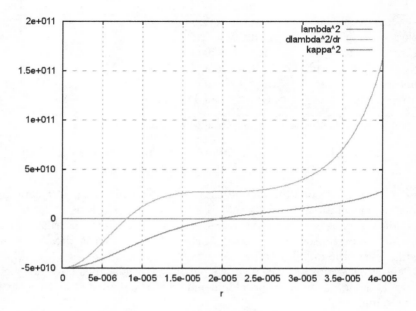

Figure 3.11: Solution functions of constraint equation (38) for $\kappa^2(0) < 0$.

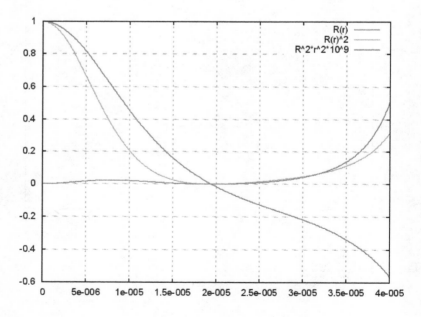

Figure 3.12: Parton solution of the Schroedinger equation.

CHAPTER 3. ECE THEORY AND BELTRAMI FIELDS

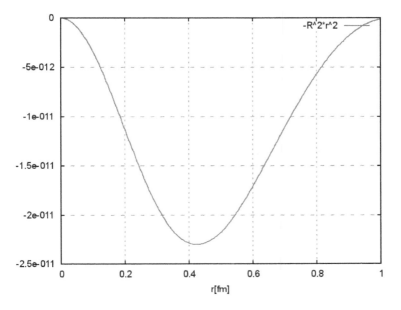

Figure 3.13: Radial wave function $-R^2 \cdot r^2$.

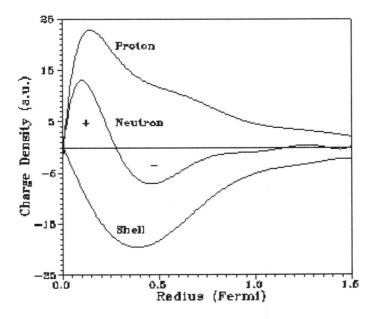

Figure 3.14: Experimental charge densities of elementary particles [30].

3.6. PARTON STRUCTURE OF ELEMENTARY PARTICLES

Chapter 4

Photon Mass and the B(3) Field

4.1 Introduction

The B(3) field was inferred in November 1991 [1]- [10] from a consideration of the conjugate product of nonlinear optics in the inverse Faraday effect. In physics before the great paradigm shift of ECE theory the conjugate product was thought to exist in free space only in a plane of two dimensions. This was absurd dogma necessitated by the need for a massless photon and the U(1) gauge invariance of the old theory [24]. The lagrangian had to be invariant under a certain type of gauge transformation. Therefore there could be no longitudinal components of the free electromagnetic field, meaning that the vector cross product known as the conjugate product could have no longitudinal component in free space, but as soon as it interacted with matter it produced an experimentally observable longitudinal magnetization. In retrospect this is grossly absurd, it defies basic geometry, the basic definition of the vector cross product in three dimensional space, or the space part of four dimensional spacetime.

The first papers on B(3) appeared in Physica B in 1992 and 1993 and can be seen in the Omnia Opera of www.aias.us. The discovery of B(3) was not immediately realized to be linked to the mass of the photon, an idea that goes back to the corpuscular theory of Newton and earlier. It was revived by Einstein as he developed the old quantum theory and special relativity, and with the inference of wave particle duality it became part of de Broglie's school of thought in the Institut Henri Poincaré in Paris. Members of this school included Proca and Vigier, whose life work was dedicated largely to the theory of photon mass and a type of quantum mechanics that rejected the Copenhagen indeterminacy. This is usually known as causal or determinist quantum mechanics. The ECE theory has clearly refuted indeterminacy in favour of causal determinism, because ECE has shown that essentially all the valid equations of physics have their origin in geometry. Indeterminism asserts that some aspects of nature are absolutely unknowable, and that there is no

4.1. INTRODUCTION

cause to an effect, and that a particle for example can do anything it likes, go forward or backward in time. To the causal determinists this is absurd and anti Baconian dogma, so they have rejected it since it was proposed, about ninety years ago. This was the first great schism in physics. The second great schism follows the emergence of ECE theory, which has split physics into dogma (the standard model) and a perfectly logical development based on geometry (ECE theory). Every effect has a cause, and the wave equations of physics are derived from geometry in a rigorously logical manner. Many aspects of the standard model have been refuted with astonishing ease. This suggests that the standard model was "not even wrong" in the words of Pauli, it was a plethora of ridiculous abstraction that could never be tested experimentally and which very few could understand. This plethora of nonsense is blasted out over the media as propaganda, doing immense harm to Baconian science. This book tries to redress some of that harm.

Vigier immediately accepted the B(3) field and in late 1992 suggested in a letter to M. W. Evans, the discoverer of B(3), that it implied photon mass because it was an experimentally observable longitudinal component of the free field and so refuted the dogma of U(1) gauge transformation. Vigier was well aware of the fact that the Proca lagrangian is not U(1) gauge invariant because of photon mass, and by 1992 had developed the subject in many directions. The subject of photon mass was as highly developed as anything in the standard physics. The two types of physics developed side by side, one being as valid as the other, but one (the standard model) being much better known. The de Broglie School of Thought was of course well known to Einstein, who invited Vigier to become his assistant, so by implication Einstein favoured the determinist school of quantum mechanics as is well known. So did Schrödinger, who worked on photon mass for many years. One of Schrödinger's last papers, with Bass, is on photon mass, from the Dublin Institute for Advanced Studies in the mid fifties. So by implication, Einstein, de Broglie and Schrödinger all rejected the standard model's U(1) gauge invariance, so they would have rejected the Higgs boson today.

The B(3) field was also accepted by protagonists of higher topology electrodynamics, three or four of whose books appear in this World Scientific series "Contemporary Chemical Physics". For example books by Lehnert and Roy, Barrett, Harmuth et al., and Crowell, and it was also accepted by Kielich, a pioneer of nonlinear optics. Other articles, notably by Reed [27] on the Beltrami fields and higher topology electrodynamics, appear in "Modern Nonlinear Optics", published in two editions and six volumes form 1992 to 2001. Piekara also worked in Paris and with Kielich, inferred the inverse Faraday effect (IFE). The latter was re-inferred by Pershan at Harvard in the early sixties and first observed experimentally in the Bloembergen School at Harvard in about 1964. The first observation used a visible frequency laser, and the IFE was confirmed at microwave frequencies by Deschamps et al. [35] in Paris in 1970 in electron plasma. So it was shown to be an ubiquitous effect that depended for its description on the conjugate product. The B(3) field was widely accepted as being a natural description of the longitudinal magnetization of the IFE.

CHAPTER 4. PHOTON MASS AND THE B(3) FIELD

Following upon the suggestion by Vigier that B(3) implied the existence of photon mass, the first attempts were made to develop O(3) electrodynamics [1]-[10], in which the indices of the complex circular basis, (1), (2) and (3), were incorporated into electrodynamics as described in earlier chapters of this book. Many aspects of U(1) gauge invariance were rejected, as described in the Omnia Opera on www.aias.us from 1993 to 2003, a decade of development. During this time, five volumes were produced by Evans and Vigier [1]- [10] in the famous van der Merwe series of "The Enigmatic Photon", a title suggested by van der Merwe himself. These are available in the Omnia Opera of www.aias.us. In the mid nineties van der Merwe had published a review article on the implications of B(3) at Vigier's suggestion, in "Foundations of Physics". This was a famous journal of avantgarde physics, one of the very few to allow publication of ideas that were not those of the standard physics.

The O(3) electrodynamics was a higher topology electrodynamics that was transitional between early B(3) theory and ECE theory, in which the photon mass and B(3) were both derived from Cartan geometry.

4.2 Derivation of the Proca Equations from ECE Theory

The Proca equation as discussed briefly in Chapter Three is the fundamental equation of photon mass theory and in this section it is derived from the tetrad postulate. The latter always gives finite photon mass in ECE theory and consider it in the format:

$$D_\mu q^a{}_\nu = \partial_\mu q^a{}_\nu + \omega^a{}_{\mu b} q^b{}_\nu - \Gamma^\lambda{}_{\mu\nu} q^a{}_\lambda = 0 \tag{4.1}$$

where $q^a{}_\nu$ is the Cartan tetrad, where $\omega^a{}_{\mu b}$ is the spin connection and $\Gamma^\lambda{}_{\mu\nu}$ is the gamma connection. Define:

$$\omega^a{}_{\mu\nu} = \omega^a{}_{\mu b} q^b{}_\nu, \tag{4.2}$$

$$\Gamma^a{}_{\mu\nu} = \Gamma^\lambda{}_{\mu\nu} q^a{}_\lambda, \tag{4.3}$$

then:

$$\partial_\mu q^a{}_\nu = \Gamma^a{}_{\mu\nu} - \omega^a{}_{\mu\nu} := \Omega^a{}_{\mu\nu}. \tag{4.4}$$

Differentiate both sides:

$$\partial^\mu \partial_\mu q^a{}_\nu = \Box q^a{}_\nu = \partial^\mu \Omega^a{}_{\mu\nu} \tag{4.5}$$

and define:

$$\partial^\mu \Omega^a{}_{\mu\nu} := -R q^a{}_\nu \tag{4.6}$$

to find the ECE wave equation:

$$(\Box + R) q^a{}_\nu = 0 \tag{4.7}$$

4.2. DERIVATION OF THE PROCA EQUATIONS FROM ECE THEORY

and the equation:

$$\partial^\mu \Omega^a{}_{\mu\nu} + R q^a{}_\nu = 0, \tag{4.8}$$

where the curvature is:

$$R = -q^\nu{}_a \partial^\mu \Omega^a{}_{\mu\nu}. \tag{4.9}$$

Now use the ECE postulate and define an electromagnetic field:

$$F^a{}_{\mu\nu} := A^{(0)} \Omega^a{}_{\mu\nu} \tag{4.10}$$

to find:

$$(\Box + R) A^a{}_\mu = 0 \tag{4.11}$$

and

$$\partial^\mu F^a{}_{\mu\nu} + R A^a{}_\nu = 0. \tag{4.12}$$

These are the Proca wave and field equations, Q. E. D.

The photon mass is defined by the curvature:

$$R = \left(\frac{mc}{\hbar}\right)^2. \tag{4.13}$$

Therefore:

$$\left(\Box + \left(\frac{mc}{\hbar}\right)^2\right) A^a{}_\mu = 0 \tag{4.14}$$

and

$$\partial^\mu F^a{}_{\mu\nu} + \left(\frac{mc}{\hbar}\right)^2 A^a{}_\nu = 0. \tag{4.15}$$

For each state of polarization a these are the Proca equations of the mid thirties. They are not U(1) gauge invariant and refute Higgs boson theory immediately, because Higgs boson theory is U(1) gauge invariant. Eq. (4.10) can be regarded as a postulate of ECE theory in which the electromagnetic field is defined by the connection $\Omega^a{}_{\mu\nu}$. By antisymmetry:

$$F^a{}_{\mu\nu} = -F^a{}_{\mu\nu} \tag{4.16}$$

and from the first Cartan structure equation:

$$T^a{}_{\mu\nu} = \partial_\mu q^a{}_\nu - \partial_\nu q^a{}_\mu + \omega^a{}_{\mu\nu} - \omega^a{}_{\nu\mu}. \tag{4.17}$$

The fundamental postulates of ECE theory are:

$$A^a{}_\mu = A^{(0)} q^a{}_\mu, \tag{4.18}$$

$$F^a{}_{\mu\nu} = A^{(0)} T^a{}_{\mu\nu}, \tag{4.19}$$

CHAPTER 4. PHOTON MASS AND THE B(3) FIELD

so:

$$F^a{}_{\mu\nu} = \partial_\mu A^a{}_\nu - \partial_\nu A^a{}_\mu + A^{(0)}\left(\omega^a{}_{\mu\nu} - \omega^a{}_{\nu\mu}\right) \quad (4.20)$$
$$= A^{(0)}\left(\Gamma^a{}_{\mu\nu} - \Gamma^a{}_{\nu\mu}\right).$$

By antisymmetry:

$$F^a{}_{\mu\nu} = 2\left(\partial_\mu A^a{}_\nu + A^{(0)}\omega^a{}_{\mu\nu}\right) \quad (4.21)$$

so:

$$F^a{}_{\mu\nu}\text{ (original)} = 2\left(F^a{}_{\mu\nu}\text{ (new)} + A^{(0)}\omega^a{}_{\mu\nu}\right). \quad (4.22)$$

The postulate (4.10) is a convenient way of deriving the two Proca equations from the tetrad postulate. In so doing:

$$R_0 = \left(\frac{m_0 c}{\hbar}\right)^2 \quad (4.23)$$

where m_0 is the rest mass of the photon. More generally define:

$$R = \left(\frac{mc}{\hbar}\right)^2 \quad (4.24)$$

where:

$$m = \gamma m_0 \quad (4.25)$$

then the de Broglie equation is generalized to:

$$E = \hbar\omega = mc^2 = \hbar c R^{1/2} \quad (4.26)$$

and the square of the mass of the moving photon is defined by the curvature:

$$m^2 = \left(\frac{\hbar}{c}\right)^2 R = \left(\frac{\hbar}{c}\right)^2 q^\nu{}_a \partial^\mu \left(\omega^a{}_{\mu\nu} - \Gamma^a{}_{\mu\nu}\right). \quad (4.27)$$

The Proca equations are discussed further in Chapter 3. The dogmatic U(1) gauge transformation of the standard physics is:

$$A^\mu \to A^\mu + \partial^\mu \chi \quad (4.28)$$

but the Proca Lagrangian in the usual standard model units is:

$$\mathcal{L} = -\frac{1}{4}F_{\mu\nu}F^{\mu\nu} + \frac{1}{2}m_0^2 A_\mu A^\mu \quad (4.29)$$

and this lagrangian is not U(1) gauge invariant because the transformation (4.28) changes it.

This fundamental problem for U(1) gauge invariance has never been resolved, and the current theory behind the Higgs boson still uses U(1) gauge invariance after many logical refutations. The result is a deep schism in physics between the scientific ECE theory and the dogmatic standard theory.

4.3 Link between Photon Mass and B(3)

The complete electromagnetic field tensor of ECE theory can be defined by:

$$F^a{}_{\mu\nu} = f^a{}_{\mu\nu} - f^a{}_{\nu\mu} + \omega^a{}_{\mu b} A^b{}_\nu - \omega^a{}_{\nu b} A^b{}_\mu \tag{4.30}$$

where:

$$A^a{}_\mu = A^{(0)} q^a{}_\mu, \quad f^a{}_{\mu\nu} = \partial_\mu A^a{}_\nu. \tag{4.31}$$

Consider now the tetrad postulate in the format:

$$\partial_\mu q^a{}_\nu = \Gamma^a{}_{\mu\nu} - \omega^a{}_{\mu\nu} := \Omega^a{}_{\mu\nu}. \tag{4.32}$$

Eq. (4.31) follows directly from the subsidiary postulate:

$$f^a{}_{\mu\nu} = A^{(0)} \Omega^a{}_{\mu\nu} \tag{4.33}$$

and as shown already in this chapter gives the Proca wave and field equations in generally covariant format. It is seen that the Proca equations are subsidiary structures of the more general nonlinear structure (4.30).

The B(3) field that is the basis of unified field theory is defined by:

$$B^a{}_{\mu\nu} = -ig\left(A^c{}_\mu A^b{}_\nu - A^c{}_\nu A^b{}_\mu\right) = \omega^a{}_{\mu b} A^b{}_\nu - \omega^a{}_{\nu b} A^b{}_\mu \tag{4.34}$$

and is derived from the non linear part of the complete field tensor (4.30). In the B(3) theory:

$$\omega^a{}_{\mu b} = -ig A^c{}_\mu \epsilon^a{}_{bc}. \tag{4.35}$$

Now define for each polarization index a:

$$g^{\mu\nu} = \partial^\mu A^\nu - \partial^\nu A^\mu. \tag{4.36}$$

It follows that:

$$\partial^\rho g^{\mu\nu} + \partial^\nu g^{\rho\mu} + \partial^\mu g^{\nu\rho} = 0. \tag{4.37}$$

This equation is the same as:

$$\partial^\mu \widetilde{g}_{\mu\nu} = 0 \tag{4.38}$$

where the tilde denotes the Hodge dual. It follows that:

$$\partial^\mu \widetilde{f}_{\mu\nu} = 0 \tag{4.39}$$

which is the homogenous field equation of the Proca structure. Eq. (4.32) allows the description of the Aharonov Bohm effects [1]- [10] with the assumption:

$$\Gamma^a{}_{\mu\nu} = \omega^a{}_{\mu\nu}. \tag{4.40}$$

CHAPTER 4. PHOTON MASS AND THE B(3) FIELD

With this assumption the potential is non zero when the field is zero. In UFT 157 on www.aias.us the following relation was derived for each polarization index a:

$$j^\mu = -\frac{R}{\mu_0} A^\mu \tag{4.41}$$

where the charge current density is:

$$j^\mu = (c\rho, \mathbf{J}) \tag{4.42}$$

and where:

$$A^\mu = \left(\frac{\phi}{c}, \mathbf{A}\right). \tag{4.43}$$

Here μ_0 is the vacuum permeability and ϵ_0 is the vacuum permittivity. So:

$$\rho = -\epsilon_0 R \phi \tag{4.44}$$

and:

$$\mathbf{J} = -\frac{R}{\mu_0} \mathbf{A} \tag{4.45}$$

where ρ is the charge density, ϕ is the scalar potential, $\mathbf{J}$ is the current density and $\mathbf{A}$ is the vector potential. A list of S. I. Units was given earlier in this book, and the units of the vacuum permeability are:

$$[\mu_0] = \mathrm{J\, s^2 c^{-2}\, m^{-1}}. \tag{4.46}$$

The complete set of equations of the Proca structure is therefore:

$$f^a{}_{\mu\nu} = A^{(0)}\left(\Gamma^a{}_{\mu\nu} - \omega^a{}_{\mu\nu}\right) \tag{4.47}$$

$$\partial^\mu f^a{}_{\mu\nu} + R A^a{}_\nu = 0 \tag{4.48}$$

$$(\Box + R) q^a{}_\mu = 0 \tag{4.49}$$

$$\partial^\mu F^a{}_{\mu\nu} = \Box A^a{}_\nu = -R A^a{}_\nu = \mu_0 j^a{}_\nu \tag{4.50}$$

$$\partial^\mu \tilde{f}^a{}_{\mu\nu} = 0 \tag{4.51}$$

$$j^\mu = -\frac{R}{\mu_0} A^\mu. \tag{4.52}$$

Now define the field tensor and its Hodge dual as:

$$f_{\mu\nu} = \begin{bmatrix} 0 & E_X/c & E_Y/c & E_Z/c \\ -E_X/c & 0 & -B_Z & B_Y \\ -E_Y/c & B_Z & 0 & -B_X \\ -E_Z/c & -B_Y & B_X & 0 \end{bmatrix}, \tag{4.53}$$

$$\tilde{f}_{\mu\nu} = \begin{bmatrix} 0 & B_X & B_Y & B_Z \\ -B_X & 0 & E_Z/c & -E_Y/c \\ -B_Y & -E_Z/c & 0 & E_X/c \\ -B_Z & E_Y/c & -E_X/c & 0 \end{bmatrix}.$$

4.3. LINK BETWEEN PHOTON MASS AND B(3)

These definitions give the inhomogeneous Proca field equation under all conditions, including the vacuum:

$$\nabla \cdot \mathbf{E} = \rho/\epsilon_0 = -R\phi \tag{4.54}$$

$$\nabla \times \mathbf{B} - \frac{1}{c^2}\frac{\partial \mathbf{E}}{\partial t} = \mu_0 \mathbf{J} = -R\mathbf{A} \tag{4.55}$$

and the homogenous field equations:

$$\nabla \cdot \mathbf{B} = 0 \tag{4.56}$$

$$\nabla \times \mathbf{E} + \frac{\partial \mathbf{B}}{\partial t} = 0 \tag{4.57}$$

under all conditions.

The solution of Eq. (4.54) is:

$$\phi = \frac{1}{\epsilon_0}\int \frac{\rho\, d^3\mathbf{x}'}{|\mathbf{x}-\mathbf{x}'|} \tag{4.58}$$

and from Eqs. (4.54) and (4.58):

$$\phi = -\frac{\rho}{\epsilon_0 R} = \frac{1}{\epsilon_0}\int \frac{\rho\, d^3\mathbf{x}'}{|\mathbf{x}-\mathbf{x}'|} \tag{4.59}$$

so:

$$\int \frac{\rho\, d^3\mathbf{x}'}{|\mathbf{x}-\mathbf{x}'|} = -\frac{\rho}{R} \tag{4.60}$$

where:

$$R = -q^a{}_\nu \partial^\mu \left(\Gamma^a{}_{\mu\nu} - \omega^a{}_{\mu\nu}\right). \tag{4.61}$$

Therefore:

$$\int \frac{\rho(\mathbf{x}')\, d^3\mathbf{x}'}{|\mathbf{x}-\mathbf{x}'|} = \frac{\rho}{q^\nu{}_a \partial^\mu(\omega^a{}_{\mu\nu} - \Gamma^a{}_{\mu\nu})}. \tag{4.62}$$

The original Proca equation of the thirties assumed that:

$$q^\nu{}_a \partial^\mu \left(\omega^a{}_{\mu\nu} - \Gamma^a{}_{\mu\nu}\right) = \left(\frac{m_0 c}{\hbar}\right)^2 \tag{4.63}$$

where m_0 is the rest mass. For electromagnetic fields in the vacuum this was assumed to be the photon rest mass, so the Proca equations were assumed to be equations of a boson with finite mass. More generally in particle physics this can be any boson. In Proca theory therefore the electromagnetic field is associated with a massive boson (i. e. a photon that has mass). Therefore the original Proca equations of the thirties assumed:

$$\phi = \frac{1}{\epsilon_0}\left(\frac{\hbar}{m_0 c}\right)^2 \rho. \tag{4.64}$$

It follows that:

$$\int \frac{\rho\, d^3\mathbf{x}'}{|\mathbf{x}-\mathbf{x}'|} = \left(\frac{\hbar}{m_0 c}\right)^2 \rho. \qquad (4.65)$$

From Eqs. (4.59) and (4.65):

$$\phi(\text{vac}) = \frac{1}{\epsilon_0}\left(\frac{\hbar}{m_0 c}\right)^2 \rho(\text{vac}) \qquad (4.66)$$

giving the photon rest mass as the ratio:

$$m_0^2 = \left(\frac{\hbar}{c}\right)^2 \frac{1}{\epsilon_0}\frac{\rho(\text{vac})}{\phi(\text{vac})} = 1.4 \times 10^{-74}\frac{\rho(\text{vac})}{\phi(\text{vac})}. \qquad (4.67)$$

Two independent experiments are needed to find $\rho(\text{vac})$ and $\phi(\text{vac})$. A list of experiments used to determine photon mass is given in ref. [37]. However, in this section the assumptions used in these determinations are examined carefully, and in the main, they are shown to be untenable. Later in this chapter a new method of determining photon mass, based on Compton scattering, will be given.

Conservation of charge current density for each polarization index a means that:

$$\partial_\mu j^\mu = 0. \qquad (4.68)$$

From Eqs. (4.68) and (4.52):

$$\partial_\mu A^\mu = 0. \qquad (4.69)$$

In the standard physics Eq. (4.69) is known as the Lorenz gauge, an arbitrary assumption. In the Proca photon mass theory the Lorenz gauge is derive analytically. In the Proca theory the 4-potential is physical, and the U(1) gauge invariance is refuted completely. In consequence, Higgs boson theory collapses.

From the well known radiative corrections [1]- [10] it is known experimentally that the vacuum contains charge current density. It follows directly from Eq. (4.52) that the vacuum also contains a 4-potential associated with photon mass. Therefore there are vacuum fields which in the non linear ECE theory include the B(3) field. The latter therefore also exists in the vacuum and is linked to photon mass and Proca theory. In the standard dogma the assumption of zero photon mass means that the vacuum fields only have transverse components. This is of course geometrical nonsense, and leads to the unphysical E(2) little group [24] of the Poincaré group. The vacuum 4-potential is:

$$A^\mu(\text{vac}) = \left(\frac{\phi(\text{vac})}{c}, \mathbf{A}(\text{vac})\right). \qquad (4.70)$$

It follows that a circuit can pick up the vacuum 4-potential via the inhomogeneous proca equations

$$\nabla \cdot \mathbf{E} = -R\phi(\text{vac}) \qquad (4.71)$$

4.3. LINK BETWEEN PHOTON MASS AND B(3)

and:

$$\nabla \times \mathbf{B} - \frac{1}{c^2}\frac{\partial \mathbf{E}}{\partial t} = -R\mathbf{A}(\text{vac}). \tag{4.72}$$

In this process, total energy is conserved through the relevant Poynting theorem derived as follows. Multiply Eq. (4.72) by $\mathbf{E}$:

$$\mathbf{E} \cdot (\nabla \times \mathbf{B}) - \frac{1}{c^2}\mathbf{E} \cdot \frac{\partial \mathbf{E}}{\partial t} = -R\mathbf{E} \cdot \mathbf{A}(\text{vac}). \tag{4.73}$$

Use:

$$\mathbf{E} \cdot \nabla \times \mathbf{B} = -\nabla \cdot \mathbf{E} \times \mathbf{B} + \mathbf{B} \cdot \nabla \times \mathbf{E} \tag{4.74}$$

in Eq. (4.73) to find the Poynting theorem of conservation of total energy density:

$$\frac{\partial W}{\partial t} + \nabla \cdot \mathbf{S} = \frac{R}{\mu_0}\mathbf{E} \cdot \mathbf{A}(\text{vac}). \tag{4.75}$$

The electromagnetic energy density in joules per metres cubed is:

$$W = \frac{1}{2}\left(\epsilon_0 E^2 + \frac{1}{\mu_0}B^2\right) \tag{4.76}$$

and the Poynting vector is:

$$\mathbf{S} = \frac{1}{\mu_0}\mathbf{E} \times \mathbf{B}. \tag{4.77}$$

Eq. (4.76) defines the electromagnetic energy density available from the vacuum, more accurately spacetime. This process is governed by the Poynting Theorem (4.75) and therefore there is conservation of total energy, there being electromagnetic energy density in the vacuum. The relevant electromagnetic field tensor is:

$$f^a{}_{\mu\nu} = \partial_\mu A^a{}_\nu \tag{4.78}$$

so either:

$$\mathbf{E} = -\nabla\phi \tag{4.79}$$

or:

$$\mathbf{E} = -\frac{\partial \mathbf{A}}{\partial t}. \tag{4.80}$$

The antisymmetry of the Cartan torsion means that the complete non-linear field of Eq. (4.30) is antisymmetric:

$$F^a{}_{\mu\nu} = -F^a{}_{\nu\mu} = f^a{}_{\mu\nu} - f^a{}_{\nu\mu} + \omega^a{}_{\mu\nu} - \omega^a{}_{\nu\mu}. \tag{4.81}$$

CHAPTER 4. PHOTON MASS AND THE B(3) FIELD

The Cartan torsion is defined by:

$$T^a{}_{\mu\nu} = q^a{}_\lambda T^\lambda{}_{\mu\nu} \tag{4.82}$$

where the antisymmetric torsion tensor $T^\lambda{}_{\mu\nu}$ is defined by the commutator of covariant derivatives:

$$[D_\mu, D_\nu] V^\rho = -T^\lambda{}_{\mu\nu} D_\lambda V^\rho + R^\rho{}_{\sigma\mu\nu} V^\sigma. \tag{4.83}$$

The torsion tensor is defined by the difference of antisymmetric connections:

$$T^\lambda{}_{\mu\nu} = \Gamma^\lambda{}_{\mu\nu} - \Gamma^\lambda{}_{\nu\mu} \tag{4.84}$$

and the tetrad postulate means that:

$$\Gamma^a{}_{\mu\nu} = -\Gamma^a{}_{\nu\mu} = \partial_\mu q^a{}_\nu + \omega^a{}_{\mu\nu}. \tag{4.85}$$

It follows that the antisymmetry in Eq. (4.30) is defined by:

$$f^a{}_{\mu\nu} + \omega^a{}_{\mu b} A^b{}_\nu = -\left(f^a{}_{\nu\mu} + \omega^a{}_{\nu b} A^b{}_\mu\right). \tag{4.86}$$

If Eq. (4.79) is used for the sake of argument then the Poynting Theorem becomes:

$$\frac{\partial W}{\partial t} + \boldsymbol{\nabla} \cdot \mathbf{S} = -\frac{1}{2}\frac{R}{\mu_0}\frac{\partial}{\partial t}\left(A^2(\text{vac})\right). \tag{4.87}$$

From Eq. (4.45):

$$\mathbf{A}(\text{vac}) = -\frac{\mu_0}{R}\mathbf{J}(\text{vac}) \tag{4.88}$$

so we arrive at:

$$\frac{\partial W}{\partial t} + \boldsymbol{\nabla} \cdot \mathbf{S} = -\frac{1}{2}\mu_0 R \frac{\partial}{\partial t}\left(\frac{J^2(\text{vac})}{R}\right) \tag{4.89}$$

which shows that the vacuum energy density and vacuum Poynting vector are derived from the time derivative of the vacuum current density squared divided by R.

In practical applications we are interested in transferring the electromagnetic energy density of the vacuum to a circuit which can use the energy density. In an isolated circuit consider the equation:

$$\Box A^a{}_\mu = \mu_0 j^a{}_\mu. \tag{4.90}$$

When the circuit interacts with the vacuum:

$$j^a{}_\mu \to j^a{}_\mu + j^a{}_\mu(\text{vac}) \tag{4.91}$$

so the Proca equation becomes:

$$\Box A^a{}_\mu = \mu_0 \left(j^a{}_\mu + j^a{}_\mu(\text{vac})\right) \tag{4.92}$$

4.3. LINK BETWEEN PHOTON MASS AND B(3)

and

$$\partial^\mu F^a{}_{\mu\nu} = \mu_0 \left(j^a{}_\mu + j^a{}_\mu(\text{vac})\right). \tag{4.93}$$

The Coulomb law is modified to:

$$\boldsymbol{\nabla} \cdot \mathbf{E} = \frac{1}{\epsilon_0}(\rho(\text{circuit}) + \rho(\text{vac})) \tag{4.94}$$

and the equation governing the scalar potential is:

$$(\Box + R)\phi = \frac{\rho(\text{vac})}{\epsilon_0}. \tag{4.95}$$

The d'Alembertian operator is defined by:

$$\Box = \frac{1}{c^2}\frac{\partial^2}{\partial t^2} - \nabla^2. \tag{4.96}$$

The time dependent part of ϕ of the circuit is therefore defined by:

$$\frac{1}{c^2}\frac{\partial^2 \phi}{\partial t^2} + R\phi = \frac{\rho(\text{vac})}{\epsilon_0}. \tag{4.97}$$

The most fundamental unit of mass of the circuit is the electron mass m_e, whose rest angular frequency is defined by the de Broglie wave particle dualism:

$$R_e = \left(\frac{m_e c}{\hbar}\right)^2 = \frac{\omega_e^2}{c^2} = q^\nu{}_a \partial^\mu \left(\omega^a{}_{\mu\nu} - \Gamma^a{}_{\mu\nu}\right). \tag{4.98}$$

So Eq. (4.97) becomes:

$$\frac{\partial^2 \phi}{\partial t^2} + \omega_e^2 \phi = \frac{c^2 \rho(\text{vac})}{\epsilon_0} \tag{4.99}$$

which is an Euler Bernoulli resonance equation provided that:

$$\frac{c^2 \rho(\text{vac})}{\epsilon_0} = A \cos \omega t. \tag{4.100}$$

The solution of the Euler Bernoulli equation

$$\frac{\partial^2 \phi}{\partial t^2} + \omega_e^2 \phi = A \cos \omega t \tag{4.101}$$

is well known to be:

$$\phi(t) = \frac{A \cos \omega t}{\left(\omega_e^2 - \omega^2\right)^{1/2}}. \tag{4.102}$$

At resonance:

$$\omega_e = \omega \tag{4.103}$$

CHAPTER 4. PHOTON MASS AND THE B(3) FIELD

and the circuit's scalar potential becomes infinite for all A, however tiny in magnitude. This allows the circuit design of a device to pick up practical quantities of electromagnetic radiation density from the vacuum by resonance amplification. The condenser plates used to observe the well known Casimir effect can be incorporated in the circuit design as in previous work by Eckardt, Lindstrom and others [36].

From Eqs. (4.41) and (4.44)

$$\frac{c^2 \rho(\text{vac})}{\epsilon_0} = -c^2 R \phi(\text{vac}) \tag{4.104}$$

and if we consider the space part of the scalar potential ϕ then:

$$\Box \to -\nabla^2 \tag{4.105}$$

and for each polarization index a the Proca equation reduces to:

$$\nabla^2 \phi = \left(\frac{mc}{\hbar}\right)^2 \phi. \tag{4.106}$$

The radial part of the Laplacian in polar coordinates is defined by:

$$\nabla^2 \phi = \frac{\partial^2 \phi}{\partial r^2} + \frac{2}{r} \frac{\partial \phi}{\partial r} \tag{4.107}$$

so there is a solution to Eq. (4.106) known as the Yukawa potential:

$$\phi = \frac{B}{r} \exp\left(-\left(\frac{mc}{\hbar}\right) r\right). \tag{4.108}$$

This solution was used in early particle physics but was discarded as unphysical. The early experiments to detect photon mass [1]- [10] all assume the validity of the Yukawa potential. However the basic equation:

$$\Box A_\mu = \mu_0 j_\mu \tag{4.109}$$

also has the solution:

$$\phi = \frac{e}{4\pi \epsilon_0} \left(\left(1 - \frac{\mathbf{n} \cdot \mathbf{v}}{c}\right) |\mathbf{r} - \mathbf{r}'|\right)^{-1}_{tr} \tag{4.110}$$

and

$$\mathbf{A} = \frac{\mu_0 e \mathbf{v}}{4\pi} \left(\left(1 - \frac{\mathbf{n} \cdot \mathbf{v}}{c}\right) |\mathbf{r} - \mathbf{r}'|\right)^{-1}_{tr} \tag{4.111}$$

which are the well known Liénard Wiechert solutions. Here tr is the retarded time defined by:

$$tr = t - \frac{1}{c} |\mathbf{r} - \mathbf{r}'|, \quad c = \frac{|\mathbf{r} - \mathbf{r}'|}{t - tr}. \tag{4.112}$$

Therefore the static potential of the Proca equation is given by Eq. (4.110) with:

$$\mathbf{v} = 0 \tag{4.113}$$

4.3. LINK BETWEEN PHOTON MASS AND B(3)

and the static vacuum charge density in coulombs per cubic metre is given by:

$$\rho(\text{vac}) = - \left(\frac{mc}{\hbar}\right)^2 \frac{1}{4\pi} \left(\frac{e}{|\mathbf{r} - \mathbf{r}'|}\right)_{tr} \quad (4.113a)$$

which is the Coulomb law for any photon mass.

This means that photon mass does not affect the Coulomb law, known to be one of the most precise laws in physics. Similarly the photon mass does not affect the Ampère Maxwell law or Ampère law. This is observed experimentally [1]- [10] with high precision, so it is concluded that the usual Liénard Wiechert solution is the physical solution, and that the Yukawa solution is mathematically correct but not physical. On the other hand the standard physics ignores the Liénard Wiechert solution, and other solutions, and asserts arbitrarily that the Yukawa solution must be used in photon mass theory. The use of the Yukawa potential means that there are deviations from the Coulomb and Ampère laws. These have never been observed so the standard physics concludes that the photon mass is zero for all practical purposes. This is an entirely arbitrary conclusion based on the anthropomorphic claim of zero photon mass, a circular argument that is invalid. The theory of this chapter shows that the Coulomb and Ampère laws are true for any photon mass, and the latter cannot be determined from these laws. In other words these laws are not affected by photon mass in the sense that their form remains the same. For example the inverse square dependence of the Coulomb law is the same for any photon mass. The concept of photon mass is not nearly as straightforward as it seems, for example UFT 244 on www.aias.us shows that Compton scattering when correctly developed gives a photon mass much different from Eq. (4.67). These are unresolved questions in particle physics because UFT 244 has shown violation of conservation of energy in the basic theory of particle scattering.

Before proceeding to the description of determination of photon mass by Compton scattering a mention is made of the origin of the idea of photon mass. This was by Henri Poincaré in his Palermo memoir submitted on July 23rd 1905, (Henri Poincaré, "Sur la Dynamique de l'Electron" Rendiconti del Circolo Matematico di Palermo, 21, 127–175 (1905)). This paper suggested that the photon velocity v could be less than c, which is the constant of the Lorentz transformation. Typically for Poincaré he introduced several new ideas in relativity, including new four vectors usually attributed to later papers of Einstein. So Poincaré can be regarded as a co pioneer of special relativity with many others. Einstein himself suggested a zero photon mass as a first tentative idea, simply because an object moving at c must have zero mass, otherwise the equations of special relativity become singular. Later, Einstein may have been persuaded by the de Broglie School in the Institut Henri Poincaré in Paris to consider finite photon mass, but this is not clear. It was therefore de Broglie who took up the idea of finite photon mass from Poincaré. He was influenced by the works of Henri Poincaré before inferring wave particle duality in 1923, when he suggested that particles such as the electron could be wave like. Confusion arises sometimes when it is asserted that the vacuum speed of light is c. This is not the meaning of c in special and general relativity, c

is the constant in the Lorentz transform. Lorentz and Poincaré had inferred the tensorial equations of electromagnetism much earlier than Einstein as is well known. They had shown that the Maxwell Heaviside equations obey the Lorentz transform. ECE has developed equations of electromagnetism that are generally covariant, and therefore also Lorentz covariant in a well defined limit. It is well known that Einstein and others were impressed by the work of de Broglie, Einstein described him famously as having lifted a corner of the veil.

Louis de Broglie proceeded to develop the theory of photon mass and causal quantum mechanics until the 1927 Solvay Conference, when indeterminism was proposed, mainly by Bohr, Heisenberg and Pauli. It was rejected by Einstein, Schrödinger, de Broglie and others. Later de Broglie returned to deterministic quantum mechanics at the suggestion of Vigier. A minority of physicists have continued to develop finite photon mass theory, setting upper limits on the magnitude of the photon mass. There are multiple problems with the idea of zero photon mass, as is well known [24]. These are discussed in comprehensive detail in the five volumes of "The Enigmatic Photon" (Kluwer, 1994–2002) by M. W. Evans and J.-P. Vigier. Wigner [24] for example showed that special relativity can be developed in terms of the Poincaré group, or extended Lorentz group. In this analysis the little group of the Poincaré group for a massless particle is the Euclidean E(2), the group of rotations and translations in a two dimensional plane. This is obviously incompatible with the four dimensions of spacetime or the three dimensions of space. The little group for a massive particle is three dimensional and physical, no longer two dimensional.

This is the most obvious problem for a massless particle, and one of its manifestations is that the electromagnetic field in free space must be transverse and two dimensional, despite the fact that the theory of electromagnetism is built on four dimensional spacetime. The massless photon can have only two senses of polarization, labelled the transverse conjugates (1) and (2) in the complex circular basis [1]- [10] used in earlier chapters. This absurd dogma took hold because of the prestige of Einstein, but prestige is no substitute for logic. The idea of zero photon mass developed into U(1) gauge invariance, which became embedded into the standard model of physics. The electromagnetic sector of standard physics is still based on U(1) gauge invariance, refuted by the B(3) field in 1992 and in comprehensive developments since then. The idea of U(1) gauge invariance is in fact refuted by the Poincaré paper of 1905 described already, and by the work of Wigner, so it is merely dogmatic, not scientific. It is refuted by effects of nonlinear optics, notably the inverse Faraday effect, and in many other ways. It was refuted comprehensively in chapter 3 by the fact that the Beltrami equations of free space electromagnetism have intricate longitudinal solutions in free space. According to the U(1) dogma, these do not exist, an absurd conclusion. Probably the most absurd idea of the U(1) dogma is the Gupta Bleuler condition, in which the time like (0) and longitudinal polarizations (3) are removed artificially [24]. There are also multiple well known problems of canonical quantization of the massless electromagnetic field. These are discussed in a standard text such as Ryder [24], and in great detail in "The Enigmatic Photon" [1]- [10]. Finally the electroweak theory, which can be described as U(1) × SU(2), was refuted completely in UFT 225.

The entire standard unified field theory depends on U(1) gauge invariance, so the entire theory is refuted as described above. Obviously there cannot be a Higgs boson.

4.4 Measurement of Photon Mass by Compton Scattering

The theory of particle scattering has been advanced greatly during the course of development of ECE theory in papers such as UFT 155 to UFT 171 on www.aias.us, reviewed in UFT 200. It has been shown that the idea of zero photon mass is incompatible with a rigorously correct theory of scattering, for example Compton scattering. This is because of the numerous problems discussed at the end of Section 4.3 – zero photon mass is incompatible with special relativity, a theory upon which traditional Compton scattering is based. In UFT 158 to UFT 171 it was found that the Einstein de Broglie equations are not self consistent, a careful scholarly examination of the theory showed up wildly inconsistent results, which were also present in equal mass electron positron scattering.

The theory of Compton scattering with finite photon mass was first given in UFT 158 to UFT 171 and the notation of those papers is used here. The relativistic classical conservation of energy equation is:

$$\gamma m_1 c^2 + m_2 c^2 = \gamma' m_1 c^2 + \gamma'' m_2 c^2 \tag{4.114}$$

where m_1 is the photon mass, m_2 is the electron mass, and where the Lorentz factors are defined by the velocities as usual. The photon mass is given by the equation first derived in UFT 160 on www.aias.us:

$$m_1^2 = \left(\frac{\hbar}{c^2}\right)^2 \left[\frac{1}{2a}\left(-b \pm (b^2 - 4ac')^{1/2}\right)\right] \tag{4.115}$$

$$a = 1 - \cos^2\theta,$$
$$b = (\omega'^2 + \omega^2)\cos^2\theta - 2A$$
$$A = \omega\omega' - x_2(\omega - \omega')$$
$$c' = A^2 - \omega^2\omega'^2\cos^2\theta$$

where ω' is the scattered gamma ray frequency, ω the incident gamma ray frequency, and where:

$$x_2 = \frac{m_2 c^2}{\hbar}. \tag{4.116}$$

Here $\hbar$ is the reduced Planck constant and c the speed of light in vacuo. The scattering angle is θ. Experimental data on Compton scattering can be used with the electron mass found in standards laboratories:

$$m_2 = 9.10953 \times 10^{-31} \text{ kg} \tag{4.117}$$

CHAPTER 4. PHOTON MASS AND THE B(3) FIELD

so:

$$x_2 = 7.76343 \times 10^{20} \text{ rad s}^{-1}. \tag{4.118}$$

The two solutions of Eq. (4.115) for photon mass are given later in this section. One solution is always real valued and this root is usually taken to be the physical value of the mass of the photon. It varies with scattering angle but is always close to the electron mass. The photon in this method is much heavier than thought previously. The other solution can be imaginary valued, and usually this solution would be discarded as unphysical. However R theory means that a real valued curvature can be found as follows:

$$R = mm^* \left(\frac{c}{\hbar}\right)^2 \tag{4.119}$$

where $*$ denotes complex conjugate. It is shown later that an imaginary valued mass can be interpreted in terms of superluminal propagation.

The velocity of the photon after it has been scattered from a stationary electron is given by the de Broglie equation:

$$\gamma' m_1 c^2 = \hbar \omega' \tag{4.120}$$

and is c for all practical purposes for all scattering angles (Section 4.3). A photon as heavy as the electron does not conflict therefore with the results of the Michelson Morley experiment but on a cosmological scale a photon as heavy as this would easily account for any mass discrepancy claimed at present to be due to "dark matter". Photon mass physics differs fundamentally from standard physics as explained in comprehensive detail [1]- [10] in the five volumes of "The Enigmatic Photon" in the Omnia Opera of www.aias.us. A photon as heavy as the electron would mean that previous attempts at assessing photon mass would have to be re-assessed as discussed already in this chapter. The Yukawa potential would have to be abandoned or redeveloped.

However the theory of the photoelectric effect can be made compatible with a heavy photon as follows. Consider a heavy photon colliding with a static electron. The energy conservation equation is:

$$\gamma m_0 c^2 + m_2 c^2 = \gamma' m_0 c^2 + \gamma'' m_2 c^2. \tag{4.121}$$

The de Broglie equation can be used as follows:

$$\hbar \omega = \gamma m_0 c^2 \tag{4.122}$$

$$\hbar \omega'' = \gamma'' m_2 c^2 \tag{4.123}$$

If the photon is stopped by the collision then the conservation of energy equation is:

$$\hbar \omega + m_2 c^2 = m_0 c^2 + \hbar \omega'' \tag{4.124}$$

where m_0 is the rest mass of the photon. This concept does not exist in the standard model because a massless photon is never at rest. So:

$$m_0 = m_2 + \frac{\hbar}{c^2} \left(\omega - \omega''\right). \tag{4.125}$$

4.4. MEASUREMENT OF PHOTON MASS BY COMPTON SCATTERING

If for the sake of argument the masses of the photon and electron are the same, then:

$$m_0 = m_2 \tag{4.126}$$

and:

$$\omega = \omega'' \tag{4.127}$$

i. e. all the energy of the photon is transferred to the electron.
If:

$$\omega \neq \omega'' \tag{4.128}$$

then:

$$\hbar(\omega - \omega'') = \Phi + (m_0 - m_2)c^2 = \Phi \tag{4.129}$$

where Φ is the binding energy of the photoelectric effect. From Eq. (4.129):

$$\hbar\omega + m_2 c^2 = m_0 c^2 + \hbar\omega'' + \Phi \tag{4.130}$$

i. e.:

$$\hbar\omega = \hbar\omega'' + \Phi = E + \Phi \tag{4.131}$$

or:

$$E = \hbar\omega - \Phi \tag{4.132}$$

which is the usual equation of the photoelectric effect, Q. E. D. The heavy photon does not disappear and transfers its energy to the electron, and the heavy photon is compatible with the photoelectric effect.

It is interesting to inspect the result for photon mass for a wider range of parameters and to see if there are "islands of stability" [33]. Starting from Eq. (4.115), there are in general four solutions for m_1, appearing in two pairs with positive and negative sign. We sorted out the negative solutions and plotted the results in a surface plot for the range of ω' and θ as obtained from an experiment [34]. The graphs are shown in Figs. 4.1 and 4.2. The areas having zero values (black) are those of imaginary mass. It can be seen that both solutions have continuous regions of well-defined values. There is even a symmetry in the angle dependence. One solution rises for increasing scattering angles while the other decreases correspondingly. There are plateaus for $m_1 \approx 1$ which is the electron mass. However a true region of constant mass does not exist, leading the de Broglie Einstein theory ad absurdum.

A major and fundamental problem for standard physics emerges from consideration of equal mass Compton scattering as described in UFT 160 on www.aias.us. It can be argued as follows that equal mass Compton scattering violates conservation of energy. Consider a particle of mass m colliding with an initially static particle of mass m. If the equations of conservation

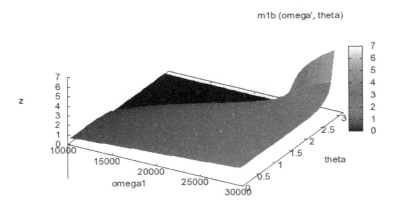

Figure 4.1: Surface plot for photon mass $m_1(\omega', \theta)$, first solution.

of energy and momentum are assumed to be true initially, they can be solved simultaneously to give:

$$x^2 + (\omega^2 - x^2)^{1/2} (\omega'^2 - x^2)^{1/2} \cos\theta = \omega\omega' - (\omega - \omega')x \quad (4.133)$$

where:

$$x = \omega_0 = \frac{mc^2}{\hbar} \quad (4.134)$$

is the rest frequency of the particle of mass m, ω' is the scattered frequency, and ω the incoming frequency of particle m colliding with an initially static particle of mass m. The scattering angle is θ and from Eq. (4.133):

$$\cos^2\theta = \frac{\omega_0^2 + \omega_0(\omega - \omega') - \omega\omega'}{\omega_0^2 - \omega_0(\omega - \omega') - \omega\omega'}. \quad (4.135)$$

In order that

$$0 \leq \cos^2\theta \leq 1 \quad (4.136)$$

then:

$$\omega < \omega'. \quad (4.137)$$

The de Broglie equation means that the collision can be described by:

$$\hbar\omega + \hbar\omega_0 = \hbar\omega' + \hbar\omega'' \quad (4.138)$$

4.4. MEASUREMENT OF PHOTON MASS BY COMPTON SCATTERING

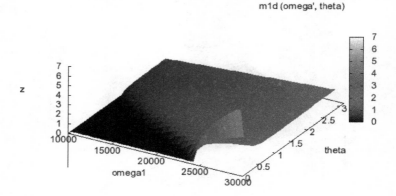

Figure 4.2: Surface plot for photon mass $m_1(\omega', \theta)$, second solution.

so:
$$\omega + \omega_0 = \omega' + \omega'' \tag{4.139}$$

and:
$$\omega - \omega' = \omega'' - \omega_0. \tag{4.140}$$

Therefore:
$$\omega'' < \omega_0. \tag{4.141}$$

From Eqs. (4.137) and (4.141):
$$\omega + \omega_0 < \omega' + \omega''. \tag{4.142}$$

However the initial conservation of energy equation is (4.139), so the theory violates conservation of energy and contradicts itself. This is a disaster for particle scattering theory because violation of conservation of energy occurs at the fundamental level. Quantum electrodynamics and string theory, or Higgs boson theory of particle scattering are invalidated.

If two particles of mass m_1 and m_2 collide and both are moving, the initial conservation of energy equation is:
$$\gamma m_1 c^2 + \gamma_2 m_2 c^2 = \gamma' m_1 c^2 + \gamma'' m_2 c^2 \tag{4.143}$$

i.e.
$$\hbar \omega + \gamma_2 m_2 c^2 = \hbar \omega' + \hbar \omega''. \tag{4.144}$$

CHAPTER 4. PHOTON MASS AND THE B(3) FIELD

Define

$$x_2 = \gamma_2 m_2 c^2 / \hbar \qquad (4.145)$$

then:

$$x_2 := \omega_2 = \omega' + \omega'' - \omega. \qquad (4.146)$$

The equation of conservation of momentum is:

$$\mathbf{p} = \mathbf{p}_1 + \mathbf{p}_2 = \mathbf{p}' + \mathbf{p}''. \qquad (4.147)$$

Solving Eqs. (4.143) and (4.147) simultaneously leads to:

$$x_2 (\omega - \omega') = \omega \omega' - \left(x_1^2 + (\omega^2 - x_1^2)^{1/2} (\omega'^2 - x_1^2)^{1/2} \cos \theta \right). \qquad (4.148)$$

For equal mass scattering:

$$\gamma_2 x (\omega - \omega') = \omega \omega' - \left(x^2 + (\omega^2 - x^2)^{1/2} (\omega'^2 - x^2)^{1/2} \cos \theta \right) \qquad (4.149)$$

where

$$x = mc^2 / \hbar. \qquad (4.150)$$

By definition:

$$\gamma_2 = \left(1 - \frac{v^2}{c^2} \right)^{-1/2} \qquad (4.151)$$

so:

$$(\omega^2 - x^2)^{1/2} (\omega'^2 - x^2)^{1/2} \cos \theta = \omega \omega' - (\omega - \omega') \left(1 - \frac{v^2}{c^2} \right)^{-1/2} x - x^2. \qquad (4.152)$$

For

$$v \ll c \qquad (4.153)$$

then:

$$\left(1 - \frac{v^2}{c^2} \right)^{-1/2} \sim 1 + \frac{1}{2} \frac{v^2}{c^2} \qquad (4.154)$$

so Eq. (4.152) is approximated by:

$$(\omega^2 - x^2)^{1/2} (\omega'^2 - x^2)^{1/2} \cos \theta = - \left((x - \omega')(x + \omega) + \frac{1}{2} \frac{v^2}{c^2} x (\omega - \omega') \right). \qquad (4.155)$$

Therefore:
$$(\omega - x)(\omega + x)(\omega' - x)(\omega' + x)\cos^2\theta = \\ (x - \omega')^2(x + \omega)^2 + \frac{v^2}{c^2}x(\omega - \omega')(x - \omega')(x + \omega) + \frac{1}{4}\frac{v^4}{c^4}x^2(\omega - \omega')^2. \quad (4.156)$$

To order (v/c):
$$\cos^2\theta = \frac{x^2 + x(\omega - \omega')(1 + v^2/c^2) - \omega\omega'}{x^2 - x(\omega - \omega') - \omega\omega'}. \quad (4.157)$$

However:
$$0 \leq \cos^2\theta \leq 1 \quad (4.158)$$

so:
$$(\omega - \omega')\left(1 + \frac{v^2}{c^2}\right) < -(\omega - \omega') \quad (4.159)$$

i.e.:
$$\omega < \omega'. \quad (4.160)$$

The conservation of energy equation (4.143) is:
$$\omega + \omega_2 = \omega' + \omega'' \quad (4.161)$$

so:
$$\omega' - \omega = \omega_2 - \omega''. \quad (4.162)$$

From Eqs. (4.160) and (4.162):
$$\omega_2 > \omega''. \quad (4.163)$$

Add Eqs. (4.160) and (4.163):
$$\omega + \omega'' < \omega' + \omega_2 \quad (4.164)$$

so conservation of energy is again violated at the fundamental level and the whole of particle scattering theory is refuted, including Higgs boson theory.

4.5 Photon Mass and Light Deflection due to Gravitation

In papers of 1923 and 1924 (L. de Broglie, Comptes Rendues, 77, 507 (1923) and Phil. Mag., 47, 446 (1924)) Louis de Broglie used the concept of photon mass to lock together the Planck theory of the photon as quantum of energy

CHAPTER 4. PHOTON MASS AND THE B(3) FIELD

and the theory of special relativity. He derived equations which are referred to as the de Broglie Einstein equations in this book. He quantized the photon momentum, producing wave particle dualism, and these papers led directly to the inference of the Schrödinger equation. In UFT 150B and UFT 155 on www.aias.us, photon mass was shown to be responsible for light deflection and time change due to gravitation and the obsolete methods of calculating these phenomena were shown to be incorrect in many ways. This is an example of a pattern in which the ECE theory as it developed made the old physics entirely obsolete. Photon mass emerged as one of the main counter examples to standard physics – the Higgs boson does not exist because of finite photon mass, which also implies that there is a cosmological red shift without an expanding universe. Therefore photon mass also refutes Big Bang, as does spacetime torsion [1]- [10]. The red shift can be derived from the original 1924 de Broglie Einstein equations without any further assumption and the de Broglie Einstein equations can be derived from Cartan geometry (chapter 1).

The existence of photon mass can be proven as in UFT 157 on www.aias.us with light deflection due to gravitation using the Planck distribution for one photon. The result is consistent with a photon mass of about 10^{-51} kg for a light beam heated to 2,500 K as it grazes the sun and this result is one of the ways of proving photon mass, inferred by the B(3) field. Prior to this result, estimates of photon mass had been given as less than an upper bound of about 10^{-52} kg, and many methods assumed the validity of the Yukawa potential. These methods have been criticized earlier in this chapter. The Einsteinian theory of light deflection due to gravitation used zero photon mass and is riddled with errors as shown in UFT 150B and UFT 155. Therefore the experimental data on light deflection due to gravitation were thoroughly re-interpreted in UFT 157 to give a reasonable estimate of photon mass. Once photon mass is accepted it works its way through in to all the experiments that originally signalled the onset of quantum theory in the late nineteenth century: black body radiation, specific heats, the photoelectric effect, atomic and molecular spectra, and in the nineteen twenties, Compton scattering. As already argued in the context of the Proca equation, photon mass indicates the existence of a vacuum potential, which can be amplified by spin connection resonance to produce energy from spacetime.

The de Broglie Einstein equations are valid in the classical limit of the Proca wave equation of special relativistic quantum mechanics. It has already been shown that the Proca equation is a limit of the ECE wave equation obtained from the tetrad postulate of Cartan geometry and the development of wave equations from the tetrad postulate provides the long sought for unification of gravitational theory and quantum mechanics. The ECE equation of quantum electrodynamics is:

$$(\Box + R) A^a{}_\mu = 0 \tag{4.165}$$

where R is a well defined scalar curvature and where:

$$A^a{}_\mu = A^{(0)} q^a{}_\mu. \tag{4.166}$$

Here $A^{(0)}$ is the scalar potential magnitude and $q^a{}_\mu$ is the Cartan tetrad defined

4.5. PHOTON MASS AND LIGHT DEFLECTION DUE TO...

in chapter one. Eq. (4.165) reduces to the 1934 Proca equation in the limit:

$$R \to \left(\frac{mc}{\hbar}\right)^2 \tag{4.167}$$

where m is the mass of the photon, c is a universal constant, and $\hbar$ is the reduced Planck constant. Note carefully that c is not the velocity of the photon of mass m, and following upon the Palermo memoir of Poincaré, de Broglie interpreted c as the maximum velocity available in special relativity.

Eq. (4.165) in the classical limit is the Einstein energy equation:

$$p^\mu p_\mu = m^2 c^2 \tag{4.168}$$

where:

$$p^\mu = \left(\frac{E}{c}, \mathbf{p}\right) \tag{4.169}$$

and where m is the mass of the photon. Here E is the relativistic energy:

$$E = \gamma m c^2 \tag{4.170}$$

and $\mathbf{p}$ is the relativistic momentum:

$$\mathbf{p} = \gamma m \mathbf{v}_g. \tag{4.171}$$

The factor γ is the result of the Lorentz transformation and was denoted by de Broglie as:

$$\gamma = \left(1 - \frac{v_g^2}{c^2}\right)^{-1/2} \tag{4.172}$$

where v_g is the group velocity:

$$v_g = \frac{\partial \omega}{\partial \kappa}. \tag{4.173}$$

The de Broglie Einstein equations are:

$$p_\mu = \hbar \kappa^\mu \tag{4.174}$$

where the 4-wavenumber is:

$$\kappa^\mu = \left(\frac{\omega}{c}, \boldsymbol{\kappa}\right). \tag{4.175}$$

Eq. (4.174) is a logically inevitable consequence of the Planck theory of the energy quantum of light later called "the photon", published in 1901, and the theory of special relativity. The standard model has attempted to reject the inexorable logic of Eq. (4.174) by rejecting m. Eq. (4.174) can be written out as:

$$E = \hbar \omega = \gamma m c^2 \tag{4.176}$$

CHAPTER 4. PHOTON MASS AND THE B(3) FIELD

and:

$$\mathbf{p} = \hbar \boldsymbol{\kappa} = \gamma m \mathbf{v}_g. \tag{4.177}$$

In his original papers of 1923 and 1924 de Broglie defined the velocity in the Lorentz transformation as the group velocity, which is the velocity of the envelope of two or more waves:

$$v_g = \frac{\Delta \omega}{\Delta \kappa} = \frac{\omega_2 - \omega_1}{\kappa_2 - \kappa_1}, \tag{4.178}$$

and for many waves Eq. (4.173) applies. The phase velocity v_p was defined by de Broglie as:

$$v_p = \frac{E}{p} = \frac{\omega}{\kappa}, \tag{4.179}$$

$$v_g v_p = c^2,$$

which is an equation independent of the Lorentz factor γ and universally valid. The standard model makes the arbitrary and fundamentally erroneous assumptions:

$$m =? \ 0, \ v_g = v_p =? \ c. \tag{4.180}$$

In physical optics the phase velocity is defined by:

$$v_p = \frac{\omega}{\kappa} = \frac{c}{n} \tag{4.181}$$

where $n(\omega)$ is the frequency dependent refractive index, in general a complex quantity (UFT 49, UFT 118 and OO 108 in the Omnia Opera on www.aias.us). The group velocity in physical optics is:

$$v_g = c \left(n + \omega \frac{dn}{d\omega} \right)^{-1} \tag{4.182}$$

and it follows that:

$$v_p v_g = c^2 = \frac{c^2}{n \left(n + \omega \frac{dn}{d\omega} \right)} \tag{4.183}$$

giving the differential equation:

$$\frac{dn}{d\omega} = -\frac{n}{2\omega}. \tag{4.184}$$

A solution of this equation is

$$n = \frac{D}{\omega^{1/2}} \tag{4.185}$$

where D^2 is a constant of integration with the units of angular frequency. So:

$$n = \left(\frac{\omega_0}{\omega} \right)^{1/2} \tag{4.186}$$

103

where ω_0 is a characteristic angular frequency of the electromagnetic radiation. Eq. (4.186) has been derived directly from the original papers of de Broglie [1]-[10] using only the equations (4.181) and (4.182) of physical optics or wave physics. The photon mass does not appear in the final Eq. (4.186) but the photon mass is basic to the meaning of the calculation. If ω_0 is interpreted as the emitted angular frequency of light in a far distant star, then ω is the angular frequency of light reaching the observer. If:

$$n > 1 \tag{4.187}$$

then:

$$\omega < \omega_0. \tag{4.188}$$

and the light has been red shifted, meaning that its observable angular frequency (ω) is lower than its emitted angular frequency (ω_0), and this is due to photon mass, not an expanding universe. The refractive index $n(\omega)$ is that of the spacetime between star and observer. Therefore in 1924 de Broglie effectively explained the cosmological red shift in terms of photon mass. "Big Bang" (a joke coined by Hoyle) is now known to be erroneous in many ways, and was the result of imposed and muddy pathology supplanting the clear science of de Broglie.

In 1924 de Broglie also introduced the concept of least (or "rest") angular frequency:

$$\hbar\omega_0 = mc^2 \tag{4.189}$$

and kinetic angular frequency ω_κ. The latter can be defined in the non relativistic limit:

$$\hbar\omega = mc^2 \left(1 - \frac{v_g^2}{c^2}\right)^{-1/2} \sim mc^2 + \frac{1}{2}mv_g^2 \tag{4.190}$$

so:

$$\hbar\omega_\kappa \sim \frac{1}{2}mv_g^2. \tag{4.191}$$

Similarly, in the non relativistic limit:

$$\hbar\kappa \sim mv_g + \frac{1}{2}m\frac{v_g^3}{c^2}, \tag{4.192}$$

so the least wavenumber, κ_0, is:

$$\hbar\kappa_0 \sim mv_g \tag{4.193}$$

and the kinetic wavenumber is:

$$\hbar\kappa_\kappa \sim \frac{1}{2}m\frac{v_g^3}{c^2}. \tag{4.194}$$

The total angular frequency in this limit is:

$$\omega = \omega_0 + \omega_\kappa \qquad (4.195)$$

and the total wavenumber is:

$$\kappa = \kappa_0 + \kappa_\kappa. \qquad (4.196)$$

The kinetic energy of the photon was defined by de Broglie by omitting the least (or "rest") frequency:

$$T = \hbar\omega_\kappa \sim \frac{1}{2}mv_g^2 = \frac{p^2}{2m} \qquad (4.197)$$

where:

$$p = mv_g. \qquad (4.198)$$

Using Eqs. (4.189) and (4.193) it is found that:

$$v_p = \frac{c^2}{v_g} = \frac{\omega_0}{\kappa_0} \qquad (4.199)$$

and using Eqs. (4.191) and (4.194)

$$v_p = \frac{c^2}{v_g} = \frac{\omega_\kappa}{\kappa_\kappa}. \qquad (4.200)$$

Therefore:

$$v_p = \frac{\omega}{\kappa} = \frac{\omega_0 + \omega_\kappa}{\kappa_0 + \kappa_\kappa} \qquad (4.201)$$

a possible solution of which is:

$$\frac{\omega_\kappa}{\kappa_0} = v_p. \qquad (4.202)$$

Using Eqs. (4.193) and (4.191):

$$\frac{\omega_\kappa}{\kappa_0} = \frac{1}{2}v_g \qquad (4.203)$$

so it is found that in these limits:

$$v_g = 2v_p. \qquad (4.204)$$

The work of de Broglie has been extended in this chapter to give a simple derivation of the cosmological red shift due to the existence of photon mass, and conversely, the red shift is a cosmological proof of photon mass. In standard model texts, photon mass is rarely discussed, and the work of de Broglie is distorted and never cited properly. The current best estimate of photon mass

4.5. PHOTON MASS AND LIGHT DEFLECTION DUE TO...

is of the order of 10^{-52} kg. In UFT 150B and UFT 155 on www.aias.us the photon mass from light deflection was calculated as:

$$m = \frac{R_0}{c^2 a} E \qquad (4.205)$$

using:

$$E = \hbar\omega. \qquad (4.206)$$

This gave the result:

$$m = 3.35 \times 10^{-41} \text{ kg}. \qquad (4.207)$$

Here R_0 is the distance of closest approach, taken to be the radius of the sun:

$$R_0 = 6.955 \times 10^8 \text{ m} \qquad (4.208)$$

and a is a distance parameter computed to high accuracy:

$$a = 3.3765447822 \times 10^4 \text{ m}. \qquad (4.209)$$

In a more complete theory, given here, the photon in a light beam grazing the sun has a mean energy given by the Planck distribution [1]- [10]:

$$\langle E \rangle = \hbar\omega \left(\frac{e^{-\hbar\omega/(kT)}}{1 - e^{-\hbar\omega/(kT)}} \right) \qquad (4.210)$$

where k is Boltzmann's constant and T the temperature of the photon. It is found that a photon mass of:

$$m = 9.74 \times 10^{-52} \text{ kg} \qquad (4.211)$$

is compatible with a temperature of 2,500 K. The temperature of the photosphere at the sun's surface is 5,778 K, while the temperature of the sun's corona is 1–3 million K. Using Eq. (4.176) it is found that:

$$v_g = 2.99757 \times 10^8 \text{ m s}^{-1} \qquad (4.212)$$

which is less than the maximum speed of relativity theory:

$$c = 2.9979 \times 10^8 \text{ m s}^{-1}. \qquad (4.213)$$

As discussed in Note 157(13) the mean energy $\langle E \rangle$ is related to the beam intensity I in joules per square metre by

$$I = 8\pi \left(\frac{f}{c} \right)^2 \langle E \rangle \qquad (4.214)$$

where f is the frequency of the beam in hertz. The intensity can be expressed as:

$$I = 8\pi f^2 m \left(1 - \frac{v_g^2}{c^2} \right)^{-1/2}. \qquad (4.215)$$

CHAPTER 4. PHOTON MASS AND THE B(3) FIELD

The total energy density of the light beam in joules per cubic metre is:

$$U = \frac{f}{c} I \tag{4.216}$$

and its power density in watts per square metre (joules per second per square metre) is:

$$\Phi = cU = fI = 8\pi f^3 m \left(1 - \frac{v_g^2}{c^2}\right)^{-1/2}. \tag{4.217}$$

The power density is an easily measurable quantity, and implies finite photon mass through Eq. (4.217). In the standard model there is no photon mass, so there is no power density, an absurd result. The power density is related to the magnitude of the electric field strength (**E**) and the magnetic flux density (**B**) of the beam by:

$$\Phi = \epsilon_0 c E^2 = \frac{cB^2}{\mu_0}. \tag{4.218}$$

The units in S. I. are as follows:

$$\left.\begin{array}{rl} E = & \text{volt m}^{-1} = \text{J C}^{-1}\text{m}^{-1} \\ B = & \text{tesla} = \text{J s C}^{-1}\text{m}^{-2} \\ \epsilon_0 = & \text{J}^{-1}\text{C}^2\text{m}^{-1} \\ \mu_0 = & \text{J s}^2\text{C}^{-2}\text{m}^{-1} \end{array}\right\} \tag{4.219}$$

where ϵ_0 and μ_0 are respectively the vacuum permittivity and permeability defined by:

$$\epsilon_0 \mu_0 = \frac{1}{c^2} \tag{4.220}$$

so:

$$\Phi = 8\pi f^3 m \left(1 - \frac{v_g^2}{c^2}\right)^{-1/2} = \epsilon_0 c E^2 = \frac{cB^2}{\mu_0}. \tag{4.221}$$

4.6 Difficulties with the Einstein Theory of Light Deflection due to Gravitation

The famous Einstein theory of light deflection due to gravitation is based on the idea of zero photon mass because in 1905 Einstein inferred such an idea from the basics of special relativity, he conjectured that a particle can travel at c if and only if its mass is identically zero, and assumed that photons travelled at c. Poincaré on the other hand realized that photons can travel at less than c if they have mass, and that c is the constant in the Lorentz transform. The Einsteinian calculation of light deflection due to gravitation was therefore based on the then new general relativity applied with a massless particle. In

4.6. DIFFICULTIES WITH THE EINSTEIN THEORY OF LIGHT...

the influential UFT 150B on www.aias.us it was shown that Einstein's method contains several fundamental errors. However precisely measured, such data cannot put right these errors, and the Einstein theory is completely refuted experimentally in whirlpool galaxies, so that it cannot be used anywhere in cosmology.

The Einstein method is based on the gravitational metric:

$$ds^2 = c^2 d\tau^2 = c^2 dt^2 \left(1 - \frac{r_0}{r}\right) - dr^2 \left(1 - \frac{r_0}{r}\right)^{-1} - r^2 d\phi^2 \tag{4.222}$$

usually and incorrectly attributed to Schwarzschild. Here, cylindrical polar coordinates are used in the XY plane. In Eq. (4.222) r_0 is the so called Schwarzschild radius, the particle of mass m orbits the mass M, for example the sun. The infinitesimal of proper time is $d\tau$. The lagrangian for this calculation is:

$$\mathcal{L} = \frac{m}{2} \left(\left(\frac{dt}{d\tau}\right)^2 \left(1 - \frac{r_0}{r}\right) - \left(1 - \frac{r_0}{r}\right)^{-1} \left(\frac{dr}{d\tau}\right)^2 - r^2 \left(\frac{d\phi}{d\tau}\right)^2 \right) \tag{4.223}$$

and the total energy and momentum are given as the following constants of motion:

$$E = mc^2 \left(1 - \frac{r_0}{r}\right) \frac{dt}{d\tau}, \quad L = mr^2 \frac{d\phi}{d\tau}. \tag{4.224}$$

Since $m \ll M$ the Schwarzschild radius is:

$$r_0 = \frac{2MG}{c^2}. \tag{4.225}$$

Therefore the calculation assumes that the mass m is not zero. For light grazing the sun, this is the photon mass.

The equation of motion is obtained from Eq. (4.222) by multiplying both sides by $1 - \frac{r_0}{r}$ to give:

$$m \left(\frac{dr}{d\tau}\right)^2 = \frac{E^2}{mc^2} - \left(1 - \frac{r_0}{r}\right) \left(mc^2 + \frac{L^2}{mr^2}\right). \tag{4.226}$$

The infinitesimal of proper time is eliminated as follows:

$$\frac{dr}{d\tau} = \frac{d\phi}{d\tau} \frac{dr}{d\phi} = \left(\frac{L^2}{mr^2}\right) \frac{dr}{d\phi} \tag{4.227}$$

to give the orbital equation:

$$\left(\frac{dr}{d\phi}\right)^2 = r^4 \left(\frac{1}{b^2} - \left(1 - \frac{r_0}{r}\right) \left(\frac{1}{a^2} + \frac{1}{r^2}\right)\right) \tag{4.228}$$

where the two constant lengths a and b are defined by:

$$a = \frac{L}{mc}, \quad b = \frac{cL}{E}. \tag{4.229}$$

CHAPTER 4. PHOTON MASS AND THE B(3) FIELD

The solution of Eq. (4.228) is:

$$\phi = \int \frac{1}{r^2} \left(\frac{1}{b^2} - \left(1 - \frac{r_0}{r}\right)\left(\frac{1}{a^2} + \frac{1}{r^2}\right) \right)^{-1/2} dr \qquad (4.230)$$

and the light deflection due to gravitation is:

$$\Delta\phi = 2 \int_{R_0}^{\infty} \frac{1}{r^2} \left(\frac{1}{b^2} - \left(1 - \frac{r_0}{r}\right)\left(\frac{1}{a^2} + \frac{1}{r^2}\right) \right)^{-1/2} dr - \pi \qquad (4.231)$$

where R_0 is the distance of closest approach, essentially the radius of the sun. Using:

$$u = 1/r, \quad du = -\frac{1}{r^2} dr \qquad (4.232)$$

the integral may be rewritten as:

$$\Delta\phi = 2 \int_0^{1/R_0} \left(\frac{1}{b^2} - (1 - r_0 u)\left(\frac{1}{a^2} + u^2\right) \right)^{-1/2} du - \pi. \qquad (4.233)$$

If we are to accept the gravitational metric for the sake of argument its correct use must be to assume an identically non zero photon mass and to integrate Eq. (4.233), producing an equation for the experimentally observed deflection $\Delta\phi$ in terms of m, a and b.

However, because of his conjecture of zero photon mass, Einstein used the null geodesic condition:

$$ds^2 = 0 \qquad (4.234)$$

which means that m is identically zero. This assumption means that:

$$a = \infty. \qquad (4.235)$$

However, the angular momentum is L is a constant of motion, so Eq. (4.235) means:

$$m = 0, \quad \frac{d\phi}{d\tau} = \infty \qquad (4.236)$$

which in the obsolete physics of the standard model was known as the ultrarelativistic limit. In this Einsteinian light deflection theory Eq. (4.223) is defined to be pure kinetic in nature, but at the same time the theory sets up an effective potential:

$$V(r) = \frac{1}{2} mc^2 \left(-\frac{r_0}{r} + \frac{a^2}{r^2} - \frac{r_0 a^2}{r^3} \right) \qquad (4.237)$$

and also assumes circular orbits:

$$\frac{dr}{d\tau} = 0. \qquad (4.238)$$

4.6. DIFFICULTIES WITH THE EINSTEIN THEORY OF LIGHT...

However, this assumption means that:

$$\frac{1}{b^2} = \left(1 - \frac{r_0}{r}\right)\left(\frac{1}{a^2} + \frac{1}{r^2}\right) \qquad (4.239)$$

and the denominator of Eq. (4.230) becomes zero and the integral becomes infinite. In order to circumvent this difficulty Einstein assumed:

$$\frac{r_0}{r} \to 0 \qquad (4.240)$$

which must mean:

$$r \to \infty \qquad (4.241)$$

and

$$m \to 0, \ a \to \infty. \qquad (4.242)$$

The effective potential was therefore defined as:

$$V(r) \xrightarrow[m\to 0, \ a\to\infty, \ r\to\infty]{} mc^2 \left(\frac{a}{r}\right)^2 \left(1 - \frac{r_0}{r}\right) \qquad (4.243)$$

which is mathematically indeterminate. Einstein also assumed:

$$mc^2 \to 0 \qquad (4.244)$$

so the equation of motion (4.229) becomes:

$$\frac{E^2}{2mc^2} = \frac{L^2}{mr^2}\left(\frac{1}{2} - \frac{MG}{c^2 r}\right). \qquad (4.245)$$

He used:

$$r = R_0 \qquad (4.246)$$

in this equation, thus finding an expression for b_0:

$$\frac{1}{b_0^2} = \frac{1}{R_0^2} - \frac{r_0}{R_0^3}. \qquad (4.247)$$

Finally he used Eq. (4.247) in Eq. (4.233) with:

$$a^2 \to \infty \qquad (4.248)$$

to obtain the integral:

$$\Delta\phi = 2\int_0^{1/R_0} \left(\frac{R_0 - r_0}{R_0^3} - u^2 + r_0 u^3\right)^{-1/2} du - \pi. \qquad (4.249)$$

It was claimed by Einstein that this integral is:

$$\Delta\phi = \frac{4MG}{c^2 R_0} \qquad (4.250)$$

CHAPTER 4. PHOTON MASS AND THE B(3) FIELD

but this is doubtful for reasons described in UFT 150B, whose calculations were all carried out with computer algebra. The experimental result for light grazing the sun is given for example by NASA Cassini as

$$\Delta\phi = 1.75'' = 8.484 \times 10^{-6} \, \text{rad}, \tag{4.251}$$

but Eq. (4.250) depends on the assumption of data such as:

$$R_0 = 6.955 \times 10^8 \, \text{m}, \quad M = 1.9891 \times 10^{30} \, \text{kg},$$
$$G = 6.67428 \times 10^{-11} \, \text{m}^3 \, \text{kg}^{-1} \, \text{s}^2. \tag{4.252}$$

In fact only MG is known with precision experimentally, not M and G individually. The radius R_0 is subject to considerable uncertainty. If we accept the dubious gravitational metric for the sake of argument, the experimental data must be evaluated from Eq. (4.231) with finite photon mass, and independent methods used to evaluate a and b.

Einstein's formula (4.249) for light deflection depends on the radius parameters R_0, and r_0. R_0 represents the radius of the sun (6.955×10^8 metres) while the so called Schwarzschild radius r_0 is 2,954 metres. So:

$$r_0 \ll R_0 \tag{4.253}$$

which implies from Eq. (4.247) that:

$$b_0 \sim R_0. \tag{4.254}$$

This gives the integral:

$$\Delta\phi = 2 \int_0^{1/R_0} \left(\frac{R_0 - r_0}{R_0^3} - u^2 + r_0 u^3 \right)^{-1/2} du - \pi \tag{4.255}$$

which has no analytical solution. Its numerical integration is also difficult, even with contemporary methods. The square root in the integral has zero crossings, leading to infinite values of the integrand and as discussed in Section 3 of UFT 150B there is a discrepancy between the experimental data, Einstein's claim and the numerical evaluation of the integral.

The correct method of evaluating the light deflection is obviously to use a finite mass m in Eq. (4.231). In a first rough approximation, UFT 150B used:

$$E = \hbar\omega \tag{4.256}$$

for one photon. More accurately a Planck distribution can be used. However Eq. (4.256) gives:

$$a = \frac{\hbar\omega}{mc^2} b. \tag{4.257}$$

The parameter b is a constant of motion, and is determined by the need for zero deflection when the mass M of the sun is absent. This gives:

$$\Delta\phi = 2 \int_0^{1/R_0} \left(\frac{1}{b^2} - u^2 \right)^{-1/2} du - \pi = 0 \tag{4.258}$$

4.6. DIFFICULTIES WITH THE EINSTEIN THEORY OF LIGHT...

and as described in UFT 150B this gives a photon mass of:

$$m = 3.35 \times 10^{-41} \, \text{kg} \tag{4.259}$$

which again a lot heavier than the estimates in the standard literature.

So in summary of these sections, the B(3) field implies a finite photon mass which can be estimated by Compton scattering and by light deflection due to gravitation. The photon mass is not zero, but an accurate estimate of its value needs refined calculations. These are simple first attempts only. There are multiple problems with the claim that light deflection by the sun is twice the Newtonian value, because the latter is itself heuristic, and because Einstein's methods are dubious, as described in UFT 150B and UFT 155. The entire Einstein method is refuted by its neglect of torsion, as explained in great detail in the two hundred and sixty UFT papers available to date.

Chapter 5

The Unification of Quantum Mechanics and General Relativity

5.1 Introduction

The standard physics has completely failed to unify quantum mechanics and general relativity, notably because of indeterminacy, a non Baconian idea introduced at the Solvay Conference of 1927. The current attempts of the standard physics at unification revolve around hugely expensive particle colliders, and these attempts are limited to the unification of the electromagnetic and weak and strong nuclear fields, leaving out gravitation completely. So it is reasonable to infer that the standard physics will never be able to produce a unified field theory. In great contrast ECE theory has succeeded with unifying all four fundamental fields with a well known geometry due to Cartan as described in foregoing chapters of this book.

Towards the end of the nineteenth century the classical physics evolved gradually into special relativity and the old quantum theory. The experiments that led to this great paradigm shift in natural philosophy are very well known, so need only a brief description here. There were experiments on the nature of broadband (black body) radiation leading to the Rayleigh Jeans law, the Stefan Boltzmann distribution and similar. The failure of the Rayleigh Jeans law led to the Planck distribution and his inference of what was later named the photon. The photoelectric effect could not be explained using the classical physics, the Brownian motion needed a new type of stochastic physics indicating the existence of molecules, first proposed by Dalton. The specific heats of solids could not be explained adequately with classical nineteenth century physics. Atomic and molecular spectra could not be explained with classical methods, notably the anomalous Zeeman effect.

The Michelson Morley experiment gave results that could not be explained with the classical Newtonian physics, so that Fitzgerald in correspondence

5.1. INTRODUCTION

with Heaviside suggested a radically new physics that came to be known as special relativity. The mathematical framework for special relativity was very nearly inferred by Heaviside but was developed by Lorentz and Poincare. Einstein later made contributions of his own. The subjects of special relativity and quantum theory began to develop rapidly. The many contributions of Sommerfeld are typically underestimated in the history of science, those of his students and post doctorals are better known. The old quantum theory evolved into the Schroedinger equation after the inference by de Broglie of wave particle dualism. Peter Debye asked his student Schroedinger to try to solve the puzzle posed by the fact that a particle could be a wave and vice versa, and during this era Compton gave an impetus to the idea of photon as particle by scattering high frequency electromagnetic radiation from a metal foil - Compton scattering.

The Schroedinger equation proved to be an accurate description of for example spectral phenomena in the non relativistic limit. In the simplest instance the Schroedinger equation quantizes the classical kinetic energy of the free particle, and does not attempt to incorporate special relativity into quantum mechanics. Sommerfeld had made earlier attempts but the main problem remained, how to quantize the Einstein energy equation of special relativity. The initial attempts by Klein and Gordon resulted in negative probability, so were abandoned for this reason. Pauli had applied his algebra to the Schroedinger equation, but none of these methods were successful in describing the g factor, Landé factor or Thomas precession in one unified framework of relativistic quantum mechanics.

Dirac famously solved the problem with the use of four by four matrices and Pauli algebra but in so doing ran into the problem of negative energies. Dirac suggested tentatively that negative energies could be eliminated with the Dirac sea, but this introduced an unobservable, the Dirac sea still has not been observed experimentally. Unobservables began to proliferate in twentieth century physics, reducing it to dogma. However, Dirac was famously successful in explaining within one framework the g factor of the electron, the Landé factor, the Thomas factor and the Darwin term, and in producing a theory free of negative probabilities. The Dirac sea seemed to give rise to antiparticles which were observed. The Dirac sea itself cannot be observed, and the problem of negative energies was not solved by Dirac.

It is not clear whether Dirac ever accepted indeterminacy, a notion introduced by Bohr and Heisenberg and immediately rejected by Einstein, Schroedinger, de Broglie and others as anti Baconian and unphysical. The Dirac equation reduces to the Schroedinger and Heisenberg equations in well defined limits, but indeterminacy is pure dogma. It is easily disproven experimentally and has taken on a life of its own that cannot be described as science. Heisenberg described the Dirac equation as an all time low in physics, but many would describe indeterminacy in the same way. In this chapter, indeterminacy is disproven straightforwardly with the use of higher order commutators. Heisenberg's own methods are used to disprove the Heisenberg Uncertainty Principle, a source of infinite confusion for nearly ninety years. One of the major outcomes of ECE theory is the rejection of the Heisenberg Uncertainty

Principle in favour of a quantum mechanics based on geometry.

The negative energy problem that plagued the Dirac equation is removed in this chapter by producing the fermion equation of relativistic quantum mechanics. This equation is not only Lorentz covariant but also generally covariant because it is derived from the tetrad postulate of a generally covariant geometry – Cartan geometry. All the equations of ECE theory are automatically generally covariant and Lorentz covariant in a well defined limit of general covariance. So the fermion equation is the first equation of quantum mechanics unified with general relativity. It has the major advantages of producing rigorously positive energy levels and of being able to express the theory in terms of two by two matrices. The fermion equation produces everything that the Dirac equation does, but with major advantages. So it should be viewed as an improvement on the deservedly famous Dirac equation, an improvement based on geometry and the ECE unified field theory. The latter also produces the d'Alembert and Klein Gordon equations, and indeed all of the valid wave equations of physics. Some of these are discussed in this chapter.

5.2 The Fermion Equation

The structure of ECE theory is the most fundamental one known in physics at present, simply because it is based directly on a rigorously correct geometry. The fermion equation can be expressed as in UFT 173 on www.aias.us in a succinct way:

$$\pi_\mu \psi \sigma^\mu = m c \sigma^1 \psi \tag{5.1}$$

where the fermion operator in covariant representation is defined as:

$$\pi_\mu = (\pi_0, \pi_1, \pi_2, \pi_3). \tag{5.2}$$

Here:

$$\pi_0 = \sigma^0 p_0, \quad \pi_i = \sigma^3 p_i \tag{5.3}$$

where p_μ is the energy momentum four vector:

$$p_\mu = (p_0, p_1, p_2, p_3). \tag{5.4}$$

The Pauli matrices are defined by:

$$\sigma^\mu = (\sigma^0, \sigma^1, \sigma^2, \sigma^3) \tag{5.5}$$

where:

$$\sigma^0 = \begin{bmatrix} 1 & 0 \\ 0 & 1 \end{bmatrix}, \quad \sigma^1 = \begin{bmatrix} 0 & 1 \\ 1 & 0 \end{bmatrix}, \quad \sigma^2 = \begin{bmatrix} 0 & -i \\ i & 0 \end{bmatrix}, \quad \sigma^3 = \begin{bmatrix} 1 & 0 \\ 0 & -1 \end{bmatrix}. \tag{5.6}$$

The eigenfunction of Eq.(5.1) is the tetrad [1]- [10]:

$$\psi = \begin{bmatrix} \psi_1^R & \psi_2^R \\ \psi_1^L & \psi_2^L \end{bmatrix} \tag{5.7}$$

5.2. THE FERMION EQUATION

whose entries are defined by the right and left Pauli spinors:

$$\phi^R = \begin{bmatrix} \psi_1^R \\ \psi_2^R \end{bmatrix}, \quad \phi^L = \begin{bmatrix} \psi_1^L \\ \psi_2^L \end{bmatrix}. \tag{5.8}$$

This eigenfunction is referred to as "the fermion spinor".

The position representation of the fermion operator is defined by the symbol δ and is:

$$\delta^\mu = -\frac{i}{\hbar} \pi_\mu. \tag{5.9}$$

Therefore the fermion equation is the first order differential equation:

$$i\hbar \delta_\mu \psi \sigma^\mu = mc\sigma^1 \psi. \tag{5.10}$$

For purposes of comparison, the covariant format of the Dirac equation in chiral representation [24] is:

$$\gamma^\mu \delta_\mu \psi_D = mc\psi_D. \tag{5.11}$$

where:

$$\psi_D = \begin{bmatrix} \phi^R \\ \phi^L \end{bmatrix} \tag{5.12}$$

is a column vector with four entries, and where the Dirac matrices in chiral representation [24] are:

$$\gamma^\mu = \left(\gamma^0, \gamma^1, \gamma^2, \gamma^3\right). \tag{5.13}$$

The complete details of the development of Eq. (5.1) are given in Note 172(8) accompanying UFT 172 on www.aias.us The ordering of terms in Eq. (5.1) is important because matrices do not commute and ψ is a 2 x 2 matrix. The energy eigenvalue of Eq.(5.1) is rigorously positive, never negative. The complex conjugate of the adjoint matrix of the fermion spinor is referred to as the "adjoint spinor" of the fermion equation, and is defined by:

$$\psi^+ = \begin{bmatrix} \psi_1^{R*} & \psi_1^{L*} \\ \psi_2^{R*} & \psi_2^{L*} \end{bmatrix}. \tag{5.14}$$

The adjoint equation of Eq. (5.1) is defined as:

$$-i\hbar \delta_\mu \psi^+ \sigma^\mu = mc\sigma^1 \psi^+ \tag{5.15}$$

where the complex conjugate of ψ has been used. These equations have well known counterparts in the Dirac theory [1]- [10], [24] but in that theory the 4 x 4 gamma matrices are used and the definition of the adjoint spinor is more complicated.

The probability four-current of the fermion equation is defined as:

$$j^\mu = \frac{1}{2} \text{Tr} \left(\psi \sigma^\mu \psi^+ + \psi^+ \sigma^\mu \psi \right) \tag{5.16}$$

CHAPTER 5. THE UNIFICATION OF QUANTUM MECHANICS AND...

and its Born probability is:

$$j^0 = \psi_1^R \psi_1^{R*} + \psi_2^R \psi_2^{R*} + \psi_1^L \psi_1^{L*} + \psi_2^L \psi_2^{L*} \tag{5.17}$$

which is rigorously positive as required of a probability. It is the same as the Born probability of the chiral representation [1]- [10], [24] of the Dirac equation. In the latter the four current is defined as:

$$j_D^\mu = \overline{\psi}_D \gamma^\mu \psi_D \tag{5.18}$$

and the adjoint Dirac spinor is a four entry row vector defined by:

$$\overline{\psi}^D = \psi_D^+ \gamma^0. \tag{5.19}$$

It is shown as follows that the probability four-current of the fermion equation is conserved:

$$\delta_\mu j^\mu = 0. \tag{5.20}$$

To prove this result multiply both sides of Eq.(5.1) from the right with ψ^+:

$$i\hbar \delta_\mu \psi \sigma^\mu \psi^+ = mc\sigma^1 \psi\psi^+. \tag{5.21}$$

Multiply both sides of Eq.(5.15) from the right with ψ:

$$-i\hbar \delta_\mu \psi^+ \sigma^\mu \psi = mc\sigma^1 \psi^+\psi \tag{5.22}$$

and subtract Eq.(5.22) from Eq.(5.21):

$$i\hbar \delta_\mu \left(\psi \sigma^\mu \psi^+ + \psi^+ \sigma^\mu \psi \right) = mc\sigma^1 \left(\psi\psi^+ - \psi^+\psi \right). \tag{5.23}$$

By definition:

$$\psi\psi^+ - \psi^+\psi = \begin{bmatrix} \psi_1^R & \psi_2^R \\ \psi_1^L & \psi_2^L \end{bmatrix} \begin{bmatrix} \psi_1^{R*} & \psi_1^{L*} \\ \psi_2^{R*} & \psi_2^{L*} \end{bmatrix} - \begin{bmatrix} \psi_1^{R*} & \psi_1^{L*} \\ \psi_2^{R*} & \psi_2^{L*} \end{bmatrix} \begin{bmatrix} \psi_1^R & \psi_2^R \\ \psi_1^L & \psi_2^L \end{bmatrix} \tag{5.24}$$

so

$$\text{Trace}(\psi\psi^+ - \psi^+\psi) = 0. \tag{5.25}$$

Therefore:

$$\text{Trace}\left(\delta_\mu(\psi\sigma^\mu\psi^+ - \psi^+\sigma^\mu\psi) \right) = 0 \tag{5.26}$$

and

$$\delta_\mu j^\mu = 0. \tag{5.27}$$

Q. E. D.

The fermion equation (5.1) may be expanded into two simultaneous equations:

$$(E + c\boldsymbol{\sigma} \cdot \mathbf{p}) \phi^L = mc^2 \phi^R \tag{5.28}$$

$$(E - c\boldsymbol{\sigma} \cdot \mathbf{p}) \phi^R = mc^2 \phi^L \tag{5.29}$$

5.2. THE FERMION EQUATION

in which E and $\mathbf{p}$ are the operators of quantum mechanics:

$$E = i\hbar \frac{\partial}{\partial t}, \quad \mathbf{p} = -i\hbar \boldsymbol{\nabla}. \tag{5.30}$$

Eqs.(5.28) and (5.29) may be developed as:

$$(E - c\boldsymbol{\sigma} \cdot \mathbf{p})(E + c\boldsymbol{\sigma} \cdot \mathbf{p}) \phi^L = m^2 c^4 \phi^L \tag{5.31}$$
$$(E + c\boldsymbol{\sigma} \cdot \mathbf{p})(E - c\boldsymbol{\sigma} \cdot \mathbf{p}) \phi^R = m^2 c^4 \phi^R \tag{5.32}$$

from which there emerge equations such as:

$$(E^2 - c^2 \boldsymbol{\sigma} \cdot \mathbf{p}\, \boldsymbol{\sigma} \cdot \mathbf{p}) \phi^R = m^2 c^4 \phi^R. \tag{5.33}$$

Using the quantum postulates this becomes the wave equation:

$$\left(\Box + \left(\frac{mc}{\hbar}\right)^2\right) \phi^R = 0 \tag{5.34}$$

and it becomes clear that the fermion equation is a factorization of the ECE wave equation:

$$\left(\Box + \left(\frac{mc}{\hbar}\right)^2\right) \psi = 0 \tag{5.35}$$

whose eigenfunction is the tetrad (ψ).

Therefore the fermion equation is obtained from the tetrad postulate and Cartan geometry. The tetrad is defined by:

$$\begin{bmatrix} V^R \\ V^L \end{bmatrix} = \begin{bmatrix} \psi_1^R & \psi_2^R \\ \psi_1^L & \psi_2^L \end{bmatrix} \begin{bmatrix} V^1 \\ V^2 \end{bmatrix} \tag{5.36}$$

i.e. as a matrix relating two column vectors.

The parity operator P acts on the fermion spinor as follows:

$$P\psi = \begin{bmatrix} \psi_1^L & \psi_2^L \\ \psi_1^R & \psi_2^R \end{bmatrix} \tag{5.37}$$

and the anti fermion is obtained straightforwardly from the fermion equation by operating on each term with P as follows:

$$P(E) = E, \; P(\mathbf{p}) = -\mathbf{p}, \; P\begin{bmatrix} \psi_1^R & \psi_2^R \\ \psi_1^L & \psi_2^L \end{bmatrix} = \begin{bmatrix} \psi_1^L & \psi_2^L \\ \psi_1^R & \psi_2^R \end{bmatrix}. \tag{5.38}$$

Note carefully that the eigenstates of energy are always positive, both in the fermion and anti fermion equations. The anti fermion is obtained from the fermion by reversing helicity:

$$P(\boldsymbol{\sigma} \cdot \mathbf{p}) = -\boldsymbol{\sigma} \cdot \mathbf{p} \tag{5.39}$$

and has opposite parity to the fermion, the same mass as the fermion, and the opposite electric charge. The static fermion is indistinguishable from the static

CHAPTER 5. THE UNIFICATION OF QUANTUM MECHANICS AND...

anti fermion [24]. So CPT symmetry is conserved as follows form fermion to anti fermion:

$$CPT \to (-C)(-P)T \tag{5.40}$$

where C is the charge conjugation operator and T the motion reversal operator. Note carefully that there is no negative energy anywhere in the analysis.

The pair of simultaneous equations (5.28) and (5.29) can be written as:

$$(E - c\boldsymbol{\sigma} \cdot \mathbf{p})(E + \boldsymbol{\sigma} \cdot \mathbf{p})\phi^L, = m^2 c^4 \phi^L \tag{5.41}$$

an equation which can be re arranged as:

$$(E^2 - m^2 c^4)\phi^L = c^2 \boldsymbol{\sigma} \cdot \mathbf{p}\, \boldsymbol{\sigma} \cdot \mathbf{p}\, \phi^L \tag{5.42}$$

and factorized to give:

$$(E - mc^2)(E + mc^2)\phi^L = c^2 \boldsymbol{\sigma} \cdot \mathbf{p}\, \boldsymbol{\sigma} \cdot \mathbf{p}\, \phi^L. \tag{5.43}$$

If **p** is real valued, Pauli algebra means that:

$$\boldsymbol{\sigma} \cdot \mathbf{p}\, \boldsymbol{\sigma} \cdot \mathbf{p} = p^2 \tag{5.44}$$

so if E and **p** are regarded as functions, not operators, Eq.(5.43) becomes the Einstein energy equation:

$$E^2 - m^2 c^4 = c^2 p^2 \tag{5.45}$$

multiplied by ϕ^L on both sides. It is well known [1]-[10] that the Einstein energy equation is a way of writing the relativistic energy and momentum:

$$E = \gamma m c^2, \tag{5.46}$$

$$\mathbf{p} = \gamma m \mathbf{v}. \tag{5.47}$$

Realizing this, Eq.(5.43) can be linearized as follows. First, express it as:

$$(E - mc^2)\phi^L = \frac{c^2 \boldsymbol{\sigma} \cdot \mathbf{p}\, \boldsymbol{\sigma} \cdot \mathbf{p}}{E + mc^2} \phi^L \tag{5.48}$$

and approximate the total energy:

$$E = \gamma m c^2 \tag{5.49}$$

by the rest energy:

$$E \approx mc^2, \tag{5.50}$$

then Eq.(5.48) becomes:

$$(E - mc^2)\phi^L = \frac{1}{2m} c^2 \boldsymbol{\sigma} \cdot \mathbf{p}\, \boldsymbol{\sigma} \cdot \mathbf{p}\, \phi^L \tag{5.51}$$

which has the structure of the free particle Schroedinger equation:

$$E_{NR}\,\phi^L = \frac{p^2}{2m}\,\phi^L \tag{5.52}$$

in which the non relativistic limit of the kinetic energy is defined in the limit $v \ll c$ by:

$$E_{NR} = E - m\,c^2 = (\gamma - 1)m\,c^2 \to \frac{p^2}{2m}. \tag{5.53}$$

So the fermion equation reduces correctly to the non relativistic Schroedinger equation for the free particle, Q. E. D.

The great importance of the fermion equation to chemical physics emerges from the fact that it can describe the phenomena for which the Dirac equation is justly famous while at the same time eliminating the problem of negative energy as we have just seen. In quantum field theory this leads to a free fermion quantum field theory. This aim is very difficult to achieve [24] in the standard quantum field theory because methods have to be devised to deal with the negative energy. The latter is due simply to Dirac's choice of gamma matrices.

The way in which the fermion equation describes the g factor of the electron, the Landé factor, the Thomas factor and Darwin term is described in the following section.

5.3 Interaction of the ECE Fermion with the Electromagnetic Field

The simplest and most powerful way of describing this interaction for each polarization index a of ECE theory is through the minimal prescription

$$p^\mu \to p^\mu - eA^\mu \tag{5.54}$$

where a negative sign is used [24] because the charge on the electron is -e. Eq.(5.54) can be written as:

$$E \to E - e\phi \tag{5.55}$$

and:

$$\mathbf{p} \to \mathbf{p} - e\mathbf{A}. \tag{5.56}$$

Using Eqs.(5.55) and (5.56) in the Einstein energy equation (5.45) gives:

$$(E - e\phi)^2 = c^2(\mathbf{p} - e\mathbf{A})^2 + m^2\,c^4 \tag{5.57}$$

which can be factorized as follows:

$$(E - e\phi - m\,c^2)(E - e\phi + m\,c^2) = c^2(\mathbf{p} - e\mathbf{A})^2 \tag{5.58}$$

CHAPTER 5. THE UNIFICATION OF QUANTUM MECHANICS AND...

and written as:

$$E = mc^2 + e\phi + c^2 \frac{(\mathbf{p} - e\mathbf{A})}{(E - e\phi + mc^2)}(\mathbf{p} - e\mathbf{A}) \tag{5.59}$$

in a form ready for quantization. The latter is carried out with:

$$\mathbf{p} \to -i\hbar\boldsymbol{\nabla} \tag{5.60}$$

and produces many well known effects and new effects of spin orbit coupling described in papers of ECE theory such as UFT 248 ff on www.aias.us.

The most famous result of the Dirac equation, and its improved version, the ECE fermion equation, is electron spin resonance, which depends on the use of the Pauli matrices as is very well known. In this section the various intricacies of this famous derivation are explained systematically. Electron spin resonance occurs in the presence of a static magnetic field, so the scalar potential can be omitted from consideration leaving hamiltonians such as:

$$H_2\psi = \frac{1}{2m}\left(\boldsymbol{\sigma}\cdot(-i\hbar\boldsymbol{\nabla} - e\mathbf{A})\ \boldsymbol{\sigma}\cdot(-i\hbar\boldsymbol{\nabla} - e\mathbf{A})\right)\psi. \tag{5.61}$$

Note carefully that the operator $\boldsymbol{\nabla}$ acts on the wave function, which is denoted ψ for ease of notation. The following type of Pauli algebra:

$$\boldsymbol{\sigma}\cdot\mathbf{V}\ \boldsymbol{\sigma}\cdot\mathbf{W} = \mathbf{V}\cdot\mathbf{W} + i\boldsymbol{\sigma}\cdot\mathbf{V}\times\mathbf{W} \tag{5.62}$$

leads to:

$$H_2\psi = \frac{1}{2m}\left(ie\hbar\left(\boldsymbol{\nabla}\cdot\mathbf{A} + i\boldsymbol{\sigma}\cdot\boldsymbol{\nabla}\times\mathbf{A}\right) - \hbar^2\left(\nabla^2 + i\boldsymbol{\sigma}\cdot\boldsymbol{\nabla}\times\boldsymbol{\nabla}\right)\right. \tag{5.63}$$
$$\left. + e^2\left(A^2 + i\boldsymbol{\sigma}\cdot\mathbf{A}\times\mathbf{A}\right) + ie\hbar(\mathbf{A}\cdot\boldsymbol{\nabla} + i\boldsymbol{\sigma}\cdot(\mathbf{A}\times\boldsymbol{\nabla}))\right)\psi.$$

Assuming that $\mathbf{A}$ is real valued, then:

$$\mathbf{A}\times\mathbf{A} = 0. \tag{5.64}$$

Also:

$$\boldsymbol{\nabla}\times\boldsymbol{\nabla} = 0 \tag{5.65}$$

so:

$$H_2\psi = \frac{1}{2m}\left(-\hbar^2\nabla^2\psi + e^2A^2\psi + ie\hbar\boldsymbol{\nabla}\cdot(\mathbf{A}\psi)\right. \tag{5.66}$$
$$\left. - e\hbar\boldsymbol{\sigma}\cdot\boldsymbol{\nabla}\times(\mathbf{A}\psi) + ie\hbar\mathbf{A}\cdot\boldsymbol{\nabla}\psi - e\hbar\boldsymbol{\sigma}\cdot\mathbf{A}\times\boldsymbol{\nabla}\psi\right).$$

It can be seen that the fermion equation produces many effects in general, all of which are experimentally observable. So it is a very powerful result of geometry and ECE unified field theory. Gravitational effects can be considered through the appropriate minimal prescription as in papers such as UFT 248 ff. Many of these effects remain to be observed.

5.3. INTERACTION OF THE ECE FERMION WITH THE...

Electron spin resonance is given by the term:

$$H_2\psi = -\frac{e\hbar}{2m}\boldsymbol{\sigma}\cdot(\boldsymbol{\nabla}\times(\mathbf{A}\psi) + \mathbf{A}\times\boldsymbol{\nabla}\psi) + ... \tag{5.67}$$

$$= -\frac{e\hbar}{2m}\boldsymbol{\sigma}\cdot\mathbf{B} + ...$$

where the standard relation between $\mathbf{B}$ and $\mathbf{A}$ has been used to illustrate the argument:

$$\mathbf{B} = \boldsymbol{\nabla}\times\mathbf{A}. \tag{5.68}$$

In the rigorous ECE theory the spin connection enters into the analysis. A vast new subject area of chemical physics emerges because electron spin resonance (ESR) and nuclear magnetic resonance (NMR) dominate the subjects of chemical physics and analytical chemistry.

Use of a complex valued potential such as that in an electromagnetic field rather than a static magnetic field produces many more effects through the equation:

$$((E - e\phi) + c\boldsymbol{\sigma}\cdot(\mathbf{p} - e\mathbf{A}))((E - e\phi) - c\boldsymbol{\sigma}\cdot(\mathbf{p} - e\mathbf{A}^*))\phi^R \tag{5.69}$$
$$= m^2 c^4 \phi^R,$$
$$\psi := \phi^R,$$

i.e.

$$(E - e\phi - mc^2)(E - e\phi + mc^2)\psi$$
$$= c^2\boldsymbol{\sigma}\cdot(\mathbf{p} - e\mathbf{A})\,\boldsymbol{\sigma}\cdot(\mathbf{p} - e\mathbf{A}^*)\psi + ec(E - e\phi)\boldsymbol{\sigma}\cdot(\mathbf{A}^* - \mathbf{A})\psi$$

where * denotes "complex conjugate". Eq.(5.69) can be linearized as:

$$(E - e\phi - mc^2)\psi = \frac{c^2\boldsymbol{\sigma}\cdot(\mathbf{p} - e\mathbf{A})\,\boldsymbol{\sigma}\cdot(\mathbf{p} - e\mathbf{A}^*)}{E - e\phi + mc^2}\psi \tag{5.70}$$
$$+ \frac{ec(E - e\phi)}{E - e\phi + mc^2}\boldsymbol{\sigma}\cdot(\mathbf{A}^* - \mathbf{A})\psi$$

and re-arranged as follows:

$$E\psi = (e\phi + mc^2)\psi \tag{5.71}$$
$$+ \frac{1}{2m}\boldsymbol{\sigma}\cdot(\mathbf{p} - e\mathbf{A})\left(1 - \frac{e\phi}{2mc^2}\right)^{-1}\boldsymbol{\sigma}\cdot(\mathbf{p} - e\mathbf{A}^*)\psi$$
$$+ \frac{e}{2mc}(mc^2 - e\phi)\left(1 - \frac{e\phi}{2mc^2}\right)^{-1}\boldsymbol{\sigma}\cdot(\mathbf{A}^* - \mathbf{A})\psi.$$

In the approximation:

$$e\phi << mc^2 \tag{5.72}$$

Eq.(5.71) gives:

$$E\psi = (H_1 + H_2 + H_3)\psi \tag{5.73}$$

where the three hamiltonians are defined as follows:

$$H_1 = E\phi + mc^2, \tag{5.74}$$

$$H_2 = \frac{1}{2m}\sigma \cdot (\mathbf{p} - e\mathbf{A})\left(1 + \frac{e\phi}{2mc^2}\right)\sigma \cdot (\mathbf{p} - e\mathbf{A}^*), \tag{5.75}$$

$$H_3 = \frac{1}{2}ec\left(1 + \frac{e\phi}{2mc^2}\right)\sigma \cdot (\mathbf{A}^* - \mathbf{A}), \tag{5.76}$$

leading to many new fermion resonance effects using the electromagnetic field rather than the static magnetic field.

For example the H_2 hamiltonian can be developed as:

$$\begin{aligned}H_{21}\psi = \frac{1}{2m}\,(ie\hbar(\boldsymbol{\nabla} \cdot \mathbf{A}^* + i\sigma \cdot \boldsymbol{\nabla} \times \mathbf{A}^* \\ -\hbar^2(\nabla^2 + i\sigma \cdot \boldsymbol{\nabla} \times \boldsymbol{\nabla}) \\ +e^2(\mathbf{A} \cdot \mathbf{A}^* + i\sigma \cdot \mathbf{A} \times \mathbf{A}^*) \\ +ie\hbar(\mathbf{A} \cdot \boldsymbol{\nabla} + i\sigma \cdot \mathbf{A} \times \boldsymbol{\nabla})))\,\psi,\end{aligned} \tag{5.77}$$

an equation that can be written as:

$$\begin{aligned}H_{21}\psi = \frac{1}{2m}\,\big(ie^2\sigma \cdot \mathbf{A} \times \mathbf{A}^*\psi - e\hbar\sigma \cdot \mathbf{A} \times \boldsymbol{\nabla}\psi \\ -e\hbar\sigma \cdot \boldsymbol{\nabla}\psi \times \mathbf{A}^* - e\hbar\sigma \cdot (\boldsymbol{\nabla} \times \mathbf{A}^*)\psi + ...\big),\end{aligned} \tag{5.78}$$

giving four out of many terms that can give novel fermion resonance effects. Using for the sake of argument:

$$\mathbf{B}^* = \boldsymbol{\nabla} \times \mathbf{A}^* \tag{5.79}$$

then the hamiltonian reduces to:

$$H_{211} = -\frac{e\hbar}{2m}\sigma \cdot \mathbf{B}^* \tag{5.80}$$

and a term due to the conjugate product of the electromagnetic field:

$$H_{212} = i\frac{e^2}{2m}\sigma \cdot \mathbf{A} \times \mathbf{A}^* \tag{5.81}$$

which defines the B(3) field introduced in previous chapters:

$$\mathbf{B}^{(3)*} = -ig\mathbf{A} \times \mathbf{A}^* = -ig\mathbf{A}^{(1)} \times \mathbf{A}^{(2)}. \tag{5.82}$$

Eq.(5.81) is the hamiltonian that defines radiatively induced fermion resonance (RFR), extensively discussed elsewhere [1]- [10] but derived here in a rigorous way from the fermion equation or chiral representation of the Dirac equation.

Spin orbit coupling and the Thomas factor can be derived from the H_{22} hamiltonian defined as follows:

$$H_{22}\psi = \frac{e}{4m^2c^2}\left(\sigma \cdot (\mathbf{p} - e\mathbf{A})\phi\,\sigma \cdot (\mathbf{p} - e\mathbf{A})\right)\psi. \tag{5.83}$$

5.3. INTERACTION OF THE ECE FERMION WITH THE...

This hamiltonian has its origins in the following equation:

$$E\psi = \left(\frac{mc^2 + e\phi + c^2(\mathbf{p} - e\mathbf{A})}{E - e\phi + mc^2} \cdot (\mathbf{p} - e\mathbf{A}) \right) \psi \tag{5.84}$$

in the approximation:

$$E = \gamma mc^2 \approx mc^2. \tag{5.85}$$

In this approximation. Eq.(5.84) becomes:

$$E\psi = \left(mc^2 + e\phi + \frac{1}{2m}(\mathbf{p} - e\mathbf{A}) \left(1 - \frac{e\phi}{2mc^2} \right)^{-1} \cdot (\mathbf{p} - e\mathbf{A}) \right) \psi \tag{5.86}$$

and in the approximation:

$$e\phi << 2mc^2 \tag{5.87}$$

the H_{22} hamiltonian is recovered as the last term on the right hand side.

In the derivation of the spin orbit coupling term several assumptions are made, but not always made clear in textbooks. The vector potential $\mathbf{A}$ is not considered in the derivation of spin orbit interaction, so that only electric field effects are considered. Therefore the relevant hamiltonian reduces to:

$$H_{22}\psi = \frac{e}{4m^2c^2} \boldsymbol{\sigma} \cdot \mathbf{p}\phi \, \boldsymbol{\sigma} \cdot \mathbf{p}\psi. \tag{5.88}$$

It is assumed that the first $\mathbf{p}$ is the operator:

$$\mathbf{p} = -i\hbar \boldsymbol{\nabla} \tag{5.89}$$

but that the second $\mathbf{p}$ is a function. This point is rarely if ever made clear in the textbooks. This assumption can be justified only on the grounds that it seems to succeed in describing the experimental data. When this assumption is made Eq.(5.88) reduces to:

$$H_{22}\psi = -\frac{ie\hbar}{4m^2c^2} \boldsymbol{\sigma} \cdot \boldsymbol{\nabla}\phi \, \boldsymbol{\sigma} \cdot \mathbf{p}\psi. \tag{5.90}$$

The $\boldsymbol{\nabla}$ operator acts on $\phi \, \boldsymbol{\sigma} \cdot \mathbf{p}\psi$, so by the Leibnitz Theorem :

$$\boldsymbol{\nabla}(\phi \, \boldsymbol{\sigma} \cdot \mathbf{p}\psi) = \boldsymbol{\nabla}(\boldsymbol{\sigma} \cdot \mathbf{p})\phi\psi + \boldsymbol{\sigma} \cdot \mathbf{p}\,\boldsymbol{\nabla}(\phi\psi) \tag{5.91}$$

and the spin orbit interaction term emerges from:

$$H_{22}\psi = -\frac{ie\hbar}{4m^2c^2}(\boldsymbol{\sigma} \cdot \boldsymbol{\nabla}(\phi\psi)\, \boldsymbol{\sigma} \cdot \mathbf{p}). \tag{5.92}$$

In this equation the Leibnitz Theorem asserts that:

$$\boldsymbol{\nabla}(\phi \, \psi) = (\boldsymbol{\nabla}\phi)\psi + \phi \, (\boldsymbol{\nabla}\psi) \tag{5.93}$$

CHAPTER 5. THE UNIFICATION OF QUANTUM MECHANICS AND...

so the spin orbit interaction term is:

$$H_{22}\psi = -\frac{ie\hbar}{4m^2c^2}(\boldsymbol{\sigma}\cdot\boldsymbol{\nabla}\phi\,\boldsymbol{\sigma}\cdot\mathbf{p})\psi + ... \tag{5.94}$$

It is seen Eq. (5.94) is only one out of many possible effects that emerge from the fermion equation and which should be systematically investigated experimentally.

In the development of the spin orbit term the obsolete standard physics is used as follows:

$$\mathbf{E} = -\boldsymbol{\nabla}\phi \tag{5.95}$$

so the spin orbit hamiltonian becomes:

$$H_{22}\psi = -\frac{ie\hbar}{4m^2c^2}\boldsymbol{\sigma}\cdot\mathbf{E}\,\boldsymbol{\sigma}\cdot\mathbf{p}\psi. \tag{5.96}$$

Now use the Pauli algebra:

$$\boldsymbol{\sigma}\cdot\mathbf{E}\,\boldsymbol{\sigma}\cdot\mathbf{p} = \mathbf{E}\cdot\mathbf{p} + i\boldsymbol{\sigma}\cdot\mathbf{E}\times\mathbf{p} \tag{5.97}$$

so the real part of the hamiltonian from these equations becomes:

$$H_{22}\psi = \frac{e\hbar}{4m^2c^2}\boldsymbol{\sigma}\cdot\mathbf{E}\times\mathbf{p}\psi \tag{5.98}$$

in which $\mathbf{p}$ is regarded as a function, and not an operator. If this second $\mathbf{p}$ is regarded as an operator, then new effects appear.

Note carefully that in the derivation of the Zeeman effect, ESR, NMR and the g factor of the electron, both $\mathbf{p}$'s are regarded as operators, but in the derivation of spin orbit interaction, only the first $\mathbf{p}$ is regarded as an operator, the second $\mathbf{p}$ is regarded as a function.

Finally in the standard derivation of spin orbit interaction, the Coulomb potential of electrostatics is chosen for the scalar potential:

$$\phi = -\frac{e}{4\pi\epsilon_0 r} \tag{5.99}$$

so the electric field strength is:

$$\mathbf{E} = -\boldsymbol{\nabla}\phi = -\frac{e}{4\pi\epsilon_0 r^3}\mathbf{r}. \tag{5.100}$$

The relevant spin orbit hamiltonian becomes:

$$H_{22}\psi = -\frac{e^2\hbar}{8\pi c^2\epsilon_0 m^2 r^3}\boldsymbol{\sigma}\cdot\mathbf{r}\times\mathbf{p}\psi \tag{5.101}$$

in which the orbital angular momentum is:

$$\mathbf{L} = \mathbf{r}\times\mathbf{p}. \tag{5.102}$$

5.3. INTERACTION OF THE ECE FERMION WITH THE...

Therefore the spin orbit hamiltonian is:

$$H_{22}\psi = -\frac{e^2\hbar}{8\pi c^2\epsilon_0 m^2 r^3}\boldsymbol{\sigma}\cdot\mathbf{L}\psi. \tag{5.103}$$

In the description of atomic and molecular spectra, the spin angular momentum operator is defined as:

$$\mathbf{S} = \frac{1}{2}\hbar\boldsymbol{\sigma} \tag{5.104}$$

and the orbital angular momentum also becomes an operator. So:

$$H_{22}\psi = -\xi\,\mathbf{S}\cdot\mathbf{L} = -\frac{e^2}{8\pi c^2\epsilon_0 m^2 r^3}\mathbf{S}\cdot\mathbf{L}\psi \tag{5.105}$$

and the Thomas factor of two is contained in Eq. (5.105) as part of the denominator. The derivation of the Thomas factor is one of the strengths of the fermion equation, which as we have argued does not suffer from the negative energy problem of the Dirac equation.

Consider again the H_{22} hamiltonian:

$$H_{22}\psi = \frac{e}{4m^2c^2}\boldsymbol{\sigma}\cdot(\mathbf{p}-e\mathbf{A})\phi\,\boldsymbol{\sigma}\cdot(\mathbf{p}-e\mathbf{A})\psi \tag{5.106}$$

and assume that:

$$\mathbf{A} = \mathbf{0} \tag{5.107}$$

so:

$$H_{22}\psi = \frac{e}{4m^2c^2}\boldsymbol{\sigma}\cdot\mathbf{p}\phi\,\boldsymbol{\sigma}\cdot\mathbf{p}\psi. \tag{5.108}$$

In the derivation of spin orbit coupling and the Thomas factor the first $\mathbf{p}$ is regarded as an operator and the second $\mathbf{p}$ as a function. In the derivation of the Darwin term both $\mathbf{p}$'s are regarded as operators, defined by:

$$-i\hbar\boldsymbol{\nabla}\psi = \mathbf{p}\psi \tag{5.109}$$

with expectation value:

$$\langle\mathbf{p}\rangle = \int\psi^*\mathbf{p}\psi\,d\tau. \tag{5.110}$$

Therefore the Darwin term is obtained from:

$$H_{22}\psi = \frac{e}{4m^2c^2}\boldsymbol{\sigma}\cdot(-i\hbar\boldsymbol{\nabla})\phi\,\boldsymbol{\sigma}\cdot(-i\hbar\boldsymbol{\nabla})\psi \tag{5.111}$$

and is a quantum mechanical phenomenon with no classical counterpart.

From Eq. (5.111):

$$H_{22}\psi = -\frac{e\hbar^2}{4m^2c^2}\left(\boldsymbol{\sigma}\cdot\boldsymbol{\nabla}\phi\,\boldsymbol{\sigma}\cdot\boldsymbol{\nabla}\right)\psi \tag{5.112}$$

and the fist del operator $\boldsymbol{\nabla}$ operates on all that follows it, so:

$$H_{22}\psi = -\frac{e\hbar^2}{4m^2c^2}\boldsymbol{\sigma}\cdot\boldsymbol{\nabla}\left(\phi\,\boldsymbol{\sigma}\cdot\boldsymbol{\nabla}\psi\right). \tag{5.113}$$

The Leibnitz Theorem is used as follows:

$$\boldsymbol{\nabla}\left(\phi\boldsymbol{\sigma}\cdot\boldsymbol{\nabla}\right) = (\boldsymbol{\nabla}\phi)(\boldsymbol{\sigma}\cdot\boldsymbol{\nabla}\psi) + \phi\boldsymbol{\nabla}\left(\boldsymbol{\sigma}\cdot\boldsymbol{\nabla}\psi\right). \tag{5.114}$$

Therefore:

$$H_{22}\psi = -\frac{e\hbar^2}{4m^2c^2}\left(\boldsymbol{\sigma}\cdot\boldsymbol{\nabla}\phi\boldsymbol{\sigma}\cdot\boldsymbol{\nabla}\psi + \boldsymbol{\sigma}\cdot\phi\boldsymbol{\nabla}(\boldsymbol{\sigma}\cdot\boldsymbol{\nabla}\psi)\right). \tag{5.115}$$

Usually the Darwin term is considered to be:

$$H_{\text{Darwin}}\psi = -\frac{e\hbar^2}{4m^2c^2}\boldsymbol{\sigma}\cdot\boldsymbol{\nabla}\phi\,\boldsymbol{\sigma}\cdot\boldsymbol{\nabla}\psi. \tag{5.116}$$

and the second term in Eq. (5.115) can be developed as:

$$\boldsymbol{\sigma}\cdot\boldsymbol{\nabla}(\boldsymbol{\sigma}\cdot\boldsymbol{\nabla}\psi) = (\boldsymbol{\sigma}\cdot\boldsymbol{\nabla})(\boldsymbol{\sigma}\cdot\boldsymbol{\nabla})\psi \tag{5.117}$$

so:

$$H_{22}\psi = -\frac{e\hbar^2}{4m^2c^2}\left(\boldsymbol{\nabla}\phi\cdot\boldsymbol{\nabla}\psi + \phi\nabla^2\psi\right). \tag{5.118}$$

5.4 New Electron Spin Orbit Effects from the Fermion Equation

On the classical standard level consider the kinetic energy of an electron of mass m and linear momentum $\mathbf{p}$:

$$H = \frac{p^2}{2m} \tag{5.119}$$

and use the minimal prescription (5.56) to describe the interaction of an electron with a vector potential $\mathbf{A}$. The interaction hamiltonian is defined by:

$$H = \frac{1}{2m}(\mathbf{p} - e\mathbf{A})\cdot(\mathbf{p} - e\mathbf{A}) \tag{5.120}$$
$$= \frac{p^2}{2m} - \frac{e}{2m}(\mathbf{p}\cdot\mathbf{A} + \mathbf{A}\cdot\mathbf{p}) + \frac{e^2A^2}{2m}.$$

As discussed in earlier chapters the vector potential can be defined by:

$$\mathbf{A} = \frac{1}{2}\mathbf{B}\times\mathbf{r}. \tag{5.121}$$

Now consider the following term of the hamiltonian:

$$H_1 = -\frac{e}{2m}(\mathbf{p}\cdot\mathbf{A} + \mathbf{A}\cdot\mathbf{p}) = -\frac{e}{4m}(\mathbf{p}\cdot\mathbf{B}\times\mathbf{r} + \mathbf{B}\times\mathbf{r}\cdot\mathbf{p}) \tag{5.122}$$

5.4. NEW ELECTRON SPIN ORBIT EFFECTS FROM THE FERMION...

where the orbital angular momentum can be defined as follows:

$$\mathbf{p} \cdot (\mathbf{B} \times \mathbf{r}) = \mathbf{B} \cdot \mathbf{r} \times \mathbf{p} = \mathbf{B} \cdot \mathbf{L}. \tag{5.123}$$

This analysis gives the well known hamiltonian for the interaction of a magnetic dipole moment with the magnetic flux density:

$$H_1 = -\frac{e}{2m} \mathbf{L} \cdot \mathbf{B} = -\mathbf{m}_D \cdot \mathbf{B}. \tag{5.124}$$

The classical hamiltonian responsible for Eq. (5.124) is:

$$H_1 = -\frac{e}{2m} (\mathbf{p} \cdot \mathbf{A} + \mathbf{A} \cdot \mathbf{p}) \tag{5.125}$$

which can be written in the SU(2) basis as:

$$H_1 = -\frac{e}{2m} (\boldsymbol{\sigma} \cdot \mathbf{p}\, \boldsymbol{\sigma} \cdot \mathbf{A} + \boldsymbol{\sigma} \cdot \mathbf{A}\, \boldsymbol{\sigma} \cdot \mathbf{p}). \tag{5.126}$$

Using Pauli algebra:

$$\boldsymbol{\sigma} \cdot \mathbf{p}\, \boldsymbol{\sigma} \cdot \mathbf{A} = \mathbf{p} \cdot \mathbf{A} + i\boldsymbol{\sigma} \cdot \mathbf{p} \times \mathbf{A} \tag{5.127}$$

$$\boldsymbol{\sigma} \cdot \mathbf{A}\, \boldsymbol{\sigma} \cdot \mathbf{p} = \mathbf{A} \cdot \mathbf{p} + i\boldsymbol{\sigma} \cdot \mathbf{A} \times \mathbf{p} \tag{5.128}$$

and the same result is obtained because:

$$i\boldsymbol{\sigma} \cdot (\mathbf{p} \times \mathbf{A} + \mathbf{A} \times \mathbf{p}) = 0. \tag{5.129}$$

However, as discussed for example by H. Merzbacher in "Quantum Mechanics" (Wiley, 1970):

$$\boldsymbol{\sigma} \cdot \mathbf{p} = \frac{1}{r^2} \boldsymbol{\sigma} \cdot \mathbf{r}\, (\mathbf{r} \cdot \mathbf{p} + i\boldsymbol{\sigma} \cdot \mathbf{L}) \tag{5.130}$$

$$\boldsymbol{\sigma} \cdot \mathbf{A} = \frac{1}{r^2} \boldsymbol{\sigma} \cdot \mathbf{r}\, (\mathbf{r} \cdot \mathbf{A} + i\boldsymbol{\sigma} \cdot \mathbf{r} \times \mathbf{A}) \tag{5.131}$$

in which:

$$\frac{1}{r^2} \boldsymbol{\sigma} \cdot \mathbf{r}\, \boldsymbol{\sigma} \cdot \mathbf{r} = 1. \tag{5.132}$$

Therefore:

$$\boldsymbol{\sigma} \cdot \mathbf{p}\, \boldsymbol{\sigma} \cdot \mathbf{A} = \frac{1}{r^2} (\mathbf{r} \cdot \mathbf{p}\, \mathbf{r} \cdot \mathbf{A} + i\boldsymbol{\sigma} \cdot \mathbf{L}\, \mathbf{r} \cdot \mathbf{A} \tag{5.133}$$
$$+ i\mathbf{r} \cdot \mathbf{p}\, \boldsymbol{\sigma} \cdot \mathbf{r} \times \mathbf{A} - \boldsymbol{\sigma} \cdot \mathbf{L}\, \boldsymbol{\sigma} \cdot \mathbf{r} \times \mathbf{A}).$$

From comparison of the real and imaginary parts of Eqs. (5.127) and (5.133):

$$\mathbf{p} \cdot \mathbf{A} = \frac{1}{r^2} (\mathbf{r} \cdot \mathbf{p}\, \mathbf{r} \cdot \mathbf{A} - \boldsymbol{\sigma} \cdot \mathbf{L}\, \boldsymbol{\sigma} \cdot \mathbf{r} \times \mathbf{A}) \tag{5.134}$$

CHAPTER 5. THE UNIFICATION OF QUANTUM MECHANICS AND...

in which:

$$\boldsymbol{\sigma} \cdot \mathbf{p} \times \mathbf{A} = \boldsymbol{\sigma} \cdot \mathbf{L}\, \mathbf{r} \cdot \mathbf{A} + \mathbf{r} \cdot \mathbf{p}\, \boldsymbol{\sigma} \cdot \mathbf{r} \times \mathbf{A}, \tag{5.135}$$

$$\mathbf{r} \cdot \mathbf{A} = \frac{1}{2}\mathbf{r} \cdot \mathbf{B} \times \mathbf{r} = \frac{1}{2}\mathbf{B} \cdot \mathbf{r} \times \mathbf{r} = 0. \tag{5.136}$$

Therefore we obtain the important identities:

$$\mathbf{p} \cdot \mathbf{A} = -\frac{1}{r^2}\boldsymbol{\sigma} \cdot \mathbf{L}\, \boldsymbol{\sigma} \cdot \mathbf{r} \times \mathbf{A}, \tag{5.137}$$

$$\boldsymbol{\sigma} \cdot \mathbf{p} \times \mathbf{A} = \mathbf{r} \cdot \mathbf{p}\, \boldsymbol{\sigma} \cdot \mathbf{r} \times \mathbf{A}. \tag{5.138}$$

The hamiltonian (5.125) can therefore be written as:

$$H_1 = -\frac{e}{m}\mathbf{p} \cdot \mathbf{A} = \frac{e}{mr^2}\boldsymbol{\sigma} \cdot \mathbf{L}\, \boldsymbol{\sigma} \cdot \mathbf{r} \times \mathbf{A} = -\mathbf{m}_B \cdot \mathbf{B}. \tag{5.139}$$

Finally use eqs. (5.121) and (5.139) to find:

$$H_1 = \frac{e}{2m}\boldsymbol{\sigma} \cdot \mathbf{L}\left(\boldsymbol{\sigma} \cdot \mathbf{B} - \frac{\boldsymbol{\sigma} \cdot \mathbf{r}}{r^2}\mathbf{B} \cdot \mathbf{r}\right) = -\mathbf{m}_D \cdot \mathbf{B}. \tag{5.140}$$

It can be seen that the well known hamiltonian responsible for the Zeeman effect has been developed into a hamiltonian that gives electron spin resonance of a new type, a resonance that arises from the interaction of the Pauli matrix with the magnetic field as in Eq. (5.140). If the magnetic field is aligned in the Z axis then:

$$\sigma_Z = \begin{bmatrix} 1 & 0 \\ 0 & -1 \end{bmatrix} \tag{5.141}$$

and the electron spin orbit (ESOR) resonance frequency is:

$$\omega = \frac{eB}{m\hbar}\boldsymbol{\sigma} \cdot \mathbf{L}. \tag{5.142}$$

This compares with the usual ESR frequency:

$$\omega = \frac{eB}{m} \tag{5.143}$$

from the hamiltonian derived already in this chapter from the fermion equation.

The ESOR hamiltonian contains a novel spin orbit coupling when quantized:

$$H_1\psi = \frac{e}{2m}\boldsymbol{\sigma} \cdot \mathbf{B}\, \boldsymbol{\sigma} \cdot \mathbf{L}\, \psi. \tag{5.144}$$

Defining the spin angular momentum as:

$$\mathbf{S} = \frac{1}{2}\hbar\boldsymbol{\sigma} \tag{5.145}$$

5.4. NEW ELECTRON SPIN ORBIT EFFECTS FROM THE FERMION...

gives [1]- [10]:

$$\mathbf{L} \cdot \mathbf{S}\psi = \frac{1}{2}(J^2 - L^2 - S^2)\psi \qquad (5.146)$$
$$= \frac{1}{2}\hbar^2 \left(J(J+1) - L(L+1) - S(S+1)\right)\psi$$

so the energy levels of the ESOR hamiltonian operator are:

$$E = \frac{e\hbar}{2m}\left(J(J+1) - L(L+1) - S(S+1)\right)\boldsymbol{\sigma} \cdot \mathbf{B} \qquad (5.147)$$

giving the ESOR frequency:

$$\omega = \frac{e\hbar}{2m}\left(J(J+1) - L(L+1) - S(S+1)\right) \qquad (5.148)$$

in which the total angular momentum J is defined by the Clebsch Gordan series:

$$J = L + S, L + S - 1, \ldots, |L - S|. \qquad (5.149)$$

Eq. (5.144) was first derived in UFT 249 and is different from the well known ESR spin hamiltonian:

$$H_{\text{ESR}} = -\frac{e}{2m}\mathbf{L} \cdot \mathbf{B} + \lambda \mathbf{S} \cdot \mathbf{L} - \frac{e\hbar}{2m}\boldsymbol{\sigma} \cdot \mathbf{B} = -g_{\text{Spin}}\boldsymbol{\sigma} \cdot \mathbf{B}. \qquad (5.150)$$

It was derived using well known Pauli algebra together with the fermion equation and potentially gives rise to many useful spectral effects.

For chemical physicists and analytical chemists therefore the most useful format of the fermion equation is:

$$E\psi = \left(mc^2 + e\phi + \frac{1}{2m}\boldsymbol{\sigma} \cdot (\mathbf{p} - e\mathbf{A})(1 + \frac{e\phi}{2mc^2})\boldsymbol{\sigma} \cdot (\mathbf{p} - e\mathbf{A})\right)\psi \qquad (5.151)$$

and a few examples have been given in this chapter of its usefulness. In ECE theory Eq. (5.151) has been derived from Cartan geometry and by using the minimal prescription. The fermion equation as argued is the chiral Dirac equation without the problem of negative energy, which to chemists was never of much interest. In chemistry the subject is approached as follows. Consider one term of the complete equation (5.151):

$$H_1\psi = -\frac{e}{2m}\left(\boldsymbol{\sigma} \cdot \mathbf{A}\,\boldsymbol{\sigma} \cdot \mathbf{p} + \boldsymbol{\sigma} \cdot \mathbf{p}\,\boldsymbol{\sigma} \cdot \mathbf{A}\right)\psi. \qquad (5.152)$$

By regarding $\boldsymbol{\sigma}$ as a function rather than an operator this term can be developed using Pauli algebra as follows:

$$H_1\psi = -\frac{e}{2m}\left(\mathbf{A} \cdot \mathbf{p} + \mathbf{p} \cdot \mathbf{A} + i\boldsymbol{\sigma} \cdot (\mathbf{A} \times \mathbf{p} + \mathbf{p} \times \mathbf{A})\right)\psi. \qquad (5.153)$$

CHAPTER 5. THE UNIFICATION OF QUANTUM MECHANICS AND...

For a uniform magnetic field:

$$\mathbf{A} = \frac{1}{2}\mathbf{B} \times \mathbf{r} \tag{5.154}$$

so:

$$H_1\psi = -\frac{e}{4m}(\mathbf{B} \times \mathbf{r} \cdot \mathbf{p} + \mathbf{p} \cdot \mathbf{B} \times \mathbf{r}) \tag{5.155}$$
$$+i\boldsymbol{\sigma} \cdot ((\mathbf{B} \times \mathbf{r}) \times \mathbf{p} + \mathbf{p} \times (\mathbf{B} \times \mathbf{r}))\,\psi.$$

By regarding $\mathbf{p}$ as a function:

$$\mathbf{B} \times \mathbf{r} \cdot \mathbf{p} = \mathbf{B} \cdot \mathbf{r} \times \mathbf{p} = \mathbf{B} \cdot \mathbf{L} \tag{5.156}$$

so the hamiltonian becomes:

$$H_1\psi = -\frac{e}{2m}\mathbf{L} \cdot \mathbf{B}\psi + (i\boldsymbol{\sigma} \cdot \mathbf{A} \times \mathbf{p} + i\boldsymbol{\sigma} \cdot \mathbf{p} \times \mathbf{A})\,\psi. \tag{5.157}$$

At this stage $\mathbf{p}$ is regarded as an operator so the second term on the right hand side of eq. (5.157) does not vanish. The use of $\mathbf{p}$ and $\boldsymbol{\sigma}$ as functions or operators is arbitrary, and justified only by the final comparison with experimental data. From Eqs. (5.157) and (5.154) the hamiltonian can be written in the format used in chemistry

$$H_1\psi = (-\frac{e}{2m}\mathbf{L} \cdot \mathbf{B} - \frac{e\hbar}{2m}\boldsymbol{\sigma} \cdot \mathbf{B})\psi \tag{5.158}$$
$$= -\frac{e}{2m}(\mathbf{L} + 2\mathbf{S})\psi$$

The total angular momentum is conserved so Eq. (5.158) can be written as:

$$H_1\psi = -\frac{e}{2m}g_L\mathbf{J} \cdot \mathbf{B}\,\psi \tag{5.159}$$

where:

$$J = L + S, \ldots, |L - S| \tag{5.160}$$

from the Clebsch Gordan series.

The conventional spin orbit term emerges as described earlier in this chapter from another term of the hamiltonian:

$$H_{so}\psi = -\frac{e}{4mc^2}\boldsymbol{\sigma} \cdot (\mathbf{p} - e\mathbf{A})\phi\,\boldsymbol{\sigma} \cdot (\mathbf{p} - e\mathbf{A})\psi \tag{5.161}$$

in which the first $\mathbf{p}$ is described as an operator but in which the second $\mathbf{p}$ is a function, giving the spin orbit term:

$$H_{so}\psi = -\frac{ie\hbar}{4m^2c^2}\boldsymbol{\sigma} \cdot \boldsymbol{\nabla}\phi\,\boldsymbol{\sigma} \cdot \mathbf{p}\,\psi. \tag{5.162}$$

So the complete ESR hamiltonian is:

$$H\psi = (-\frac{e}{2m}\mathbf{L} \cdot \mathbf{B} - \frac{e\hbar}{2m}\boldsymbol{\sigma} \cdot \mathbf{B} - \xi\mathbf{S} \cdot \mathbf{L})\psi \tag{5.163}$$

5.4. NEW ELECTRON SPIN ORBIT EFFECTS FROM THE FERMION...

in which the spin orbit coupling constant is:

$$\xi = \frac{e}{4\pi c^2 \epsilon_0 m^2 r^3}. \tag{5.164}$$

Finally both **S** and **L** are operators, so:

$$\mathbf{S} \cdot \mathbf{L}\,\psi = \frac{\hbar^2}{2}\left(J(J+1) - L(L+1) - S(S+1)\right)\psi. \tag{5.165}$$

The above is the very well known conventional description of ESR in the language used by chemists, and is a description based in ECE theory on geometry. In ECE theory it can be developed in many ways because it is generally covariant while the obsolete standard description is Lorentz covariant.

However, several new spectroscopies can be developed using a well known Pauli algebra but one which seems never to have been applied to fermion resonance spectroscopies:

$$\boldsymbol{\sigma} \cdot \mathbf{p} = \frac{1}{r^2}\boldsymbol{\sigma} \cdot \mathbf{r}\left(\mathbf{r} \cdot \mathbf{p} + i\boldsymbol{\sigma} \cdot \mathbf{L}\right), \tag{5.166}$$

$$\boldsymbol{\sigma} \cdot \mathbf{A} = \frac{1}{r^2}\boldsymbol{\sigma} \cdot \mathbf{r}\left(\mathbf{r} \cdot \mathbf{A} + i\boldsymbol{\sigma} \cdot \mathbf{r} \times \mathbf{A}\right). \tag{5.167}$$

For a uniform magnetic field:

$$\mathbf{r} \cdot \mathbf{A} = 0 \tag{5.168}$$

so:

$$\mathbf{p} \cdot \mathbf{A} = \frac{1}{r^2}\boldsymbol{\sigma} \cdot \mathbf{L}\,\boldsymbol{\sigma} \cdot \mathbf{A} \times \mathbf{r} \tag{5.169}$$

and

$$\boldsymbol{\sigma} \cdot \mathbf{p} \times \mathbf{A} = \frac{1}{r^2}\mathbf{r} \cdot \mathbf{p}\,\boldsymbol{\sigma} \cdot \mathbf{r} \times \mathbf{A} \tag{5.170}$$

as in note 250(7) accompanying UFT 250 on www.aias.us. Using these results it is found that:

$$H_1\psi = -\frac{e}{2m}\left(\mathbf{p} \cdot \mathbf{A} + \mathbf{A} \cdot \mathbf{p}\right)\psi \tag{5.171}$$
$$= -\frac{e}{mr^2}\boldsymbol{\sigma} \cdot \mathbf{A} \times \mathbf{r}\,\boldsymbol{\sigma} \cdot \mathbf{L}\,\psi.$$

Using Eq. (5.171) for a uniform magnetic field gives:

$$\mathbf{A} \times \mathbf{r} = \frac{1}{2}(\mathbf{B} \times \mathbf{r}) \times \mathbf{r} = \frac{1}{2}\left(\mathbf{r}(\mathbf{r} \cdot \mathbf{B}) - r^2\mathbf{B}\right) \tag{5.172}$$

giving a novel spin orbit hamiltonian in the useful form:

$$H_1\psi = \frac{e}{\hbar m}\boldsymbol{\sigma} \cdot \left(\mathbf{B} - \frac{\mathbf{r}}{r^2}(\mathbf{r} \cdot \mathbf{B})\right)\mathbf{S} \cdot \mathbf{L}\,\psi. \tag{5.173}$$

CHAPTER 5. THE UNIFICATION OF QUANTUM MECHANICS AND...

Its expectation value is:

$$\langle H_1 \rangle = \frac{e}{\hbar m} \int \psi^* H_1 \psi \, d\tau \tag{5.174}$$

with the normalization:

$$\int \psi^* \psi \, d\tau = 1. \tag{5.175}$$

Using the result:

$$\mathbf{S} \cdot \mathbf{L} \psi = \frac{\hbar^2}{2} \left(J(J+1) - L(L+1) - S(S+1) \right) \psi \tag{5.176}$$

the energy eigenvalues of the hamiltonian are:

$$E = \frac{e\hbar}{2m} \left(J(J+1) - L(L+1) - S(S+1) \right) \\ \left(\boldsymbol{\sigma} \cdot \mathbf{B} - \int \psi^* \frac{\boldsymbol{\sigma} \cdot \mathbf{r}}{r^2} \mathbf{r} \cdot \mathbf{B} \psi \, d\tau \right) \tag{5.177}$$

as in note 250(9) accompanying UFT 250 on www.aias.us.

In spherical polar coordinates:

$$\begin{aligned} X &= r \sin\theta \cos\phi \\ Y &= r \sin\theta \sin\phi \\ X &= r \cos\theta \end{aligned} \tag{5.178}$$

and integration of a function over all space means:

$$\int f \, d\tau = \int_{\phi=0}^{2\pi} \int_{\theta=0}^{\pi} \int_0^\infty f r^2 \sin\theta \, dr \, d\theta \, d\phi. \tag{5.179}$$

If the magnetic field is aligned in the Z axis then in Cartesian coordinates:

$$\frac{\boldsymbol{\sigma} \cdot \mathbf{r}}{r^2} \mathbf{r} \cdot \mathbf{B} = \sigma_Z B_Z \frac{Z^2}{X^2 + Y^2 + Z^2} \tag{5.180}$$

and if it is assumed on average that:

$$\left\langle \frac{Z^2}{X^2 + Y^2 + Z^2} \right\rangle = \frac{1}{3} \tag{5.181}$$

the Eq. (5.177) reduces to:

$$E = \frac{1}{3} \frac{e\hbar}{m} \sigma_Z B_Z \left(J(J+1) - L(L+1) - S(S+1) \right) \tag{5.182}$$

and electron spin orbit resonance occurs at:

$$\omega = \frac{2}{3} \frac{e}{m} B_Z \left(J(J+1) - L(L+1) - S(S+1) \right). \tag{5.183}$$

5.4. NEW ELECTRON SPIN ORBIT EFFECTS FROM THE FERMION...

In spherical coordinates:

$$\frac{Z^2}{X^2 + Y^2 + Z^2} = \cos^2 \theta \tag{5.184}$$

so:

$$\int \psi^* \frac{\boldsymbol{\sigma} \cdot \mathbf{r}}{r^2} \mathbf{r} \cdot \mathbf{B} \psi \, d\tau \tag{5.185}$$

$$= \sigma_Z B_Z \int_{\phi=0}^{2\pi} \int_{\theta=0}^{\pi} \int_0^{\infty} \psi^* \cos^2 \theta \, \psi \, r^2 \sin \theta \, dr \, d\theta \, d\phi.$$

It is seen that this part of the hamiltonian is r dependent and must be evaluated for each wave function ψ. The only analytical wave functions are those of atomic H, so computational methods can be used to evaluate the energy levels of Eq. (5.185) for the H atom. The results are given in UFT 250 on www.aias.us As an example we show the two contributions of Eq. (5.177) in Table 5.1. These results are obtained for quantum numbers of atomic H. The column E_1 gives the contribution of the $\boldsymbol{\sigma} \cdot \mathbf{B}$ term and column E_2 the contribution of the integral in (5.177). F_j is the factor of quantum numbers J, L, etc. It can be seen that there are no contributions for s sates and the contributions of E_1 are always larger than those of E_2.

Consider now the hamiltonian:

$$H = -\frac{e}{2m} \left(\boldsymbol{\sigma} \cdot \mathbf{p} \, \boldsymbol{\sigma} \cdot \mathbf{A} + \boldsymbol{\sigma} \cdot \mathbf{A} \, \boldsymbol{\sigma} \cdot \mathbf{p} \right) \tag{5.186}$$

in its quantized form:

$$H\psi = -\frac{e}{2m} \frac{\hbar}{i} \left(\boldsymbol{\sigma} \cdot \boldsymbol{\nabla} \, \boldsymbol{\sigma} \cdot \mathbf{A} + \boldsymbol{\sigma} \cdot \mathbf{A} \, \boldsymbol{\sigma} \cdot \boldsymbol{\nabla} \right) \psi. \tag{5.187}$$

Note that:

$$\mathbf{r} \cdot \mathbf{p} = \frac{\hbar}{i} r \, \mathbf{e}_r \cdot \boldsymbol{\nabla} = \frac{\hbar}{i} r \frac{\partial}{\partial r} \tag{5.188}$$

where the radial unit vector is defined as:

$$\mathbf{e}_r = \frac{\mathbf{r}}{r}. \tag{5.189}$$

From Pauli algebra:

$$\boldsymbol{\sigma} \cdot \mathbf{A} = \frac{\boldsymbol{\sigma} \cdot \mathbf{r}}{r^2} (\mathbf{r} \cdot \mathbf{A} + i\boldsymbol{\sigma} \cdot \mathbf{r} \times \mathbf{A}) \tag{5.190}$$

and for a uniform magnetic field

$$\mathbf{A} = \frac{1}{2} \mathbf{B} \times \mathbf{r} \tag{5.191}$$

in which:

$$\mathbf{r} \cdot \mathbf{A} = 0 \tag{5.192}$$

CHAPTER 5. THE UNIFICATION OF QUANTUM MECHANICS AND ...

n	L	M_L	J	S	M_S	M_J	F_j	E_1	E_2	$E_1 + E_2$
1	0	0	1/2	1/2	-1/2	-1/2	0	0	0	0
1	0	0	1/2	1/2	1/2	-1/2	0	0	0	0
2	0	0	1/2	1/2	-1/2	-1/2	0	0	0	0
2	0	0	1/2	1/2	1/2	-1/2	0	0	0	0
2	1	-1	3/2	1/2	-1/2	-3/2	1	1	$\frac{1}{5}$	$\frac{6}{5}$
2	1	-1	3/2	1/2	1/2	-1/2	1	1	$\frac{1}{5}$	$\frac{6}{5}$
2	1	0	1/2	1/2	-1/2	-1/2	-1	-1	$-\frac{3}{5}$	$-\frac{8}{5}$
2	1	0	3/2	1/2	1/2	1/2	1	1	$\frac{3}{5}$	$\frac{8}{5}$
2	1	1	1/2	1/2	-1/2	1/2	-1	-1	$-\frac{1}{5}$	$-\frac{6}{5}$
2	1	1	3/2	1/2	1/2	3/2	1	1	$\frac{1}{5}$	$\frac{6}{5}$
3	0	0	1/2	1/2	-1/2	-1/2	0	0	0	0
3	0	0	1/2	1/2	1/2	-1/2	0	0	0	0
3	1	-1	3/2	1/2	-1/2	-3/2	1	1	$\frac{1}{5}$	$\frac{6}{5}$
3	1	-1	3/2	1/2	1/2	-1/2	1	1	$\frac{1}{5}$	$\frac{6}{5}$
3	1	0	1/2	1/2	-1/2	-1/2	-1	-1	$-\frac{3}{5}$	$-\frac{8}{5}$
3	1	0	3/2	1/2	1/2	1/2	1	1	$\frac{3}{5}$	$\frac{8}{5}$
3	1	1	1/2	1/2	-1/2	3/2	-1	-1	$-\frac{1}{5}$	$-\frac{6}{5}$
3	1	1	5/2	1/2	1/2	-5/2	1	1	$\frac{1}{5}$	$\frac{6}{5}$
3	2	-2	5/2	1/2	-1/2	-5/2	2	2	$\frac{2}{7}$	$\frac{16}{7}$
3	2	-2	5/2	1/2	1/2	-3/2	2	2	$\frac{2}{7}$	$\frac{16}{7}$
3	2	-1	3/2	1/2	-1/2	-3/2	-2	-2	$-\frac{6}{7}$	$-\frac{20}{7}$
3	2	-1	5/2	1/2	1/2	-1/2	2	2	$\frac{6}{7}$	$\frac{20}{7}$
3	2	0	3/2	1/2	-1/2	-1/2	-2	-2	$-\frac{22}{21}$	$-\frac{64}{21}$
3	2	0	5/2	1/2	1/2	1/2	2	2	$\frac{22}{21}$	$\frac{64}{21}$
3	2	1	3/2	1/2	-1/2	1/2	-2	-2	$-\frac{6}{7}$	$-\frac{20}{7}$
3	2	1	5/2	1/2	1/2	3/2	2	2	$\frac{6}{7}$	$\frac{20}{7}$
3	2	2	3/2	1/2	-1/2	3/2	-2	-2	$-\frac{2}{7}$	$-\frac{16}{7}$
3	2	2	5/2	1/2	1/2	5/2	2	2	$\frac{2}{7}$	$\frac{16}{7}$

Table 5.1: Energies E_1, E_2 and $E_1 + E_2$ in units of $e\hbar/(2m)$.

5.4. NEW ELECTRON SPIN ORBIT EFFECTS FROM THE FERMION...

it follows that:

$$\boldsymbol{\sigma} \cdot \mathbf{A} = i \frac{\boldsymbol{\sigma} \cdot \mathbf{r}}{r^2} \boldsymbol{\sigma} \cdot \mathbf{r} \times \mathbf{A}. \tag{5.193}$$

As in note 251(1) accompanying UFT 251 on www.aias.us it follows that:

$$\boldsymbol{\sigma} \cdot \mathbf{p} \, \boldsymbol{\sigma} \cdot \mathbf{A} \psi = \frac{\hbar}{r} \left(\boldsymbol{\sigma} \cdot \mathbf{r} \times \mathbf{A} \frac{\partial \psi}{\partial r} + \frac{\partial}{\partial r} (\boldsymbol{\sigma} \cdot \mathbf{r} \times \mathbf{A}) \right) \psi \tag{5.194}$$
$$- \frac{1}{r^2} \boldsymbol{\sigma} \cdot \mathbf{r} \times \mathbf{A} \, \boldsymbol{\sigma} \cdot \mathbf{L} \psi.$$

Using Eq. (5.191) it follows that:

$$\frac{1}{r} \mathbf{r} \times \mathbf{A} = \frac{r}{2} \left(\mathbf{B} - \mathbf{e}_r (\mathbf{B} - \mathbf{e}_r) \right) \tag{5.195}$$

and that:

$$\frac{1}{r} \frac{\partial}{\partial r} \mathbf{r} \times \mathbf{A} = \mathbf{B} - \frac{1}{2r} \frac{\partial}{\partial r} (r^2 \mathbf{e}_r (\mathbf{e}_r \cdot \mathbf{B})) \tag{5.196}$$
$$= \mathbf{B} - \mathbf{e}_r (\mathbf{e}_r \cdot \mathbf{B})$$

so:

$$\boldsymbol{\sigma} \cdot \mathbf{p} \, \boldsymbol{\sigma} \cdot \mathbf{A} \psi = \boldsymbol{\sigma} \cdot \mathbf{B}_1 (\hbar \psi + \frac{1}{2} r \frac{\partial \psi}{\partial r} - \frac{1}{2} \boldsymbol{\sigma} \cdot \mathbf{L} \psi) \tag{5.197}$$

in which the modified magnetic flux density is:

$$\mathbf{B}_1 = \mathbf{B} - \mathbf{e}_r (\mathbf{e}_r \cdot \mathbf{B}). \tag{5.198}$$

The hamiltonian (5.187) can therefore be developed as:

$$H\psi = -\frac{e\hbar}{2m} \boldsymbol{\sigma} \cdot \mathbf{B}_1 (\psi + r \frac{\partial \psi}{\partial r}) + \frac{e}{2m} \boldsymbol{\sigma} \cdot \mathbf{B}_1 \, \boldsymbol{\sigma} \cdot \mathbf{L} \psi. \tag{5.199}$$

Recall that the conventional development of the hamiltonian is well known:

$$H\psi = i \frac{e\hbar}{2m} (\boldsymbol{\nabla} \cdot (\mathbf{A}\psi) + \mathbf{A} \cdot \boldsymbol{\nabla}\psi) \tag{5.200}$$
$$- \frac{e\hbar}{2m} (\boldsymbol{\sigma} \cdot \boldsymbol{\nabla} \times (\mathbf{A}\psi) + \mathbf{A} \times \boldsymbol{\nabla}\psi)$$
$$= -\frac{e\hbar}{2m} \boldsymbol{\sigma} \cdot \mathbf{B} + i \frac{e\hbar}{2m} ((\boldsymbol{\nabla} \cdot \mathbf{A})\psi + 2\boldsymbol{\nabla}\psi \cdot \mathbf{A})$$

and misses the information given in Eq. (5.199).

As in note 251(2) on www.aias.us it is possible to define three novel types of hamiltonian:

$$H_1 \psi = -\frac{e\hbar}{2m} \boldsymbol{\sigma} \cdot \mathbf{B}_1 \psi \tag{5.201}$$

$$H_2 \psi = -\frac{e\hbar}{2m} \boldsymbol{\sigma} \cdot \mathbf{B}_1 r \frac{\partial \psi}{\partial r} \tag{5.202}$$

$$H_3 \psi = \frac{e}{2m} \boldsymbol{\sigma} \cdot \mathbf{B}_1 \, \boldsymbol{\sigma} \cdot \mathbf{L} \psi \tag{5.203}$$

CHAPTER 5. THE UNIFICATION OF QUANTUM MECHANICS AND...

whose energy expectation values are:

$$E_1 = -\frac{e\hbar}{2m} \int \psi^* \boldsymbol{\sigma} \cdot \mathbf{B}_1 \psi \, d\tau \tag{5.204}$$

$$E_2 = -\frac{e\hbar}{2m} \int \psi^* \boldsymbol{\sigma} \cdot \mathbf{B}_1 r \frac{\partial \psi}{\partial r} \, d\tau \tag{5.205}$$

$$E_3 = \frac{e}{2m} \int \psi^* \boldsymbol{\sigma} \cdot \mathbf{B}_1 \, \boldsymbol{\sigma} \cdot \mathbf{L} \psi \, d\tau \tag{5.206}$$

with the Born normalization:

$$\int \psi^* \psi \, d\tau = 1. \tag{5.207}$$

These are developed in UFT 251 for the hydrogenic wavefunctions, giving many novel results of usefulness to analytical chemisty.

The use of well known Pauli algebra in a new way is illustrated on the simplest level in UFT 252 with the kinetic energy hamiltonian itself:

$$H\psi = \frac{1}{2m} \boldsymbol{\sigma} \cdot \mathbf{p} \, \boldsymbol{\sigma} \cdot \mathbf{p} \, \psi \tag{5.208}$$

in which the Pauli algebra is:

$$\boldsymbol{\sigma} \cdot \mathbf{p} = \frac{1}{r^2} (\mathbf{r} \cdot \mathbf{p} + i\boldsymbol{\sigma} \cdot \mathbf{L}). \tag{5.209}$$

Therefore:

$$\boldsymbol{\sigma} \cdot \mathbf{p} \, \boldsymbol{\sigma} \cdot \mathbf{p} = \frac{1}{r^2} (\mathbf{r} \cdot \mathbf{p} + i\boldsymbol{\sigma} \cdot \mathbf{L})(\mathbf{r} \cdot \mathbf{p} + i\boldsymbol{\sigma} \cdot \mathbf{L}) \tag{5.210}$$

$$= \frac{1}{r^2} \left(\mathbf{r} \cdot \mathbf{p} \, \mathbf{r} \cdot \mathbf{p} + i(\mathbf{r} \cdot \mathbf{p} \, \boldsymbol{\sigma} \cdot \mathbf{L} + \boldsymbol{\sigma} \cdot \mathbf{L} \, \mathbf{r} \cdot \mathbf{p}) - L^2 - i\boldsymbol{\sigma} \cdot \mathbf{L} \times \mathbf{L} \right)$$

which can be quantized using:

$$\mathbf{r} \cdot \mathbf{p} \psi = \frac{\hbar}{i} r \frac{\partial \psi}{\partial r},$$

$$L^2 \psi = \hbar^2 l(l+1)\psi,$$

$$\mathbf{L} \times \mathbf{L} \psi = i\hbar \psi,$$

$$\mathbf{S} \cdot \mathbf{L} \psi = \frac{\hbar^2}{2} (j(j+1) - l(l+1) - s(s+1))\psi.$$

Therefore there are results such as the following which are instructive in the use of operators in quantum mechanics:

$$\mathbf{r} \cdot \mathbf{p}(\mathbf{r} \cdot \mathbf{p}\psi) = \frac{\hbar}{i} r \frac{\partial}{\partial r} \left((\frac{\hbar}{i} r \frac{\partial}{\partial r}) \psi \right). \tag{5.211}$$

As shown in detail in UFT 252 the hamiltonian (5.208) can be developed as:

$$-\frac{\hbar^2}{2m} \boldsymbol{\sigma} \cdot \boldsymbol{\nabla} \, \boldsymbol{\sigma} \cdot \boldsymbol{\nabla} \psi = -\frac{\hbar^2}{2m} \nabla^2 \psi + \frac{1}{m} \frac{\mathbf{S} \cdot \mathbf{L}}{r} \left(2 \frac{\partial \psi}{\partial r} + \frac{\psi}{r} \right) \tag{5.212}$$

5.4. NEW ELECTRON SPIN ORBIT EFFECTS FROM THE FERMION...

where the wavefunctions are the spherical harmonics:

$$\psi = Y_l^m. \tag{5.213}$$

The analysis gives two novel classes of energy expectation values:

$$E_1 = \frac{\hbar^2}{m}(j(j+1) - l(l+1) - s(s+1))\int \psi^* \frac{1}{r}\frac{\partial \psi}{\partial r} d\tau \tag{5.214}$$

and

$$E_2 = \frac{\hbar^2}{2m}(j(j+1) - l(l+1) - s(s+1))\int \psi^* \frac{1}{r^2}\psi \, d\tau \tag{5.215}$$

which are evaluated by computer in UFT 252.

Similarly the hamiltonian quadratic in the potential:

$$H_5 = \frac{e^2}{2m}\boldsymbol{\sigma} \cdot \mathbf{A}\, \boldsymbol{\sigma} \cdot \mathbf{A}\psi \tag{5.216}$$

can be developed as in UFT 252 using Eq. (5.191) as:

$$H_5 \psi = \frac{e^2 B_Z^2}{8m} r^2 (1 - \cos^2\theta)\psi \tag{5.217}$$

again giving novel types of spectroscopy.

The hamiltonian:

$$H_7 \psi = \frac{1}{2m}\boldsymbol{\sigma} \cdot \mathbf{p}(1 + \frac{e\phi}{2mc^2})\boldsymbol{\sigma} \cdot \mathbf{p}\psi \tag{5.218}$$

from the fermion equation gives the spin orbit component:

$$H_8 \psi = \frac{e}{4m^2 c^2}\boldsymbol{\sigma} \cdot \mathbf{p}\phi\, \boldsymbol{\sigma} \cdot \mathbf{p}\psi \tag{5.219}$$

as we have seen and Eq. (5.219) can also be developed using Eq. (5.209) to give:

$$H_8 \psi = \frac{e}{4m^2 c^2}(\mathbf{r} \cdot \mathbf{p} + i\boldsymbol{\sigma} \cdot \mathbf{L})\frac{\phi}{r^2}(\mathbf{r} \cdot \mathbf{p} + i\boldsymbol{\sigma} \cdot \mathbf{L})\psi. \tag{5.220}$$

There are several terms in this equation that can be developed as in UFT 252. For example:

$$H_9 \psi = \frac{e}{4m^2 c^2}\mathbf{r} \cdot \mathbf{p}(\frac{\phi}{r^2}\mathbf{r} \cdot \mathbf{p})\psi \tag{5.221}$$

in which:

$$\mathbf{r} \cdot \mathbf{p}\psi = -i\hbar r \frac{\partial \psi}{\partial r}. \tag{5.222}$$

So the hamiltonian gives:

$$H_9 \psi = \frac{e^2 \hbar^2}{16 m^2 c^2 \pi \epsilon_0}(j(j+1) - l(l+1) - s(s+1))\frac{1}{r^3}(3\psi - r\frac{\partial \psi}{\partial r}) \tag{5.223}$$

and two types of energy expectation values:

$$E_{91} = \frac{3e^2\hbar^2}{16m^2c^2\pi\epsilon_0}(j(j+1) - l(l+1) - s(s+1)) \int \frac{\psi\psi^*}{r^3} d\tau \qquad (5.224)$$

and

$$E_{92} = -\frac{e^2\hbar^2}{16m^2c^2\pi\epsilon_0}(j(j+1) - l(l+1) - s(s+1)) \int \frac{\psi\psi^*}{r^2} d\tau \qquad (5.225)$$

which give observable new fermion resonance spectra.

The main spin orbit hamiltonian (5.220) can be developed into the following four hamiltonians:

$$H_{10}\psi = \frac{e}{4m^2c^2}\mathbf{r}\cdot\mathbf{p}(\frac{\phi}{r^2}\mathbf{r}\cdot\mathbf{p}\psi) \qquad (5.226)$$

$$H_{11}\psi = \frac{ie}{4m^2c^2}\boldsymbol{\sigma}\cdot\mathbf{L}(\frac{\phi}{r^2}\mathbf{r}\cdot\mathbf{p}\psi) \qquad (5.227)$$

$$H_{12}\psi = \frac{ie}{4m^2c^2}\mathbf{r}\cdot\mathbf{p}(\frac{\phi}{r^2}\boldsymbol{\sigma}\cdot\mathbf{L}\psi) \qquad (5.228)$$

$$H_{13}\psi = -\frac{e}{4m^2c^2}\boldsymbol{\sigma}\cdot\mathbf{L}(\frac{\phi}{r^2}\boldsymbol{\sigma}\cdot\mathbf{L}\psi) \qquad (5.229)$$

and these are evaluated systematically in UFT 252 giving many new results.

Finally in this section the effect of gravitation on fermion resonance spectra can be evaluated as in UFT 253 using the gravitational minimal prescription:

$$E \rightarrow E + m\Phi \qquad (5.230)$$

where the gravitational potential is:

$$\Phi = -\frac{GM}{r} \qquad (5.231)$$

where G is Newton's constant and where Φ is the gravitational potential. Here M is a mass that is attracted to the mass of the electron m. Various effects of gravitation are developed in UFT 253.

5.5 Refutation of Indeterminacy: Quantum Hamilton and Force Equations

The methods used to derive the fermion equation can be used as in UFT 175 to UFT 177 on www.aias.us to derive the Schroedinger equation from differential geometry. The fundamental axioms of quantum mechanics can be derived from geometry and relativity. These methods can be used to infer the existence of the quantized equivalents of the Hamilton equations of motion, which Hamilton derived in about 1833 without the use of the lagrangian dynamics. It is very well known that the Hamilton equations use position (x) and momentum (p) as conjugate variables in a well defined classical sense [1]- [10] and so x and p are "specified simultaneously" in the dense Copenhagen jargon of the

5.5. REFUTATION OF INDETERMINACY: QUANTUM HAMILTON...

twentieth century. Therefore, by quantum classical equivalence, x and p are specified simultaneously in the quantum Hamilton equations, thus refuting the Copenhagen interpretation of quantum mechanics based on the commutator of operators of position and momentum. The quantum Hamilton equations were derived for the first time in UFT 175 in 2011, and are described in this section. They show that x and p are specified simultaneously in quantum mechanics, a clear illustration of the confusion caused by the Copenhagen interpretation.

The anti commutator $\{\hat{x}, \hat{p}\}$ is used in this section to derive further refutations of Copenhagen, in that $\{\hat{x}, \hat{p}\}$ acting on a wavefunctions that are exact solutions of Schroedinger's equation produces expectation values that are zero for the harmonic oscillator, and non zero for atomic H. The anti commutator $\{\hat{x}, \hat{p}\}$ is shown to be proportional to the commutator $[x^2, p^2]$, whose expectation values for the harmonic oscillator are all zero, while for atomic H they are all non-zero. For the particle on a ring, combinations can be zero, while individual commutators of this type are non-zero. For linear motion self inconsistencies in the Copenhagen interpretation are revealed, and for the particle on a sphere the commutator is again non-zero. The hand calculations in fifteen additional notes accompanying UFT 175 are checked with computer algebra, as are all calculations in UFT theory to which computer algebra may be applied. Tables were produced in UFT 175 of the relevant expectation values. The Copenhagen interpretation is completely refuted because in that interpretation it makes no sense for the expectation value of a commutator of operators to be both zero and non-zero for the same pair of operators. One of the operators would be absolutely unknowable and the other precisely knowable if the expectation value were non zero, and both precisely knowable if it were zero. These two interpretations refer respectively to non zero and zero commutator expectation values, and both interpretations cannot be true for the same pair of operators. Prior to the work in UFT 175 in 2011, commutators of a given pair of operators were thought to be zero or non zero, never both zero and non zero, so a clear refutation of Copenhagen was never realized. In ECE theory, Copenhagen and its unscientific, anti Baconian, jargon are not used, and expectation values are straightforward consequences of the fundamental operators introduced by Schroedinger. The latter immediately rejected Copenhagen, as did Einstein and de Broglie.

The Schroedinger equation is derived in ECE from the tetrad postulate of Cartan geometry, which is reformulated as the ECE wave equation:

$$(\Box + R)q^a{}_\nu = 0 \tag{5.232}$$

where:

$$R := q^\nu{}_a \partial^\mu (\omega^a{}_{\mu\nu} - \Gamma^a{}_{\mu\nu}) \tag{5.233}$$

as discussed earlier in this book. The fermion equation in its wave format is the limit:

$$R \to \left(\frac{mc}{\hbar}\right)^2 \tag{5.234}$$

CHAPTER 5. THE UNIFICATION OF QUANTUM MECHANICS AND...

and for the free particle reduces to:

$$-\frac{\hbar^2}{2m}\nabla^2\psi = (E - mc^2)\psi. \tag{5.235}$$

This equation reduces to the Schroedinger equation:

$$-\frac{\hbar^2}{2m}\nabla^2\psi = E_{NR}\psi \tag{5.236}$$

where:

$$E_{NR} = E - mc^2. \tag{5.237}$$

In the presence of potential energy the Schroedinger equation becomes:

$$-\frac{\hbar^2}{2m}\nabla^2\psi = (E_{NR} + V)\psi. \tag{5.238}$$

In this derivation, the fundamental axiom of quantum mechanics follows from the wave equation (5.232) and from the necessity that the classical equivalent of the hamiltonian operator H is the hamiltonian in classical dynamics, the sum of the kinetic and potential energies:

$$H = E_{NR} + V. \tag{5.239}$$

So in ECE physics, quantum mechanics can be derived from general relativity in a straightforward way that can be tested against experimental data at each stage. For example earlier in this chapter the method resulted in many new types of spin orbit spectroscopies.

The two quantum Hamilton equations are derived respectively using the well known position and momentum representations of quantum mechanics. In the position representation the Schroedinger axiom is:

$$\hat{p}\psi = -i\hbar\frac{\partial \psi}{\partial x}, \quad (\hat{p}\psi)^* = i\hbar\frac{\partial \psi}{\partial x}, \tag{5.240}$$

from which it follows that:

$$[\hat{x}, \hat{p}]\,\psi = i\hbar\psi. \tag{5.241}$$

So the expectation value of the commutator is:

$$\langle[\hat{x}, \hat{p}]\rangle = i\hbar. \tag{5.242}$$

In the position representation the expectation value, <x>, of x is x. It follows that:

$$\frac{d}{dx}\langle\hat{x}\rangle = -\frac{i}{\hbar}\langle[\hat{x}, \hat{p}]\rangle = 1. \tag{5.243}$$

Note that this tautology can be derived as follows from the equation:

$$\frac{d}{dx}\langle\hat{x}\rangle = \frac{d}{dx}\int \psi^*\hat{x}\psi\, d\tau \tag{5.244}$$

5.5. REFUTATION OF INDETERMINACY: QUANTUM HAMILTON...

which can be proven as follows. First use the Leibnitz Theorem to find that:

$$\frac{d}{dx}\int \psi^* \hat{x}\psi \, d\tau = \int \frac{d\psi^*}{dx}\hat{x}\psi \, d\tau + \int \psi^* \hat{x}\frac{d\psi}{dx} \, d\tau. \tag{5.245}$$

In quantum mechanics the operators are hermitian operators defined as follows:

$$\int \psi_m^* \hat{A}\psi_n \, d\tau = \left(\int \psi_n^* \hat{A}\psi_m \, d\tau\right)^* = \int \hat{A}^* \psi_m^* \psi_n \, d\tau. \tag{5.246}$$

Therefore it follows that Eq. (5.245) is:

$$\frac{d}{dx}\langle\hat{x}\rangle = -\frac{i}{\hbar}\int \psi^*(\hat{p}\hat{x} - \hat{x}\hat{p})\psi \, d\tau \tag{5.247}$$

which is Eq. (5.243), Q. E. D.

The first quantum Hamilton equation is obtained by generalizing x to any hermitian operator A of quantum mechanics:

$$\hat{x} \to \hat{A} \tag{5.248}$$

so one format of the first quantum Hamilton equation is:

$$\frac{d}{dx}\langle\hat{A}\rangle = \frac{i}{\hbar}\langle[\hat{p},\hat{A}]\rangle. \tag{5.249}$$

In the special case:

$$\hat{A} = \hat{H} \tag{5.250}$$

then:

$$\frac{d}{dx}\langle\hat{H}\rangle = \frac{i}{\hbar}\langle[\hat{p},\hat{H}]\rangle. \tag{5.251}$$

However, it is known that:

$$\frac{d}{dt}\langle\hat{p}\rangle = \frac{i}{\hbar}\langle[\hat{H},\hat{p}]\rangle \tag{5.252}$$

so from Eqs. (5.251) and (5.252) the quantum Hamilton equation is:

$$\frac{d}{dx}\langle\hat{H}\rangle = -\frac{d}{dt}\langle\hat{p}\rangle. \tag{5.253}$$

The expectation values in this equation are:

$$H = \langle\hat{H}\rangle, \quad p = \langle\hat{p}\rangle \tag{5.254}$$

so the first Hamilton equation of motion of 1833 follows, Q. E. D.:

$$\frac{dH}{dx} = -\frac{dp}{dt}. \tag{5.255}$$

CHAPTER 5. THE UNIFICATION OF QUANTUM MECHANICS AND...

The second quantum Hamilton equation follows from the momentum representation:

$$\hat{x}\psi = -\frac{\hbar}{i}\frac{\partial \psi}{\partial p}, \quad \hat{p}\psi = p\psi \tag{5.256}$$

from which the following tautology follows:

$$\frac{d}{dp}\langle\hat{p}\rangle = \frac{\hbar}{i}[\langle\hat{x}\rangle, \langle\hat{p}\rangle] = 1. \tag{5.257}$$

This tautology can be obtained from the equation:

$$\frac{d}{dp}\langle\hat{p}\rangle = \frac{d}{dp}\int \psi^*\hat{p}\psi\,d\tau. \tag{5.258}$$

Now generalize p to any operator A:

$$\hat{p} \to \hat{A} \tag{5.259}$$

and the second quantum Hamilton equation in one format is:

$$\frac{d}{dp}\langle\hat{A}\rangle = -\frac{i}{\hbar}\langle[\hat{x},\hat{A}]\rangle. \tag{5.260}$$

In the special case:

$$\hat{A} = \hat{H} \tag{5.261}$$

the second quantum Hamilton equation is:

$$\frac{d}{dp}\langle\hat{H}\rangle = -\frac{i}{\hbar}\langle[\hat{x},\hat{H}]\rangle. \tag{5.262}$$

However it is known that:

$$\langle[\hat{x},\hat{H}]\rangle = -\frac{\hbar}{i}\frac{\langle\hat{x}\rangle}{dt} \tag{5.263}$$

so the second quantum Hamilton equation is:

$$\frac{d}{dp}\langle\hat{H}\rangle = \frac{d}{dt}\langle\hat{x}\rangle \tag{5.264}$$

which reduces to its classical counterpart, the second quantum Hamilton equation of classical dynamics, Q. E. D.:

$$\frac{dH}{dp} = \frac{dx}{dt}. \tag{5.265}$$

Note carefully that both the quantum Hamilton equations derive directly from the familiar commutator (5.242) of quantum mechanics. Conversely the Hamilton equations of 1833 imply the commutator (5.242) given only the Schroedinger postulate in position and momentum representation respectively.

5.5. REFUTATION OF INDETERMINACY: QUANTUM HAMILTON...

In the Hamilton equations of classical dynamics, x and p are simultaneously observable, so they are also simultaneously observable in the quantized Hamilton equations of motion and in quantum mechanics in general. This argument refutes Copenhagen straightforwardly, and the arbitrary assertion that x and p are not simultaneously observable.

The anti commutator method of refuting Copenhagen was also developed in UFT 175 on www.aias.us and is based on the definition of the anti commutator:

$$\{\hat{x}, \hat{p}\}\psi = (\hat{x}\,\hat{p} + \hat{p}\,\hat{x})\,\psi. \tag{5.266}$$

In the position representation the anti commutator is:

$$\{\hat{x}, \hat{p}\}\psi = -i\hbar \left(x\frac{\partial \psi}{\partial x} + \frac{\partial}{\partial x}(x\psi) \right) = -i\hbar \left(\psi + 2x\frac{\partial \psi}{\partial x} \right). \tag{5.267}$$

Similarly the commutator of $\hat{p}^2$ and $\hat{x}^2$ is defined as:

$$[\hat{x}^2, \hat{p}^2]\psi = \left([\hat{x}^2, \hat{p}]\hat{p} + \hat{p}([\hat{x}^2, \hat{p}])\right)\psi. \tag{5.268}$$

Now use the quantum Hamilton equations to find that:

$$[\hat{p}, \hat{x}^2]\psi = -\,2i\hbar x\psi, \tag{5.269}$$

$$[\hat{x}^2, \hat{p}]\psi = 2i\hbar x\psi. \tag{5.270}$$

It follows that:

$$[\hat{x}^2, \hat{p}^2]\psi = 2i\hbar(\hat{p}\hat{x} + \hat{x}\hat{p})\psi. \tag{5.271}$$

so the following useful equation has been proven in one dimension:

$$[\hat{x}^2, \hat{p}^2]\psi = 2i\hbar\,\{\hat{x}, \hat{p}\}\,\psi. \tag{5.272}$$

In three dimensions the Schroedinger axiom in position representation is:

$$\hat{p}\psi = -i\hbar\boldsymbol{\nabla}\psi \tag{5.273}$$

and in three dimensions the relevant commutator is:

$$[\mathbf{r}, \mathbf{p}]\psi = -i\hbar(\mathbf{r}\cdot\boldsymbol{\nabla}\psi - \boldsymbol{\nabla}\cdot(\mathbf{r}\psi)) \tag{5.274}$$

where in Cartesian coordinates:

$$r^2 = X^2 + Y^2 + Z^2. \tag{5.275}$$

Therefore:

$$[\mathbf{r}, \mathbf{p}]\psi = -i\hbar(\mathbf{r}\cdot\boldsymbol{\nabla}\psi - \psi\boldsymbol{\nabla}\cdot\mathbf{r} - \mathbf{r}\cdot\boldsymbol{\nabla}\psi) \tag{5.276}$$

where:

$$\boldsymbol{\nabla}\cdot(\mathbf{r}\psi) = \psi\boldsymbol{\nabla}\cdot\mathbf{r} + \mathbf{r}\cdot\boldsymbol{\nabla}\psi \tag{5.277}$$

CHAPTER 5. THE UNIFICATION OF QUANTUM MECHANICS AND ...

in which:

$$\nabla \cdot \mathbf{r} = 3. \tag{5.278}$$

So:

$$[\hat{\mathbf{r}}, \hat{\mathbf{p}}]\psi = 3i\hbar\psi. \tag{5.279}$$

In three dimensions:

$$[\hat{r}^2, \hat{p}^2]\psi = \left([\hat{r}^2, \hat{\mathbf{p}}] \cdot \hat{\mathbf{p}} + \hat{\mathbf{p}} \cdot ([\hat{r}^2, \hat{\mathbf{p}}])\right)\psi \tag{5.280}$$

where:

$$[\hat{r}^2, \hat{\mathbf{p}}]\psi = r^2\hat{\mathbf{p}}\psi - \hat{\mathbf{p}}(r^2\psi) = i\hbar\nabla r^2\psi \tag{5.281}$$

and where:

$$\nabla r^2 = \frac{\partial r^2}{\partial X}\mathbf{i} + \frac{\partial r^2}{\partial Y}\mathbf{j} + \frac{\partial r^2}{\partial Z}\mathbf{k} \tag{5.282}$$

with:

$$r^2 = X^2 + Y^2 + Z^2. \tag{5.283}$$

So:

$$\nabla r^2 = 2\mathbf{r} \tag{5.284}$$

and the three dimensional equivalent of Eq. (5.272) is:

$$[\hat{r}^2, \hat{p}^2]\psi = 2i\hbar\{\hat{\mathbf{r}}, \hat{\mathbf{p}}\}\psi. \tag{5.285}$$

The anti commutator in this equation is:

$$(\hat{\mathbf{r}} \cdot \hat{\mathbf{p}} + \hat{\mathbf{p}} \cdot \hat{\mathbf{r}})\psi = \hat{\mathbf{r}} \cdot \hat{\mathbf{p}}\psi + \hat{\mathbf{p}} \cdot (\hat{\mathbf{r}}\psi) = -i\hbar(2\mathbf{r} \cdot \nabla\psi + 3\psi) \tag{5.286}$$

where:

$$\mathbf{r} \cdot \nabla\psi = X\frac{\partial\psi}{\partial X} + Y\frac{\partial\psi}{\partial Y} + Z\frac{\partial\psi}{\partial Z} \tag{5.287}$$

so in Cartesian coordinates:

$$\{\hat{\mathbf{r}}, \hat{\mathbf{p}}\}\psi = -i\hbar\left(2\left(X\frac{\partial\psi}{\partial X} + Y\frac{\partial\psi}{\partial Y} + Z\frac{\partial\psi}{\partial Z}\right) + 3\psi\right). \tag{5.288}$$

When considering the H atom the relevant anti commutator is:

$$\{\hat{r}, \hat{p}_r\}\psi = -i\hbar\{r, \frac{\partial}{\partial r}\}\psi. \tag{5.289}$$

With these definitions some expectation values:

$$\langle[\hat{r}^2, \hat{p}^2]\rangle = 2i\hbar\langle\{\hat{\mathbf{r}}, \hat{\mathbf{p}}\}\rangle \tag{5.290}$$

145

5.5. REFUTATION OF INDETERMINACY: QUANTUM HAMILTON...

l,m	$[r,p_r]$	$\{r,p_r\}$	$[r,p_r^2]$	$\{r,p_r^2\}$	$[r^2,p_r]$	$\{r^2,p_r\}$	$[r^2,p_r^2]$	$\{r^2,$
0,0	$i\hbar$	$2i\hbar$	0	$1/(a_0)\hbar^2$	$3ia_0\hbar$	$3ia_0\hbar$	0	0
1,0	$i\hbar$	$2i\hbar$	0	$1/(a_0)\hbar^2$	$12ia_0\hbar$	$12ia_0\hbar$	0	$3\hbar$
1,1	$i\hbar$	$2i\hbar$	0	$1/(2a_0)\hbar^2$	$10ia_0\hbar$	$10ia_0\hbar$	0	$\hbar^2$
2,0	$i\hbar$	$2i\hbar$	0	$1/(a_0)\hbar^2$	$27ia_0\hbar$	$27ia_0\hbar$	0	$8\hbar$
2,1	$i\hbar$	$2i\hbar$	0	$7/(9a_0)\hbar^2$	$25ia_0\hbar$	$25ia_0\hbar$	0	$6\hbar$
2,2	$i\hbar$	$2i\hbar$	0	$1/(3a_0)\hbar^2$	$21ia_0\hbar$	$21ia_0\hbar$	0	$2\hbar$

Table 5.2: Commutators for radial wave functions of the Hydrogen atom. a_0 is the Bohr radius.

are worked out for exact solutions of the Schroedinger equation in the fifteen calculational notes accompanying UFT 175 on www.aias.us. All expectation values were checked by computer algebra and tabulated. The result is a definitive refutation of Copenhagen because expectation values can be zero or non-zero depending on which solution of Schroedinger's equation is used, as discussed already. So this method reduces Copenhagen to absurdity, Q. E. D., a reductio ad absurdum refutation of the Copenhagen interpretation of quantum mechanics.

As an example we show various commutators and anti commutators for the wave functions of atomic Hydrogen in Table 5.2. For example r^2 commutes with p_r^2 but r does not commute with p. This means according to the Copenhagen interpretation that r^2 and p_r^2 are known at the same time, but r and p_r are not, a reductio ad absurdum.

The force equation of quantum mechanics was first inferred in 2011 in UFT 176 and UFT 177 on www.aias.us and has been very influential. It was derived from the two quantum Hamilton equations:

$$i\hbar \frac{d}{dq}\left\langle \hat{H} \right\rangle = \left\langle [\hat{H},\hat{p}] \right\rangle \tag{5.291}$$

and

$$i\hbar \frac{d}{dp}\left\langle \hat{H} \right\rangle = -\left\langle [\hat{H},\hat{q}] \right\rangle \tag{5.292}$$

applied to canonical operators $\hat{p}$ and $\hat{q}$. By using the well known [1]- [10]:

$$\frac{d}{dq}\left\langle \hat{H} \right\rangle = \left\langle \frac{d\hat{H}}{dq} \right\rangle, \quad \frac{d}{dp}\left\langle \hat{H} \right\rangle = \left\langle \frac{d\hat{H}}{dp} \right\rangle \tag{5.293}$$

these equations can be put into operator format as follows:

$$i\hbar \frac{d\hat{H}}{dq}\psi = [\hat{H},\hat{p}]\psi \tag{5.294}$$

CHAPTER 5. THE UNIFICATION OF QUANTUM MECHANICS AND...

and

$$i\hbar \frac{d\hat{H}}{dp}\psi = -[\hat{H}, \hat{q}]\psi \qquad (5.295)$$

where ψ is the wave function. If the hamiltonian is defined as:

$$H = \frac{p^2}{2m} + V(x) \qquad (5.296)$$

then:

$$\frac{dH}{dx} = \frac{dV}{dx} \qquad (5.297)$$

because in the Hamilton dynamics x and p are independent, canonical variables. Therefore Eq. (5.293) is satisfied automatically. Using the result:

$$[\hat{H}, \hat{p}]\psi = i\hbar \frac{dV}{dx}\psi = -i\hbar F\psi \qquad (5.298)$$

where F is force, Eq. (5.291) gives the force equation of quantum mechanics:

$$-\frac{d\hat{H}}{dx}\psi = F\psi \qquad (5.299)$$

where the eigenoperator is defined by:

$$\frac{d\hat{H}}{dx} := -\hbar^2 \frac{\partial^3}{\partial x^3} + \frac{dV(x)}{dx}. \qquad (5.300)$$

In the classical limit, the corresponding principle of quantum mechanics means that Eq. (5.299) becomes one of the Hamilton equations:

$$F = \frac{dp}{dt} = -\frac{dH}{dx}. \qquad (5.301)$$

In the momentum representation Eq. (5.295) gives a second fundamental equation of quantum mechanics:

$$\frac{d\hat{H}}{dp}\psi = v\psi \qquad (5.302)$$

where the eigenvalues are those of quantized velocity. Here:

$$\frac{dH}{dp} = \frac{p}{m} \qquad (5.303)$$

and:

$$\frac{d\hat{H}}{dp}\psi = v\psi. \qquad (5.304)$$

5.5. REFUTATION OF INDETERMINACY: QUANTUM HAMILTON...

Eq. (5.302) corresponds in the classical limit to the second Hamilton equation:

$$v = \frac{dx}{dt} = \frac{dH}{dp}. \tag{5.305}$$

The general, or canonical, formulation of Eqs. (5.299) and (5.302) is as follows:

$$-\frac{d\hat{H}}{dq}\psi = F\psi \tag{5.306}$$

and

$$\frac{d\hat{H}}{dp}\psi = v\psi \tag{5.307}$$

which reduce to the canonical Hamilton equations:

$$-\frac{dH}{dq} = \frac{dp}{dt} \tag{5.308}$$

and

$$\frac{dH}{dp} = \frac{dq}{dt}. \tag{5.309}$$

The rotational equivalent of Eq. (5.310) is:

$$i\hbar\frac{d\hat{H}}{d\phi}\psi = [\hat{H}, \hat{J}_Z]\psi \tag{5.310}$$

in which the canonical variables are:

$$q = \phi, \quad \hat{p} = \hat{J}_Z. \tag{5.311}$$

For rotational problems in the quantum mechanics of atoms and molecules, H commutes with $\hat{J}_Z$ so

$$[\hat{H}, \hat{J}_Z] = 0 \tag{5.312}$$

in which case:

$$\frac{d\hat{H}}{d\phi}\psi = 0. \tag{5.313}$$

In order for $d\hat{H}/d\phi$ to be non-zero there must be a ϕ dependent potential energy in the hamiltonian:

$$H = \frac{J^2}{2I} + V(\phi) \tag{5.314}$$

so the hamiltonian operator must be:

$$\hat{H} = -\frac{\hbar^2}{2I}\hat{\Lambda}^2 + V(\phi) \tag{5.315}$$

where $\hat{\Lambda}$ is the lagrangian operator. In this case:

$$\frac{d\hat{H}}{d\phi} = -\frac{\hbar^2}{2I}\hat{\Lambda}^2 + \frac{dV}{d\phi} \tag{5.316}$$

and Eq (5.310) gives the torque equation of quantum mechanics:

$$-\frac{d\hat{H}}{d\phi}\psi = T_q\psi = -\frac{dV}{d\phi}\psi \tag{5.317}$$

where T_q are eigenvalues of torque.

There also exist higher order quantum Hamilton equations as discussed in UFT 176, and quantum Hamilton equations for rotation in a plane.

Finally as shown in detail in the influential UFT 177 on www.aias.us the force equation of quantum mechanics can be derived from the quantum Hamilton equations and is:

$$(\hat{H} - E)\frac{d\psi}{dx} = F\psi \tag{5.318}$$

where the force is defined by:

$$\frac{d}{dx}\left\langle \hat{H} \right\rangle = \frac{dH}{dx} = \frac{dV}{dx} = -F = -\frac{dp}{dt}. \tag{5.319}$$

In the force equation the hamiltonian operator acts on the derivative of the Schroedinger wave function or in general on the derivative of a quantum mechanical wave function obtained in any way, for example in computational quantum chemistry, and this is a new method of general utility as developed in UFT 175.

As some examples we show the radial quantum force component for some orbitals of Hydrogen (Figs. 5.1 - 5.3). There are poles in the force, but only at radial values where the wave functions are zero. This inhibits divergence for the expectation values of force. In some cases there are zero crossing of the force where the wave function has a minimum or maximum. This could mean that charge is shifted to places of high probability density. For molecules this could have to do with the fact that the same type of orbitals can be binding or anti-binding, depending on the symmetry. The force eigenvalues have the potential of giving new insight into chemical bonding and stability mechanisms. There are more examples in UFT papers 177 and 178.

5.5. REFUTATION OF INDETERMINACY: QUANTUM HAMILTON...

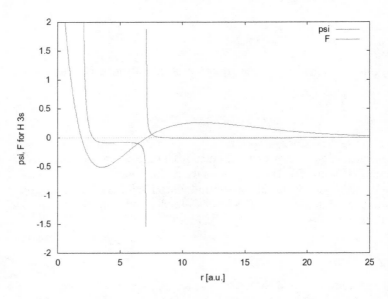

Figure 5.1: Radial wave function and quantum force for H 3s orbital.

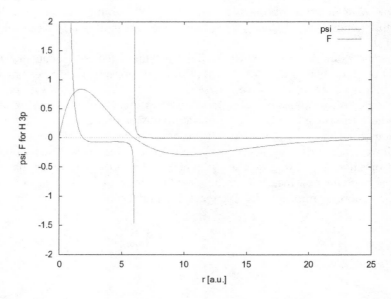

Figure 5.2: Radial wave function and quantum force for H 3p orbital.

CHAPTER 5. THE UNIFICATION OF QUANTUM MECHANICS AND...

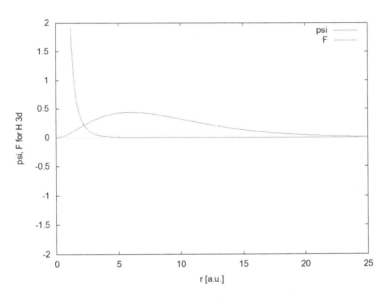

Figure 5.3: Radial wave function and quantum force for H 3d orbital.

5.5. REFUTATION OF INDETERMINACY: QUANTUM HAMILTON...

Chapter 6

Antisymmetry

6.1 Introduction

The concept of anti symmetry pervades ECE theory, and manifests itself in several important ways. The theory is based on differential forms that are anti symmetric [1]- [10] by definition, notably the torsion form. This is a vector valued two form of differential geometry, and in another language is an anti symmetric tensor with an upper a index signalling the fact that electromagnetism in ECE theory has a fundamentally different geometry that is more complete than that of the Maxwell Heaviside theory. As explained in chapter 1, the first and second Cartan Maurer structure equations define the anti symmetric torsion form and the anti symmetric curvature form, a tensor valued two form of differential geometry. In a way, the entire ECE theory is anti symmetric from the basics of geometry.

The fundamentally important achievement of Cartan geometry is to reduce everything to two fundamental objects, the torsion and curvature, which are defined in terms of the tetrad and the spin connection in a very simple way. The great elegance of the Cartan geometry is that it reduces very complicated vector and tensor equations to simple form equations. However this mathematical elegance can only be achieved at the expense of abstraction, as is always the case in mathematics. However abstract a mathematical theory, it must always reduce to well known but less elegant mathematics. If it does not, or is not comprehensible, it is either self inconsistent or effectively useless in natural philosophy. The less elegant vector format of the Cartan structure equations has proven to be the most useful in the foregoing chapters, but the structure equations show that everything is anti symmetric.

The reason for this is that the structure equations, when translated into tensorial language, are defined by the commutator of covariant derivatives. It is important to note that the structure equations are precisely the same fundamental definitions of geometry in all notations: differential form, tensor and vector. As explained already in this book the commutator is anti symmetric by definition. It is loosely referred to as a round trip in a mathematical space of any dimension. This round trip, or return journey, defines the two struc-

6.1. INTRODUCTION

ture equations of Cartan and Maurer in an elegant way and shows that the two structure equations are not independent, they are always linked by the commutator. This very fundamental property of mathematics can be looked upon, loosely writing, as a reason for the existence of the Cartan identity and the Evans identity of differential geometry. So the commutator is the "most fundamental" object in geometry. It was unknown to pioneers such as Riemann, Christoffel, Ricci, Levi-Civita and Bianchi, otherwise they would have inferred torsion, (which they obviously did not), and would have realized that the Christoffel connection is anti symmetric from the most fundamental type of reasoning in mathematics. This realization is the key to the anti symmetry laws of ECE theory developed in this chapter. They are powerful laws that refute the Maxwell Heaviside (MH) theory immediately, showing that the MH theory is lacking in information and is self insufficient and inconsistent. This is a major advance in electromagnetism that was fully realized in UFT 130 ff. on www.aias.us. It is not clear whether Cartan and Maurer inferred the commutator, it may be present in their work, but it is not made clear. The commutator is present in Lie algebra however, and is a fundamental concept there. To chemists its most well known manifestation is the commutator of Pauli matrices which gives another Pauli matrix, defining the SU(2) basis used by Dirac.

The famous role of Albert Einstein in all this was to propose that non Euclidean geometry is needed for the theory of gravitation. He finally decided in a paper published in late 1915 to use the second Bianchi identity known to him. Naturally this was the second Bianchi identity without torsion, torsion was unknown in 1915. The UFT paper 88 published about six or seven years ago has been influential in showing that the second Bianchi identity as used by Einstein is incorrect, so the Einsteinian era is over and we are entering into a post Einsteinian era of thought. One cannot make a howler in mathematics, however well intentioned, and expect to get away with it for a century – unless of course one is Einstein, who cannot be wrong. This is very familiar – human nature as distinct from nature, and human nature is almost always wrong. So people are still busy proving the precision of the Einstein theory knowing full well that it collapsed completely almost sixty years ago when the velocity curve of a whirlpool galaxy was discovered experimentally. They are dogmatists because they ignore nature, they are not Baconian scientists.

In historical fact, which is always brushed aside by dogmatists, Einstein did not get away with it at all, he was criticised severely by Schwarzschild in December 1915 in a letter which is now online and easily googled up, placed there by A. A. Vankov as discussed already in this book. Vankov has pointed out many more errors in the 1915 paper of Einstein, but UFT 88 destroyed his theory completely and replaced it with the correct second Bianchi identity. UFT88 has been studied several thousand times in about six years without any objection. So one would not like to be a dogmatist any more. If Bianchi had had the commutator at his disposal he would have inferred torsion, being the clear minded mathematician that he was. All the details of the calculation are given in UFT 99, again a heavily studied paper, again without a single objection. After Schwarzschild's untimely demise in 1916 there was a free

CHAPTER 6. ANTISYMMETRY

for all, the main critic was gone. However, Bauer and Schroedinger noted independently in 1918 that something was drastically wrong with the Einstein field equation. They were brushed aside by human nature, and the world was told that Eddington had proven general relativity. The world did not know about torsion, or in fact anything about general relativity. Eddington did not have anywhere approaching the precision to prove anything. Almost a century later people are still trying to prove that light bending is twice the Newtonian value, and their experiments are still being criticised. The critics are still being brushed aside. This chant of "twice the Newtonian value" is reminiscent of Golding's "Lord of the Flies". It is a ritual like any other. The data may or may not be precise but do not prove a mathematical howler. They can be investigated however with the post Einsteinian ECE theory and we can do our best to make sense of them. That is Baconian science.

Cartan inferred his elegant geometry in the early twenties, in the middle of the golden era of physics when profound discoveries had become commonplace. The only thing known about geometry at this time, when Einstein suddenly became famous, was that the curvature is anti symmetric in its last two indices. To the general public this meant absolutely nothing, but the same general public regarded Einstein as an Idol of the Cave. This is a metaphor, no disrespect to Einstein, who must have been intensely irritated by his new found fame, especially as he was being harassed by a bee – Elie Cartan – more irritating than any fly. Cartan had written to Einstein in the most respectful terms pointing out that Einstein's geometry had half of it missing. It contained curvature but no torsion, two wheels on the wagon, which was listing badly and about to sink. There ensued a correspondence known only to a tiny group of scholars. It was always a polite correspondence which made Einstein fully aware of torsion but the latter was not incorporated into the theory of general relativity.

There is little purpose in going in to the details of this correspondence because it was carried out at a time when the action of the commutator on a vector was not clear. The relevant contemporary equation was given in chapter 1 and is recounted here for ease of reference:

$$[D_\mu, D_\nu] V^\rho = R^\rho{}_{\mu\nu\sigma} V^\sigma - T^\lambda{}_{\mu\nu} D_\lambda V^\rho \tag{6.1}$$

Here $T^\lambda{}_{\mu\nu}$ is the torsion in tensor format [1]- [10] and $R^\rho{}_{\mu\nu\sigma}$ is the curvature in tensor format. This equation is the essence of anti symmetry in ECE theory. The commutator acts on a vector V^ρ in any dimension in any mathematical space. It is made up of the covariant derivatives defined by Christoffel in the eighteen sixties:

$$D_\mu V^\nu = \partial_\mu V^\nu + \Gamma^\nu{}_{\mu\lambda} V^\lambda \tag{6.2}$$

using the Christoffel connection $\Gamma^\nu{}_{\mu\lambda}$. It is the geometrical connection that makes the space different from that of Euclid, two thousand plus years ago. The commutator formalism is valid in n dimensions, while Euclid thought in three dimensions, without a geometrical connection.

The first thing to notice is that the commutator always produces the torsion and curvature at the same time. It makes no sense to throw away the torsion.

6.1. INTRODUCTION

This arbitrary procedure is equivalent to throwing away one of the Cartan structure equations. No expert in differential geometry would do that, only dogmatic physicists. Unfortunately, the curvature was known before Eq. (6.1) was known. The early pioneers of geometry had guessed and got it wrong, they had guessed that geometry could be described by curvature and nothing else. This guess is entirely excusable, it is how knowledge works, but it is entirely inexcusable to go on ignoring torsion once it is known. This is exactly what happened in twentieth century relativity. The latter fell flat on its face when the velocity curve of a whirlpool galaxy was discovered in about 1958.

The second thing to notice is that when the connection is made zero, or removed, the commutator of ordinary derivatives is zero:

$$[\partial_\mu, \partial_\nu] V^\rho = 0 \tag{6.3}$$

and this is a fundamental property of a space without a geometrical connection. In three dimensions such a space is that of Euclid. It has no torsion and no curvature. Notice that the curvature and torsion both vanish. It is not possible for one to exist and the other not to exist. It is becoming clear that the commutator is an elegant object of thought, it produces non Euclidean geometry and shows that this type of geometry is always described by only two types of tensor, the torsion and curvature, and that both always coexist. They both vanish in Euclidean geometry and more generally in an n dimensional space with no connection.

The most important thing to notice is that a commutator of any kind is always anti symmetric. In the case of covariant derivatives it is defined from the most fundamental principles of geometry as:

$$[D_\mu, D_\nu] V^\rho = D_\mu (D_\nu V^\rho) - D_\nu (D_\mu V^\rho) \tag{6.4}$$

so interchanging μ and ν produces the opposite sign. This is what is meant by anti symmetry. Any object with subscripts μ and ν changes sign under the action of the commutator. So it is entirely obvious and long accepted that torsion and curvature are anti symmetric:

$$T^\lambda{}_{\mu\nu} = -T^\lambda{}_{\nu\mu}, \tag{6.5}$$

$$R^\rho{}_{\mu\nu\sigma} = -R^\rho{}_{\mu\sigma\nu}. \tag{6.6}$$

If these tensors were not anti symmetric, the commutator method could not be used, and the Cartan Maurer structure equations would not be valid. In the ninety years since they were proposed, they have never been refuted logically.

The torsion tensor has been defined for ninety years by:

$$T^\lambda{}_{\mu\nu} = \Gamma^\lambda{}_{\mu\nu} - \Gamma^\lambda{}_{\nu\mu} \tag{6.7}$$

and is the difference of two Christoffel connections. In the second connection μ and ν are reversed. So the action of the commutator is:

$$[D_\mu, D_\nu] V^\rho = -\Gamma^\lambda{}_{\mu\nu} D_\lambda V^\rho + ... \tag{6.8}$$

This equation has been written in such a way as to show that there is a one to one correspondence between the commutator indices, μ and ν, and the indices μ and ν of the connection. The commutator is antisymmetric by definition, so the connection is anti symmetric from the most fundamental principles of non Euclidean geometry:

$$\Gamma^\lambda{}_{\mu\nu} = -\Gamma^\lambda{}_{\nu\mu} \tag{6.9}$$

This entirely obvious result refutes the Einsteinian general relativity immediately, so although logical to geometry it is terminally dangerous to foggy dogma or fogma. The truth is always dangerous and exciting. Argument is vulgar and often convincing.

In the development of early non Euclidean geometry the metric was inferred first by Riemann, then the connection by Christoffel, then the curvature by Ricci and Levi Civita and finally the identities known after Bianchi. This took about forty years, from the eighteen sixties to about 1902. These developments did not use the commutator, so there was no way of knowing the symmetry of the lower two indices of the connection. It could be inferred only that the connection was a matrix for each upper index λ. Clearly this pure mathematical development never considered physics, so no fact of nature was used to try to determine the symmetry of the connection. For each λ the connection is a matrix in μ and ν. A matrix in general has no symmetry, it is therefore described as asymmetric. The only thing that can be inferred logically is that the Christoffel connection is asymmetric. It is the sum of symmetric and anti symmetric components, as for any matrix. However, the commutator always produces the anti symmetric part of the connection, and at the same time produces the anti symmetric torsion and anti symmetric curvature and at the same time produces the first and second Cartan Maurer structure equations. So the entire Cartan geometry uses an anti symmetric connection and the entire Cartan geomerty is produced by the commutator. This is the essence of this chapter.

The fogma of the twentieth century ignored the commutator and asserted that Christoffel had somehow managed to prove that the connection is symmetric. If the connection is symmetric, the commutator is symmetric and vanishes. The torsion and curvature vanish, and with them the structure equations of Cartan and Maurer. So the fogma led to the darkest recesses of Plato's Cave, and we are emerging into the light with ECE theory.

6.2 Application of Antisymmetry to Electrodynamics

On the U(1) level used in the standard model the commutator of covariant derivatives acts on the gauge field [1]- [10], [24] ψ as follows:

$$[D_\mu, D_\nu,]\psi = -ig\,[\partial_\mu, A_\nu]\psi \tag{6.10}$$

6.2. APPLICATION OF ANTISYMMETRY TO ELECTRODYNAMICS

where g is a constant and where A_ν is the four potential on the U(1) level. Now let:

$$\mu \to \nu, \quad \nu \to \mu \tag{6.11}$$

then by definition:

$$[D_\mu, D_\nu]\psi = -[D_\nu, D_\mu]\psi. \tag{6.12}$$

The commutator is expanded with the Leibnitz Theorem as follows:

$$\begin{aligned}[\partial_\mu, A_\nu]\psi &= \partial_\mu (A_\nu \psi) - A_\nu (\partial_\mu \psi) \\ &= (\partial_\mu A_\nu)\psi + A_\nu (\partial_\mu \psi) - A_\nu (\partial_\mu \psi) \\ &= (\partial_\mu A_\nu)\psi. \end{aligned} \tag{6.13}$$

Therefore:

$$[\partial_\mu, A_\nu]\psi = (\partial_\mu A_\nu)\psi \tag{6.14}$$

$$[\partial_\nu, A_\mu]\psi = (\partial_\nu A_\mu)\psi \tag{6.15}$$

and Eq.(6.12) is:

$$(\partial_\mu A_\nu)\psi = -(\partial_\nu A_\mu)\psi \tag{6.16}$$

giving the antisymmetry law of ECE theory on the U(1) level in electrodynamics. It was realized in UFT 130, a heavily studied paper, that Eq. (6.16) profoundly changes the nature of electric and electronic engineering in all their aspects. They have been inexplicably missed since Heaviside's time in the late nineteenth century but are simple to derive. Eq. (6.16) immediately shows that U(1) gauge symmetry is incorrect and self inconsistent. The basic assertion of U(1) = O(2) gauge electromagnetism (flat electromagnetism) is that there are only transverse states of radiation in vacuo. This patently absurd assertion is necessitated by the early guess of Einstein that a particle moving at c must have identically zero mass. As we have seen the correct interpretation was given in July 1905 by Poincare, that c is not the speed of light in vacuo but the constant of the Lorentz transformation.

So in flat electromagnetism the transverse vector potential is:

$$\mathbf{A} = \frac{A^{(0)}}{\sqrt{2}} (i\mathbf{i} + \mathbf{j}) e^{i\phi} \tag{6.17}$$

where the electromagnetic phase is:

$$\phi = \omega t - \kappa Z. \tag{6.18}$$

Here ω is the angular frequency at instant t, κ is the wave vector magnitude at position Z. Therefore:

$$\frac{\partial A_X}{\partial Z} = -i\kappa A_X = \kappa \frac{A^{(0)}}{\sqrt{2}} e^{i\phi}, \tag{6.19}$$

CHAPTER 6. ANTISYMMETRY

$$\frac{\partial A_Y}{\partial Z} = -i\kappa A_Y = -i\kappa \frac{A^{(0)}}{\sqrt{2}} e^{i\phi}. \tag{6.20}$$

However the antisymmetry law (6.16) means that:

$$\frac{\partial A_Z}{\partial X} = -\frac{\partial A_X}{\partial Z} = -\kappa A^{(0)} e^{i\frac{\phi}{\sqrt{2}}}, \tag{6.21}$$

$$\frac{\partial A_Z}{\partial Y} = -\frac{\partial A_Y}{\partial Z} = i\kappa A^{(0)} e^{i\frac{\phi}{\sqrt{2}}},$$

showing immediately that there is a longitudinal polarization A_Z by anti symmetry. It is immediately obvious that there is no Higgs boson, which rests on flat electromagnetism, the U(1) sector symmetry of the theory behind the Higgs boson. Using the de Moivre Theorem:

$$e^{i\phi} = \cos\phi + i\sin\phi \tag{6.22}$$

so:

$$\frac{\partial A_Z}{\partial X} = -\kappa \frac{A^{(0)}}{\sqrt{2}} \cos\phi; \quad \frac{\partial A_Z}{\partial Y} = -\kappa \frac{A^{(0)}}{\sqrt{2}} \sin\phi \tag{6.23}$$

and

$$\left(\frac{\partial A_Z}{\partial X}\right)^2 + \left(\frac{\partial A_Z}{\partial Y}\right)^2 = \kappa^2 \frac{A^{(0)2}}{2}. \tag{6.24}$$

If cylindrical symmetry is used for the sake of simplicity it is found that:

$$A_Z = \pm \frac{1}{2} X \kappa A^{(0)} \tag{6.25}$$

and there are three senses of space like polarization. The Beltrami analysis of chapter 3 shows the nature of longitudinal solutions very clearly and obviously. In a sense the standard model of physics has always been a flat world fantasy. As soon as Proca developed his equations, U(1) gauge invariance collapsed. That was in 1938, and it is still being rolled out today in standard physics, but not in ECE physics.

In the obsolete flat electromagnetism, the electric field strength **E** is defined by the scalar and vector potentials by:

$$\mathbf{E} = -\nabla\phi - \frac{\partial \mathbf{A}}{\partial t} \tag{6.26}$$

and the magnetic flux density by:

$$\mathbf{B} = \nabla \times \mathbf{A}. \tag{6.27}$$

In the flat world of U(1) electromagnetism it is claimed that a static electric field is defined by:

$$\mathbf{E} = -\nabla\phi \tag{6.28}$$

6.2. APPLICATION OF ANTISYMMETRY TO ELECTRODYNAMICS

and that for a static electric field:

$$\frac{\partial \mathbf{A}}{\partial t} = \mathbf{0}. \tag{6.29}$$

The anti symmetry equations (6.16) immediately refute these assertions because:

$$\boldsymbol{\nabla}\phi = \frac{\partial \mathbf{A}}{\partial t} = \mathbf{0}. \tag{6.30}$$

The electric field is always defined by Eq. (6.30) in all situations in the natural sciences and engineering.

Similarly in gravitational theory the Newtonian acceleration due to gravity is always defined in the obsolete standard physics by:

$$\mathbf{g} = -\boldsymbol{\nabla}\Phi \tag{6.31}$$

but the anti symmetry argument shows that:

$$\mathbf{g} = -\boldsymbol{\nabla}\Phi = -\frac{1}{c}\frac{\partial \boldsymbol{\Phi}}{\partial t} \tag{6.32}$$

where Φ is the gravitational equivalent of the scalar potential ϕ and $\boldsymbol{\Phi}$ is the equivalent of the vector potential $\mathbf{A}$ in electromagnetism.

The anti symmetry law (6.16) leads to multiple difficulties for flat electromagnetism and standard physics. The law (6.16) can be expressed as two equations:

$$\boldsymbol{\nabla}\phi = \frac{\partial \mathbf{A}}{\partial t} \tag{6.33}$$

and

$$\partial_i A_j = -\partial_j A_i. \tag{6.34}$$

From Eqs. (6.27) and (6.33):

$$\boldsymbol{\nabla} \times \mathbf{E} = 0, \quad \frac{\partial \mathbf{B}}{\partial t} = \mathbf{0} \tag{6.35}$$

meaning that the magnetic field in flat electrodynamics cannot change with time, an absurdity. This is a difficulty encountered at the most basic level in the tensorial theory of electromagnetism. Apparently it was not realized by Lorentz and Poincare because they did not infer the anti symmetry law (6.16). The Faraday law of induction of the flat electromagnetism is:

$$\boldsymbol{\nabla} \times \mathbf{E} + \frac{\partial \mathbf{B}}{\partial t} = \mathbf{0} \tag{6.36}$$

so from Eq. (6.35):

$$\boldsymbol{\nabla} \times \mathbf{E} = \mathbf{0} \tag{6.37}$$

CHAPTER 6. ANTISYMMETRY

which means that the electric field strength is also static, another absurd result of assuming a zero photon mass. A static electric field on the U(1) level is defined by:

$$\mathbf{A} = 0 \tag{6.38}$$

so it follows that:

$$\mathbf{B} = \nabla \times \mathbf{A} = 0 \tag{6.39}$$

and that the magnetic flux density vanishes. From the anti symmetry equation (6.33) it follows that:

$$\nabla \phi = \frac{\partial \mathbf{A}}{\partial t} = 0 \tag{6.40}$$

and so:

$$\mathbf{E} = -\nabla \phi = 0. \tag{6.41}$$

Anti symmetry therefore results in the complete collapse of U(1) electromagnetism, both **E** and **B** vanish as a result of anti symmetry in the flat world of U(1) electromagnetism. The ship falls off the edge of the flat dogmatic world. Anti symmetry proves straightforwardly that the notion of a massless photon is empty dogma, and that the geometry used in MH theory is woefully inadequate.

Note carefully that U(1) symmetry gauge theory itself, Eq. (6.10), has been used to disprove the theory simply by using the anti symmetry of the commutator, which acts on the gauge field [1]- [10], [24] as follows:

$$[D_\mu, D_\nu] \psi = [\partial_\mu - ig A_\mu, \partial_\nu - ig A_\nu] \psi. \tag{6.42}$$

The U(1) covariant derivative is defined as:

$$D_\mu = \partial_\mu - ig A_\mu \tag{6.43}$$

where:

$$g = \frac{e}{\hbar} = \frac{\kappa}{A^{(0)}} \tag{6.44}$$

as argued in previous chapters. The photon momentum in this theory is:

$$p = \hbar \kappa = e A^{(0)}, \tag{6.45}$$

a minimal prescription. In Eq. (6.42):

$$[\partial_\mu, \partial_\nu] = 0 \tag{6.46}$$

so:

$$[D_\mu, D_\nu] \psi = -ig \left([\partial_\mu, A_\nu] - ig [A_\mu, A_\nu] \right) \psi. \tag{6.47}$$

The fundamental anti symmetry:

$$[D_\mu, D_\nu]\psi = -[D_\nu, D_\mu]\psi \tag{6.48}$$

means that:

$$[\partial_\mu, A_\nu]\psi = -[\partial_\nu, A_\mu]\psi \tag{6.49}$$

so:

$$\partial_\mu A_\nu = -\partial_\nu A_\mu \tag{6.50}$$

and we obtain Eq. (6.16) irrefutably. The only alternative is to abandon the commutator, but as argued already that means the abandonment of geometry itself.

The derivation of the anti symmetry law is so simple that it is almost trivially evident from the commutator method. Yet the law is so powerful that it can refute a century of dogma in a few lines, as we have just argued.

This catastrophe for the standard physics became evident a few years ago in UFT 132. By now it is long known that flat electromagnetism is empty dogma, and by implication the Higgs boson. The latter exists only because the media can be used to propagate the idea. As in Einstein's era the general public still has no idea of the meaning of commutator. This is an illustration of human nature rather than that of nature. The scene is now set for the entry of ECE theory and for the implementation of anti symmetry within ECE theory.

6.3 Antisymmetry in ECE Electromagnetism

In ECE electrodynamics the electromagnetic field is defined by:

$$F^a{}_{\mu\nu} = \partial_\mu A^a{}_\nu - \partial_\nu A^a{}_\mu + \omega^a{}_{\mu b} A^b{}_\nu - \omega^a{}_{\nu b} A^b{}_\mu \tag{6.51}$$

in which the antisymmetry law is determined by the antisymmetry of the Chistoffel connection:

$$\Gamma^a{}_{\mu\nu} = -\Gamma^a{}_{\nu\mu}. \tag{6.52}$$

Using the tetrad postulate the Christoffel connection becomes:

$$\Gamma^a{}_{\mu\nu} = \partial_\mu q^a{}_\mu + \omega^a{}_{\mu\nu} \tag{6.53}$$

so anti symmetry in Cartan geometry means that:

$$\partial_\mu q^a{}_\nu + \omega^a{}_{\mu\nu} + \partial_\nu q^a{}_\mu + \omega^a{}_{\nu\mu} = 0. \tag{6.54}$$

As in chapter 2 this equation translates into the following anti symmetry equation in electrodynamics:

$$\partial_\mu A^a{}_\nu + \partial_\nu A^a{}_\mu + A^{(0)}\left(\omega^a{}_{\mu\nu} + \omega^a{}_{\nu\mu}\right) = 0. \tag{6.55}$$

CHAPTER 6. ANTISYMMETRY

This was first derived in UFT 133 and UFT 134 and is a fundamental constraint on the first Cartan Maurer structure equation:

$$F^a{}_{\mu\nu} = \partial_\mu A^a{}_\nu - \partial_\nu A^a{}_\mu + A^0 \left(\omega^a{}_{\mu\nu} - \omega^a{}_{\nu\mu}\right). \tag{6.56}$$

This is known as the Lindstrom constraint and is discussed in more detail as follows, based on UFT 134.

For a single polarization the ECE theory of electromagnetism reduces to a format that is superficially similar to the Maxwell Heaviside equations:

$$\boldsymbol{\nabla} \cdot \mathbf{B} = 0 \tag{6.57}$$

$$\boldsymbol{\nabla} \times \mathbf{E} + \frac{\partial \mathbf{B}}{\partial t} = \mathbf{0} \tag{6.58}$$

$$\boldsymbol{\nabla} \cdot \mathbf{E} = \frac{\rho}{\epsilon_0} \tag{6.59}$$

$$\boldsymbol{\nabla} \times \mathbf{B} - \frac{1}{c^2}\frac{\partial \mathbf{E}}{\partial t} = \mu_0 \mathbf{J} \tag{6.60}$$

but the relation between the fields and the potentials are as follows:

$$\mathbf{E} = -\boldsymbol{\nabla}\phi - \frac{\partial \mathbf{A}}{\partial t} - \omega_0 \mathbf{A} + \boldsymbol{\omega}\phi, \tag{6.61}$$

$$\mathbf{B} = \boldsymbol{\nabla} \times \mathbf{A} - \boldsymbol{\omega} \times \mathbf{A}. \tag{6.62}$$

The electric component of the anti symmetry equation for a single polarization is:

$$\boldsymbol{\nabla}\phi - \frac{\partial \mathbf{A}}{\partial t} - \omega_0 \mathbf{A} - \boldsymbol{\omega}\phi = \mathbf{0} \tag{6.63}$$

and the magnetic anti symmetry relation restricted by the Lindstrom constraint is:

$$\boldsymbol{\nabla} \times \mathbf{A} = -\boldsymbol{\omega} \times \mathbf{A}. \tag{6.64}$$

If we apply the anti symmetry equations (6.63) and (6.64) to the field intensities $\mathbf{E}$ and $\mathbf{B}$ we see two independent definitions of $\mathbf{E}$ and a single definition of $\mathbf{B}$:

$$\mathbf{E} = -2\frac{\partial \mathbf{A}}{\partial t} - 2\omega_0 \mathbf{A} \tag{6.65}$$

or

$$\mathbf{E} = -2\boldsymbol{\nabla}\phi + 2\boldsymbol{\omega}\phi \tag{6.66}$$

and

$$\mathbf{B} = 2\boldsymbol{\nabla} \times \mathbf{A}. \tag{6.67}$$

So $\mathbf{B}$ is obviously compatible with the Gauss Law:

$$\boldsymbol{\nabla} \cdot \mathbf{B} = 0. \tag{6.68}$$

6.3. ANTISYMMETRY IN ECE ELECTROMAGNETISM

Applying the two alternative equations (6.65) and (6.66) for **E**, and (6.67) for **B**, to Faraday's Law, Eq. (6.58) gives for both cases:

$$\nabla \times \left(\phi \boldsymbol{\omega} + \frac{\partial \mathbf{A}}{\partial t} \right) = \mathbf{0} \tag{6.69}$$

and

$$\nabla \times (\omega_0 \mathbf{A}) = \mathbf{0}. \tag{6.70}$$

Take the curl of Eq. (6.63) and apply Eq. (6.70) to obtain Eq. (6.69), meaning that Eq. (6.69) contains no new information that is not already given by the electric component of the anti symmetry equations. Using the anti symmetry relations the following equations can be obtained as in UFT 134:

$$\nabla \times (\boldsymbol{\omega} \phi) - \frac{\partial}{\partial t}(\boldsymbol{\omega} \times \mathbf{A}) = \mathbf{0}, \tag{6.71}$$

$$-\nabla^2 \phi + \nabla \cdot (\boldsymbol{\omega} \phi) = \frac{\rho}{2\epsilon_0}, \tag{6.72}$$

$$-\nabla \times (\boldsymbol{\omega} \times \mathbf{A}) - \frac{1}{c^2}\frac{\partial}{\partial t}(\nabla \phi - \boldsymbol{\omega} \phi) = \frac{\mu_0}{2}\mathbf{J}. \tag{6.73}$$

Eq. (6.72) gives a resonant form of the Coulomb law which can be used to produce resonant energy from spacetime as described in the next chapter. Eqs. (6.62) to (6.65) give a set of seven equations in seven unknows as described in UFT 134. However the Coulomb and Ampère Maxwell laws are not independent. This can be shown for example by taking the divergence of Eq. (6.73):

$$\frac{1}{c^2}\frac{\partial}{\partial t}\left(-\nabla^2 \phi + \nabla \cdot (\boldsymbol{\omega} \phi)\right) = \frac{1}{2}\mu_0 \nabla \cdot \mathbf{J} \tag{6.74}$$

and integrating with respect to time to give:

$$-\nabla^2 \phi + \nabla \cdot (\boldsymbol{\omega} \phi) = \frac{\rho}{2\epsilon_0} \tag{6.75}$$

with:

$$\rho = \int \nabla \cdot \mathbf{J}\, dt. \tag{6.76}$$

Starting with Eqs. (6.65) and (6.67), Faraday's law becomes:

$$\nabla \times \left(-2\frac{\partial \mathbf{A}}{\partial t} - 2\omega_0 \mathbf{A}\right) + 2\frac{\partial}{\partial t}(\nabla \times \mathbf{A}) = \mathbf{0} \tag{6.77}$$

which can be simplified to:

$$\nabla \times (\omega_0 \mathbf{A}) = \mathbf{0} \tag{6.78}$$

CHAPTER 6. ANTISYMMETRY

and is identical with Eq. (6.70). The Coulomb and Ampère Maxwell laws take the form:

$$\nabla \cdot \frac{\partial \mathbf{A}}{\partial t} + \nabla \cdot (\omega_0 \mathbf{A}) = \frac{\rho}{2\epsilon_0}, \tag{6.79}$$

$$\nabla \times \nabla \times \mathbf{A} + \frac{1}{c^2} \frac{\partial^2 \mathbf{A}}{\partial t^2} + \frac{1}{c^2} \frac{\partial}{\partial t}(\omega_0 \mathbf{A}) = \frac{1}{2}\mu_0 \mathbf{J}. \tag{6.80}$$

Eq. (6.79) is compatible with Eq. (6.78) and shows that $\omega_0 \mathbf{A}$ represents a pure source field. Eqs. (6.79) and (6.80) represent four equations for four variables. These equations are independent if the charge and current density are chosen to be unrelated. Eq. (6.80) is a wave equation in three dimensions with transverse and longitudinal solutions that go beyond MH electrodynamics. Eq. (6.79) is a non linear diffusion equation, the non linearity being caused by the spin connection, and indicating that there is a flow of potential present in addition to MH theory. This can be considered to represent interaction with the surrounding vacuum or spacetime - the source of energy in resonance effects.

It is possible to derive a third version of the equation set using Eq. (6.70):

$$\omega_0 \mathbf{A} = -\frac{\partial}{\partial t}(\nabla \phi). \tag{6.81}$$

Substituting Eq. (6.66) and (6.68) into Eq. (6.59) and (6.60) gives:

$$\nabla \cdot \frac{\partial \mathbf{A}}{\partial t} + \nabla \cdot (\omega_0 \mathbf{A}) = -\frac{\rho}{2\epsilon_0}, \tag{6.82}$$

$$\nabla \times \nabla \times A + \frac{1}{c^2} \frac{\partial^2 \mathbf{A}}{\partial t^2} + \frac{1}{c^2} \frac{\partial}{\partial t}(\omega_0 \mathbf{A}) = \frac{1}{2}\mu_0 \mathbf{J}, \tag{6.83}$$

and using the vector identity:

$$\nabla \times \nabla \times \mathbf{A} = \nabla(\nabla \cdot \mathbf{A}) - \nabla^2 \mathbf{A} \tag{6.84}$$

time integrating Eq. (6.82) and substituting the expression for $\nabla \cdot \mathbf{A}$ Into Eq. (6.83) gives:

$$\left(-\nabla^2 + \frac{1}{c^2}\frac{\partial^2}{\partial t^2}\right)\left(\mathbf{A} + \int \omega_0 \mathbf{A}\, dt\right) = \frac{1}{2}\mu_0 \mathbf{J} + \frac{1}{2}\int \frac{\nabla \rho}{\epsilon_0}\, dt. \tag{6.85}$$

Using Eq. (6.81) this can be written more elegantly as:

$$\left(-\nabla^2 + \frac{1}{c^2}\frac{\partial^2}{\partial t^2}\right)(\mathbf{A} - \nabla \phi) = \frac{1}{2}\mu_0 \mathbf{J} + \frac{1}{2\epsilon_0}\int \nabla \rho\, dt. \tag{6.86}$$

By using Eq. (6.65):

$$\int \mathbf{E}\, dt = -2\mathbf{A} - 2\int \omega_0 \mathbf{A}\, dt = -2\mathbf{A} + 2\nabla \phi \tag{6.87}$$

6.3. ANTISYMMETRY IN ECE ELECTROMAGNETISM

which appears in Eq. (6.86). Alternatively Eq. (6.86) is according to Eq. (6.66):

$$\int \mathbf{E} \, dt = -2 \int \boldsymbol{\nabla}\phi \, dt + 2 \int \phi \boldsymbol{\omega} \, dt. \tag{6.88}$$

Substituting this alternative form of Eq. (6.88) into Eq. (6.87) we obtain:

$$\left(-\nabla^2 + \frac{1}{c^2}\frac{\partial^2}{\partial t^2}\right)\left(\int \boldsymbol{\nabla}\phi \, dt - \int \phi \boldsymbol{\omega} \, dt\right) = \frac{1}{2}\mu_0 \mathbf{J} + \frac{1}{2\epsilon_0}\int \boldsymbol{\nabla}\rho \, dt \tag{6.89}$$

and after taking the time derivative:

$$\left(-\nabla^2 + \frac{1}{c^2}\frac{\partial^2}{\partial t^2}\right)(\boldsymbol{\nabla}\phi - \boldsymbol{\omega}\phi) = \frac{1}{2}\mu_0 \frac{\partial \mathbf{J}}{\partial t} + \frac{1}{2\epsilon_0}\boldsymbol{\nabla}\rho. \tag{6.90}$$

In total, Eqs. (6.81), (6.86) and (6.90) represent nine equations in nine unknowns:

$$\omega_0 \mathbf{A} = -\frac{\partial}{\partial t}(\boldsymbol{\nabla}\phi) \tag{6.91a}$$

$$\left(-\nabla^2 + \frac{1}{c^2}\frac{\partial^2}{\partial t^2}\right)(\mathbf{A} - \boldsymbol{\nabla}\phi) = \frac{1}{2}\mu_0 \mathbf{J} + \frac{1}{2\epsilon_0}\int \boldsymbol{\nabla}\rho \, dt, \tag{6.91}$$

$$\left(-\nabla^2 + \frac{1}{c^2}\frac{\partial^2}{\partial t^2}\right)(\boldsymbol{\nabla}\phi - \boldsymbol{\omega}\phi) = \frac{1}{2}\mu_0 \frac{\partial \mathbf{J}}{\partial t} + \frac{1}{2\epsilon_0}\boldsymbol{\nabla}\rho. \tag{6.92}$$

The equations are entirely independent and represent a balanced set.

Singularities occur in the solutions, giving plenty of opportunity for resonance effects and obtaining energy from spacetime. For example if the cross product is taken of the electric portion of the anti symmetry equation (6.63) with $\mathbf{A}$:

$$\boldsymbol{\nabla}\phi \times \mathbf{A} - \frac{\partial \mathbf{A}}{\partial t} \times \mathbf{A} - \omega_0 \mathbf{A} \times \mathbf{A} - \phi\boldsymbol{\omega} \times \mathbf{A} = 0. \tag{6.93}$$

Assuming that the time derivative of $\mathbf{A}$ is parallel to $\mathbf{A}$:

$$\boldsymbol{\nabla}\phi \times \mathbf{A} = \phi\boldsymbol{\omega} \times \mathbf{A} \tag{6.94}$$

and Eq. (6.64) can be used to remove $\boldsymbol{\omega}$:

$$\boldsymbol{\nabla} \times \mathbf{A} = -\frac{1}{\phi}\boldsymbol{\nabla}\phi \times \mathbf{A}. \tag{6.95}$$

Singularities occur whenever ϕ is zero and $\boldsymbol{\nabla}\phi$ and $\mathbf{A}$ are not. Combined with the driven resonances in Eqs. (6.91) and (6.92) a rich supply of non linear solutions becomes available.

It is seen that the ECE anti symmetry equations are the only equations of electrodynamics that are self consistent and are preferred over the MH equations.

The Lindstrom magnetic constraint combined with a particular solution of the electric constraint reduces the second model described above to MH theory. Anti symmetry means that it is not possible to reduce ECE theory to MH theory simply by removing the spin connection, because that procedure produces:

$$\mathbf{E} = -\frac{\partial \mathbf{A}}{\partial t} - \nabla \phi, \tag{6.96}$$

$$\mathbf{B} = \nabla \times \mathbf{A}. \tag{6.97}$$

As shown already in this chapter these relations when used with anti symmetry generally invalidate MH theory, a major discovery of the evolution of ECE theory. However, applying the following particular solutions of the anti symmetry equations:

$$\boldsymbol{\omega}\phi = -\frac{\partial \mathbf{A}}{\partial t} \tag{6.98}$$

$$\omega_0 \mathbf{A} = \nabla \phi \tag{6.99}$$

$$\boldsymbol{\omega} \times \mathbf{A} = -\nabla \times \mathbf{A} \tag{6.100}$$

the electric and magnetic fields of the ECE theory become:

$$\mathbf{E} = -2\frac{\partial \mathbf{A}}{\partial t} - 2\nabla \phi, \tag{6.101}$$

$$\mathbf{B} = 2\nabla \times \mathbf{A}. \tag{6.102}$$

The standard MH structure is:

$$\mathbf{B} = \nabla \times \mathbf{a} \tag{6.103}$$

and comparing Eqs. (6.102) and (6.103):

$$\mathbf{a} = 2\mathbf{A}. \tag{6.104}$$

Substituting Eq. (6.103) into the Faraday Law:

$$\nabla \times \mathbf{E} + \frac{\partial \mathbf{B}}{\partial t} = 0 \tag{6.105}$$

gives:

$$\nabla \times \mathbf{E} = -\nabla \times \frac{\partial \mathbf{a}}{\partial t} \tag{6.106}$$

which has:

$$\mathbf{E} = -\frac{\partial \mathbf{a}}{\partial t} - \nabla \phi_1 \tag{6.107}$$

as the only solution. Comparing Eqs. (6.101) and (6.107) gives:

$$\phi_1 = 2\phi \tag{6.108}$$

which show that the theory designated II in the engineering model on www.aias.us reduces to the MH theory given the restrictions (6.98) to (6.100).

Note carefully that this reduction is achieved by:

$$\mathbf{B} = \nabla \times \mathbf{a} = \nabla \times \mathbf{A} - \boldsymbol{\omega} \times \mathbf{A} = 2\nabla \times \mathbf{A} \tag{6.109}$$

and not by discarding the spin connection. So the MH format achieved in this way is still a theory of general relativity, making unification with gravitation possible.

6.4 Derivation of the Equivalence Principle from Antisymmetry and Other Applications

The equivalence of inertial and gravitational mass is known as the weak equivalence principle and has been tested experimentally with great precision. In this section the equivalence principle is derived from anti symmetry. It has been shown independently [1]- [10] by Moses, Reed and Evans that any vector field in three dimensions may be expressed as the sum of three vectors:

$$\mathbf{V} = \mathbf{V}^{(1)} + \mathbf{V}^{(2)} + \mathbf{V}^{(3)} \tag{6.110}$$

in the complex circular basis defined earlier in this book. Helmholtz showed in the nineteenth century that any vector field can be written as the sum of two vectors:

$$\mathbf{V} = \mathbf{V}_s + \mathbf{V}_l \tag{6.111}$$

where:

$$\nabla \cdot \mathbf{V}_s = 0, \tag{6.112}$$

$$\nabla \times \mathbf{V}_l = \mathbf{0}. \tag{6.113}$$

The use of the complex circular basis extends the Helmholtz equation as follows:

$$\mathbf{V}_s = \mathbf{V}^{(1)} + \mathbf{V}^{(2)}, \tag{6.114}$$

$$\mathbf{V}_l = \mathbf{V}^{(3)}. \tag{6.115}$$

The most fundamental components are therefore components of $\mathbf{V}^{(1)}$, $\mathbf{V}^{(2)}$, $\mathbf{V}^{(3)}$. Examples of these fundamental components are shown below, for example a vector potential. In the first papers on ECE theory these components were identified as the objects known as tetrads in Cartan geometry. Such an identification had also been made indirectly by Reed. In Cartan's original definition of the tetrad the a index is the upper index of a four dimensional Minkowski spacetime at point P to a four dimensional manifold indexed μ. Each of the three dimensional vectors defined in Eq. (6.110) is the space like component of the following four dimensional vectors:

$$V^{(i)}_\mu = \left(V^{(i)}_0, -\mathbf{V}^{(i)}\right), \quad i = 1, 2, 3. \tag{6.116}$$

The complete four dimensional vector is the sum of these three vectors:

$$V_\mu = V^{(1)}_\mu + V^{(2)}_\mu + V^{(3)}_\mu. \tag{6.117}$$

So there exist three time like components and the complete time like component is their sum:

$$V_0 = V^{(1)}_0 + V^{(2)}_0 + V^{(3)}_0. \tag{6.118}$$

CHAPTER 6. ANTISYMMETRY

In four dimensions the a index is:

$$a = (0), (1), (2), (3) \tag{6.119}$$

so in general there also exists the component $V^{(0)}{}_0$. These fundamental elements may always be expressed as tetrad elements and defined as a 4 x 4 matrix as follows:

$$X^a = V^a{}_\mu X^\mu. \tag{6.120}$$

It follows that any four dimensional vector can be defined as a scalar valued quantity multiplied by a Cartan tetrad:

$$V^a{}_\mu = V q^a{}_\mu. \tag{6.121}$$

Therefore Cartan's differential geometry may be applied to any four dimensional vector. Normally it is applied to the tetrad and the first Cartan structure equation defines the Cartan torsion from the tetrad. The latter is the fundamental building block because it consists of fundamental components of the complete vector field. The Heaviside Gibbs vector analysis restricts consideration to **V** only, but the tetrad analysis realizes that **V** has an internal structure.

In four dimensions therefore define the fundamental vectors:

$$V^{(0)}{}_\mu = \left(V^{(0)}{}_0, \mathbf{0}\right), \tag{6.122}$$

$$V^{(i)}{}_\mu = \left(V^{(i)}{}_0, -\mathbf{V}^{(i)}\right), \quad i = 1, 2, 3. \tag{6.123}$$

Eq. (6.122) means that the space like components of $V^{(0)}{}_\mu$ are zero by definition because the superscript (0) is time like by definition. There are no space like components of a time like property. On the other hand a vector such as $V^{(1)}{}_\mu$ is a four vector, so $V^{(0)}{}_0$ in general is its non-zero time like component. In general the Cartan tetrad is defined by:

$$X^a = q^a{}_\mu X^\mu \tag{6.124}$$

where X denotes any vector field. Therefore Cartan geometry extends the Heaviside Gibbs analysis and this finding can be applied systematically to physics, notably dynamics. The Heaviside Gibbs analysis was restricted to three dimensional space with no connection, i.e. a Euclidean space. Using Cartan's differential geometry the analysis can be extended to any space of any dimension by use of the Cartan spin connection. Using this procedure all the equations of physics can be derived automatically within a unified framework, thus producing the first successful unified field theory.

Now apply this method to the concept of velocity in dynamics. The velocity tetrad is:

$$V^a{}_\mu = v q^a{}_\mu \tag{6.125}$$

6.4. DERIVATION OF THE EQUIVALENCE PRINCIPLE FROM...

where v is the scalar magnitude of velocity, i.e. the speed. The gravitational potential is defined as:

$$\Phi^a{}_\mu = cv^a{}_\mu = \Phi q^a{}_\mu. \tag{6.126}$$

In analogy the electromagnetic potential is also defined in terms of the tetrad in ECE theory:

$$A^a{}_\mu = A^{(0)} q^a{}_\mu. \tag{6.127}$$

The electromagnetic field is defined in terms of the Cartan torsion:

$$F^a{}_{\mu\nu} = A^{(0)} T^a{}_{\mu\nu} \tag{6.128}$$

and also the gravitational field:

$$g^a{}_{\mu\nu} = \Phi^{(0)} T^a{}_{\mu\nu}. \tag{6.129}$$

The acceleration due to gravity in ECE theory is therefore part of the torsion, so in general the acceleration in electrodynamics is also part of the torsion, defined conveniently as:

$$a^a{}_{\mu\nu} = cv T^a{}_{\mu\nu}. \tag{6.130}$$

In vector notation Eq. (6.129) splits in to two equations:

$$\mathbf{a}^a = -\frac{\partial \mathbf{v}^a}{\partial t} - c \boldsymbol{\nabla} v^a{}_0 - c\omega^a{}_{0b} \mathbf{v}^b + c v^b{}_0 \boldsymbol{\omega}^a{}_b \tag{6.131}$$

and

$$\boldsymbol{\Omega}^a = \boldsymbol{\nabla} \times \mathbf{v}^a - \boldsymbol{\omega}^a{}_b \times \mathbf{v}^b. \tag{6.132}$$

The spin connection is defined as:

$$\omega^a{}_{\mu b} = (\omega^a{}_{0b}, -\boldsymbol{\omega}^a{}_b). \tag{6.133}$$

In tensor notation the relation between acceleration and velocity in generally covariant dynamics is:

$$a^a{}_{\mu\nu} = c \left(\partial_\mu v^a{}_\nu - \partial_\nu v^a{}_\mu + v \left(\omega^a{}_{\mu\nu} - \omega^a{}_{\nu\mu} \right) \right). \tag{6.134}$$

So Eqs. (6.131) and (6.132) may be simplified to:

$$\mathbf{a}^a = -\frac{\partial \mathbf{v}^a}{\partial t} + c \boldsymbol{\nabla} \Phi^a + cv \boldsymbol{\omega}^a{}_{\text{orb}} \tag{6.135}$$

and:

$$\boldsymbol{\Omega}^a = \boldsymbol{\nabla} \times \mathbf{v}^a + v \boldsymbol{\omega}^a{}_{\text{spin}} \tag{6.136}$$

where:

$$\boldsymbol{\omega}^a{}_{\text{orb}} = (\omega^a{}_{01} - \omega^a{}_{10}) \mathbf{i} + (\omega^a{}_{02} - \omega^a{}_{20}) \mathbf{j} + (\omega^a{}_{03} - \omega^a{}_{30}) \mathbf{k} \tag{6.137}$$

and

$$\boldsymbol{\omega}^a{}_{\text{spin}} = (\omega^a{}_{32} - \omega^a{}_{23})\mathbf{i} + (\omega^a{}_{13} - \omega^a{}_{31})\mathbf{j} + (\omega^a{}_{21} - \omega^a{}_{12})\mathbf{k} \qquad (6.138)$$

and where:

$$v\omega^a{}_{\text{orb}} = -\omega^a{}_{ob}\mathbf{v}^b + v^b{}_0\omega^a{}_b \qquad (6.139)$$

and

$$v\omega^a{}_{\text{spin}} = -\boldsymbol{\omega}^a{}_b \times \mathbf{v}^b. \qquad (6.140)$$

Equations (6.139) and (6.140) are Coriolis type accelerations due to orbital and spin torsion. Eq. (6.135) shows that acceleration is due to the rate of change of velocity and also the gradient of the potential. If the inertial frame of Newtonian dynamics is defined as flat space time then in the inertial frame:

$$\mathbf{a}^a = -\frac{\partial \mathbf{v}^a}{\partial t} - \boldsymbol{\nabla}\Phi^a. \qquad (6.141)$$

The equivalence principle assumes that:

$$-\frac{\partial \mathbf{v}^a}{\partial t} = -\boldsymbol{\nabla}\Phi^a \qquad (6.142)$$

which is the direct result of the ECE anti symmetry law:

$$\partial_\mu v^a{}_\nu = -\partial_\nu v^a{}_\mu \qquad (6.143)$$

when

$$\mu = 0, \ \nu = 1, \qquad (6.144)$$

Q. E. D.

Force is defined by mass multiplied by acceleration, so

$$\mathbf{F}^a = -m\frac{\partial \mathbf{v}^a}{\partial t} = -m\boldsymbol{\nabla}\Phi^a \qquad (6.145)$$

which is a generalization of the weak equivalence principle assumed by Newton but not proven by him. ECE theory shows that the equivalence principle has a geometrical origin.

Chapter 7

Energy from Space Time and Low Energy Nuclear Reactions

7.1 Introduction

These phenomena when viewed as experimental data completely refute the standard model of physics, which is still unable to deal with them. There are many devices available that take energy from space time (www.et3m.net) in a reproducible and repeatable manner. These devices are being used routinely in the best industry. Low energy nuclear reactors (LENR) are about to be mass produced, but the old physics still cannot explain them. A plausible qualitative explanation for such devices has been given by ECE theory through the use of Euler Bernoulli resonance [1]- [10] in equations containing the spin connection. The first example found was spin connection resonance (SCR) in the Coulomb Law, and after that several other mechanisms were found. The theory has been greatly developed independently by Eckardt and Lindstrom. This chapter aims to explain the simple basics of spin connection resonance.

For over a hundred years there have been many reports of devices producing more electric power than inputted to a given device. Many of these reports were not reproducible and repeatable, but in the past thirty years or so the subject has become more scientific, with more details becoming available of circuit design. Some of the reports were of surges or spikes of power which could not be explained conventionally. Some of these were too large to be artifacts. The subject has been hampered greatly by pseudoscience and charlatans, so from the beginning ECE set out to give a rigorous explanation of such phenomena. A qualitative or plausible explanation was sought based on data that were likely to be reproducible and repeatable and to be free of artifact. Conventional electric resonance must be eliminated carefully before a source of energy from space time can be considered as a possible explanation.

In addition to these requirements of Baconian science the circuit design

7.1. INTRODUCTION

must preferably be made available as the scientific apparatus, in the usual manner of a scientific experiment, but very often no details of apparatus were available. Possibly this may have been due to inventors who were careful to protect patent rights. So scientists have been reluctant to approach these important subject areas in an open minded, scientific, manner. This is a pity because they are of great potential importance to humankind. If there is any chance whatsoever of obtaining energy from spacetime, then that chance should be exploited to the hilt. A coherent theory for such phenomena was not formulated until spin connection resonance was proposed. The Maxwell Heaviside (MH) theory has no explanation for energy from space time, so there has been a historical tendency to dismiss all such data as artifact, or being indicative of a lack of knowledge of basic principles such as conservation of energy. In the past there has been a widespread belief that energy from space time means energy from nothing. This absurd lack of understanding delayed the acceptance of the subject for many years.

In about 2005 one of the authors of this book (MWE) was asked to give an explanation of a very intense resonance peak in apparatus demonstrated to the U. S. Navy by Alex Hill and colleagues (www.aias.us) whose work was first drawn to the attention of MWE by Albert Collins. John Shelburne, a civilian working for the Navy in Florida, asked MWE to give a plausible explanation in terms of the then new ECE theory. The resonance peak was demonstrated to the U. S. Navy by the Alex Hill group, and the Naval civilian staff were satisfied that the effect was free of artifact. There was an intense resonance of electric power which could not be explained by conventional electric resonance theory, based on Euler Bernoulli theory. Subsequently the Alex Hill group developed devices which are now used in Fortune Fifty industry. Observers are allowed to see the devices in operation in Fortune Fifty industry.

7.2 Spin Connection Resonance from the Coulomb Law

In the simplest instance the Coulomb law in ECE theory is given by:

$$\nabla \cdot \mathbf{E} = \frac{\rho}{\epsilon_0} \qquad (7.1)$$

where:

$$\mathbf{E} = -\left(\nabla + \boldsymbol{\omega}\right)\phi \qquad (7.2)$$

where ϕ is the scalar potential in volts, $\boldsymbol{\omega}$ is the spin connection vector in inverse metres, $\mathbf{E}$ is the electric field strength in volts m^{-1}, ρ is the charge density in Cm^{-3} and ϵ_0 is the S. I. vacuum pemittivity:

$$\epsilon_0 = 8 \cdot 854 \times 10^{-12} \mathrm{J}^{-1}\mathrm{C}^2\mathrm{m}^{-1}. \qquad (7.3)$$

Thus:

$$\nabla \cdot \left(\left(\nabla + \boldsymbol{\omega}\right)\phi\right) = -\frac{\rho}{\epsilon_0} \qquad (7.4)$$

i.e.:

$$\nabla^2 \phi + \boldsymbol{\omega} \cdot \nabla \phi + \left(\nabla \cdot \boldsymbol{\omega}\right)\phi = -\frac{\rho}{\epsilon_0} \qquad (7.5)$$

which is an equation capable of giving resonant solutions from the spin connection vector. The Poisson equation does not give resonant solutions. In one Z dimension Eq. (7.5) becomes:

$$\frac{\partial^2 \phi}{\partial Z^2} + \omega_Z \frac{\partial \phi}{\partial Z} + \left(\frac{\partial \omega_Z}{\partial Z}\right)\phi = -\frac{\rho}{\epsilon_0}. \qquad (7.6)$$

The spin connection in Eq. (7.6) must be:

$$\omega_Z = \frac{2}{Z} \qquad (7.7)$$

in order to recover the standard Coulomb law off resonance. This is because:

$$\phi = -\frac{e}{4\pi\epsilon_0 Z}, \quad \frac{\partial \phi}{\partial Z} = \frac{e}{4\pi\epsilon_0 Z^2} = -\frac{\omega_Z}{2}\phi \qquad (7.8)$$

in the off resonant condition, giving Eq. (7.7). In the off resonant condition the role of the spin connection is to change the sign of the electric field according to Eq. (7.8). The way in which the field and potential are related is changed, but this has no experimental effect because $\mathbf{E}$ is effectively changed by $-\mathbf{E}$. With the spin connection (7.7), Eq. (7.6) becomes:

$$\frac{\partial^2 \phi}{\partial Z^2} + \frac{2}{Z}\frac{\partial \phi}{\partial Z} - \frac{2}{Z^2}\phi = -\frac{\rho}{\epsilon_0}. \qquad (7.9)$$

7.2. SPIN CONNECTION RESONANCE FROM THE COULOMB LAW

Now assume that the charge density is initially oscillatory:

$$\rho = \rho^{(0)} \cos(\kappa Z) \tag{7.10}$$

where κ is a wave number. Thus:

$$\frac{\partial^2 \phi}{\partial Z^2} + \frac{2}{Z}\frac{\partial \phi}{\partial Z} - \frac{2}{Z^2}\phi = -\rho^{(0)} \cos \kappa Z. \tag{7.11}$$

The partial derivatives can be changed to total derivatives to give an ordinary differential equation:

$$\frac{d^2 \phi}{dZ^2} + \frac{2}{Z}\frac{d\phi}{dZ} - \frac{2}{Z^2}\phi = -\rho^{(0)} \cos \kappa Z. \tag{7.12}$$

Using the well known Euler method this equation can be reduced to an undamped oscillator equation that has resonant solutions, and this was the earliest attempt at developing the theory of spin connection resonance in UFT63.

This was the first plausible explanation of the Alex Hill devices (www.et3m.net) which have been observed over the years by invited experts, the types of device used by Fortune Fifty companies are power saving devices in induction motors, described on the www.et3m.net site, and energy saving devices in lighting. These types of devices can be mass marketed so no better proof of the presence of energy from space time can be given. Initially, this type of energy was known as energy from the vacuum, but such a nomenclature lent itself to misrepresentation and misunderstanding, notably to absurd allegations of perpetual motion. These came about because the vacuum was confused with "nothingness", so that presumably these advocates of perpetual motion thought that no energy can be transferred from nothing to a device. On the contrary, the vacuum of general relativity contains energy, defined by the infinitesimal of proper time and the dynamic metric. This has been known for a century. So transfer of energy occurs from space time to a device. Total energy is conserved.

Therefore the nomenclature of "energy from space time" was adopted and when the request came in from the U. S. Navy to devise an explanation, one was found by using the spin connection and looking for equations with the structure of an Euler Bernoulli equation. It would then be possible for a small driving force to produce a large resonance in output electric power. This theory is the same in structure as conventional electric resonance theory, but the driving force originates in spacetime. The vacuum structure of spacetime has been greatly developed during the evolution of ECE theory by Eckardt and Lindstrom. When first asked to devise a theory the relevant author (MWE) had no details of circuit design, and was given only a qualitative account of the results. So spin connection resonance was devised to provide a qualitative description.

Subsequently it was found that spin connection resonance occurs in magnetostatics (UFT 65). The ECE equations of magnetostatics can be written as:

$$\boldsymbol{\nabla} \cdot \mathbf{B}^a = 0 \tag{7.13}$$

CHAPTER 7. ENERGY FROM SPACE TIME AND LOW ENERGY...

$$\nabla \times \mathbf{B}^a = \mu_0 \mathbf{J}^a \tag{7.14}$$

$$\mathbf{B}^a = \nabla \times \mathbf{A}^a - g\mathbf{A}^b \times \mathbf{A}^c \tag{7.15}$$

and in this case spin connection resonance is defined by the simultaneous equations:

$$\nabla \times \left(\nabla \times \mathbf{A}^a - g\mathbf{A}^b \times \mathbf{A}^c\right) = \mu_0 \mathbf{J}^a \tag{7.16}$$

and:

$$\nabla \cdot \mathbf{A}^b \times \mathbf{A}^c = 0 \tag{7.17}$$

Eq. (7.16) can be developed with the vector identities:

$$\nabla \times \nabla \times \mathbf{A}^a = -\nabla^2 \mathbf{A}^a + \nabla\left(\nabla \cdot \mathbf{A}^a\right) \tag{7.18}$$

and:

$$\nabla \times \left(\mathbf{A}^b \times \mathbf{A}^c\right) = \mathbf{A}^b \nabla \cdot \mathbf{A}^c - \mathbf{A}^c \nabla \cdot \mathbf{A}^b + \left(\mathbf{A}^c \cdot \nabla\right)\mathbf{A}^b - \left(\mathbf{A}^b \cdot \nabla\right)\mathbf{A}^c. \tag{7.19}$$

To simplify the problem for the sake of illustration, assume that the vector potential has no divergence:

$$\nabla \cdot \mathbf{A}^a = \nabla \cdot \mathbf{A}^b = \nabla \cdot \mathbf{A}^c = 0 \tag{7.20}$$

and assume that $\mathbf{A}^c$ is space independent so that:

$$\left(\mathbf{A}^b \cdot \nabla\right)\mathbf{A}^c = \mathbf{0}. \tag{7.21}$$

Eq. (7.16) becomes:

$$\nabla^2 \mathbf{A}^a + g\left(\mathbf{A}^c \cdot \nabla\right)\mathbf{A}^b = -\mu_0 \mathbf{J}^a \tag{7.22}$$

which can be reduced to:

$$\frac{\partial^2 A^a{}_Z}{\partial X^2} + \kappa_0^2 A^a{}_Z = \mu_0 J^a{}_Z(0) \cos(\kappa X) \tag{7.23}$$

as in UFT 65. This has the resonant solution:

$$A^a{}_Z \to \infty \tag{7.24}$$

at:

$$\kappa = \kappa_0 = \left(g\left(\frac{\partial A_Z}{\partial X}\right)\right)^{\frac{1}{2}}. \tag{7.25}$$

Spin connection resonance can also occur in the Faraday law of induction if it is assumed that there is a magnetic current density:

$$\nabla \times \mathbf{E}^a + \frac{\partial \mathbf{B}^a}{\partial t} = \mu_0 \mathbf{j}^a. \tag{7.26}$$

7.2. SPIN CONNECTION RESONANCE FROM THE COULOMB LAW

UFT 65 assumed that there was no scalar potential and that the electric field is defined by:

$$\mathbf{E}^a = -\frac{\partial \mathbf{A}^a}{\partial t} \tag{7.27}$$

leading to another example of spin connection resonance. Subsequently, UFT 74 led to spin connection resonance in magnetic motors (M. W. Evans and H. Eckardt, Physica B, 400, 175 - 179 (2007)). In UFT 92 the theory was developed for the Coulomb law in radial coordinates. The most influential of these early papers of ECE theory is UFT 107, which applied spin connection resonance to the Faraday disk generator using the concept of rotating spacetime. It was shown that at resonance the vector potential goes to infinity, and this seemed to give a plausible qualitative explanation of experimentally observed resonance in a variable frequency Faraday disk generator.

In these early papers the antisymmetry laws of ECE theory had not yet been inferred, but several types of spin connection resonance were defined. As explained already in this book the antisymmetry laws give the possibility of many more resonances and infinities, thus giving plenty of support for the experimental data of the Alex Hill group (www.et3m.net). Subsequently the subject of spin connection resonance was developed by Eckardt and Lindstrom, and an account of these developments is given later in this chapter. The essential point in all these developments is that spin connection resonance occurs only in a theory of general relativity applied to electromagnetism.

The theory continued to develop until it reached the stage described in UFT 259, in which charge current density had been given a geometrical meaning and in which the antisymmetry laws could be incorporated to give spin connection resonance in a simpler way than in the early papers. This is typical of the development of ECE theory, the theory simplified and clarified during the course of 260 papers to date. The latest stages of development are summarized conveniently in the analysis of the Coulomb law using the electric charge density defined by:

$$\rho^a = \epsilon_0 \left(\boldsymbol{\omega}^a{}_b \cdot \mathbf{E}^b - c \mathbf{A}^b \cdot \mathbf{R}^a{}_b \,(\text{orb}) \right) \tag{7.28}$$

where ϵ_0 is the vacuum permittivity, $\boldsymbol{\omega}^a{}_b$ is the spin connection vector, $\mathbf{E}^b$ is the electric field strength, c is the universal constant known as the speed of light, and $\mathbf{R}^a{}_b$ is the orbital part of the curvature vector. As explained already in this book the electric field strength is:

$$\mathbf{E}^a = -c\boldsymbol{\nabla} A^a{}_0 - \frac{\partial \mathbf{A}^a}{\partial t} - c\omega^a{}_{0b} \mathbf{A}^b + c A^b{}_0 \boldsymbol{\omega}^a{}_b \tag{7.29}$$

where the 4-potential is defined by:

$$A^a{}_\mu = (A^a{}_0, -\mathbf{A}^a) = \left(\frac{\phi^a}{c}, -\mathbf{A}^a \right) \tag{7.30}$$

where ϕ^a is the scalar potential. The electric current density is defined by:

$$\mathbf{J}^a = \epsilon_0 c \left(\omega^a{}_{0b} \mathbf{E}^b - c A^b{}_0 \mathbf{R}^a{}_b \,(\text{orb}) + c \boldsymbol{\omega}^a{}_b \times \mathbf{B}^b - c \mathbf{A}^b \times \mathbf{R}^a{}_b \,(\text{spin}) \right) \tag{7.31}$$

CHAPTER 7. ENERGY FROM SPACE TIME AND LOW ENERGY...

where $\mathbf{R}^a{}_b$ (spin) is the spin part of the curvature vector and where $\mathbf{B}^b$ is the magnetic flux density.

As discussed in UFT 259 the equations of electrostatics in ECE theory are

$$\nabla \cdot \mathbf{E}^a = \omega^a{}_b \cdot \mathbf{E}^b \tag{7.32}$$

$$\omega^a{}_{0b} \cdot \mathbf{E}^b = \phi^b \mathbf{R}^a{}_b \text{ (orb)} \tag{7.33}$$

$$\omega^a{}_b \times \mathbf{E}^b + \phi^b \mathbf{R}^a{}_b \text{ (spin)} = 0 \tag{7.34}$$

$$\mathbf{E}^a = -\nabla \phi^a + \phi^b \omega^a{}_b \tag{7.35}$$

In order to obtain spin connection resonance Eq. (7.32) must be extended to:

$$\nabla \cdot \mathbf{E}^a = \omega^a{}_b \cdot \mathbf{E}^b - c\mathbf{A}^b \text{ (vac)} \cdot \mathbf{R}^a{}_b \text{ (orb)} \tag{7.36}$$

where $\mathbf{A}^b$ is a vacuum potential of ECE theory. The static electric field is:

$$\mathbf{E}^a = -\nabla \phi^a + \phi^b \omega^a{}_b \tag{7.37}$$

so from Eqs. (7.36) and (7.37):

$$\nabla^2 \phi^a + \left(\omega^a{}_b \cdot \omega^b{}_c\right) \phi^c = \nabla \cdot \left(\phi^b \omega^a{}_b\right) + \omega^a{}_b \cdot \nabla \phi^b + c\mathbf{A}^b \text{ (vac)} \cdot \mathbf{R}^a{}_b \text{ (orb)}. \tag{7.38}$$

The ECE anti symmetry law means that:

$$-\nabla \phi^a = \phi^b \omega^a{}_b \tag{7.39}$$

leading to the Euler Bernoulli resonance equation:

$$\nabla^2 \phi^a + \left(\omega^a{}_b \cdot \omega^b{}_c\right) \phi^c = \frac{1}{2} c\mathbf{A}^b \text{ (vac)} \cdot \mathbf{R}^a{}_b \text{ (orb)} \tag{7.40}$$

and undamped spin connection resonance. The left hand side contains the Hooke's law term and the right hand side the driving term originating in a vacuum potential. However tiny this term may be it can be amplified greatly by undamped resonance, confirming the Alex hill result in another way. This is the most complete theory of Coulomb law resonance to date.

Denoting:

$$\rho^a \text{ (vac)} = \frac{\epsilon_0 c}{2} \mathbf{A}^b \text{ (vac)} \cdot \mathbf{R}^a{}_b \text{ (orb)} \tag{7.41}$$

the equation becomes:

$$\nabla^2 \phi^a + \left(\omega^a{}_b \cdot \omega^b{}_c\right) \phi^c = \frac{\rho^a \text{ (vac)}}{\epsilon_0}. \tag{7.42}$$

The left hand side is a field property and the right hand side is a property of the ECE vacuum. In the simplest case:

$$\nabla^2 \phi + \omega^2{}_0 \phi = \frac{\rho \text{ (vac)}}{\epsilon_0} \tag{7.43}$$

and produces undamped resonance if the driving term is oscillatory as already described in this book.

7.3 Low Energy Nuclear Reactions (LENR)

This is a most promising source of new energy, the most well known device being the Rossi reactor recently purchased for commercialization. Again the standard model of electromagnetism has no coherent explanation for the phenomenon, in which nuclear fusion occurs in simple apparatus with release of useful heat. Some of the devices used to produce this heat are well known and available in all detail. The technique has been subject to numerous independent assessments and checks for repeatability and reproducibility. Initially it was known as cold fusion, famously discovered by Pons and Fleischman in the University of Utah. Their discoveries were supported initially by the State of Utah. It was difficult initially to prove that cold fusion was reproducible and repeatable, so there ensued a very long debate which is still going on. However the LENR technique is being commercialized, and subject to control of the heat produced, will be available for domestic use.

LENR devices are already being used for military and other applications and have been subjected to the usual testing and certifying. Some academic departments are also dedicated to LENR, and many conferences, journals and newsletters dedicated to the subject. In the economics department in the University of Utah, models are being developed to research the effect of LENR on future economies. The availability of cheap and clean energy is a pre-requisite for economic growth. Stephen Bannister for example is currently preparing a Thesis on this topic in the University of Utah's Department of Economics, a Thesis which compares the first industrial revolution in Britain with the second industrial revolution expected to occur as the result of the energy techniques described in this chapter. During one such conference approximately a year and a half ago one of the authors of this book (MWE) was asked to devise a theoretical explanation for low energy nuclear reactions in terms of ECE theory in order to devise a solid and coherent framework for its development within the scope of a unified field theory. There are many theories of LENR but no consensus as to the origins of the energy needed to cause a nuclear reaction in simple apparatus in the laboratory.

The initial response to this request was UFT 226 on www.aias.us, in which a general theory of particle collisions was developed. This is overviewed briefly in this section. Consider two particles of 4-momenta p^μ and p^μ_1:

$$p^\mu = \left(\frac{E}{c}, \mathbf{p}\right), \quad p^\mu_1 = \left(\frac{E_1}{c}, \mathbf{p}_1\right). \tag{7.44}$$

In the minimal prescription on the semi classical level the collision of these particles is described by:

$$p^\mu \rightarrow p^\mu + p^\mu_1 \tag{7.45}$$

$$E \rightarrow E + E_1 \tag{7.46}$$

$$\mathbf{p} \rightarrow \mathbf{p} + \mathbf{p}_1 \tag{7.47}$$

where E is the relativistic energy

$$E = \gamma mc^2 \tag{7.48}$$

CHAPTER 7. ENERGY FROM SPACE TIME AND LOW ENERGY...

and **p** the relativistic momentum:

$$\mathbf{p} = \gamma m \mathbf{v}. \tag{7.49}$$

The Lorentz factor is defined by:

$$\gamma = \left(1 - \frac{v^2}{c^2}\right)^{-\frac{1}{2}} \tag{7.50}$$

where **v** is the velocity of a particle of mass m and where c is the speed of light in vacuo. Eq. (7.49) implies the Einstein field equation:

$$E^2 = p^2 c^2 + m^2 c^4 \tag{7.51}$$

which can be written as:

$$E^2 - m^2 c^4 = (E - mc^2)(E + mc^2) = c^2 p^2. \tag{7.52}$$

From Eqs. (7.45) and (7.51):

$$(E + E_1)^2 = c^2 (p + p_1)^2 + m^2 c^4 \tag{7.53}$$

which is the classical relativistic description of particle interaction in the minimal prescription. From Eq. (7.53):

$$(E + E_1)^2 - m^2 c^4 = c^2 (p + p_1)^2 \tag{7.54}$$

so the relativistic kinetic energy is:

$$T = E + E_1 - mc^2 = \frac{c^2 (p + p_1)^2}{E + E_1 + mc^2}. \tag{7.55}$$

This kinetic energy is a limit of the ECE fermion equation, which is derived from the Cartan geometry used in this book. The concepts of particle mass m and m_1 are limits of the more general R factor of the ECE wave equation described in UFT 181 and UFT 182. After a series of approximations described in UFT 226, and similar to those used in the derivation of the fermion equation described already in this book, the energy E can be expressed as:

$$E = \frac{c^2 (p + p_1)^2}{2mc^2 + E_1} + mc^2 \tag{7.56}$$

and the kinetic energy as:

$$T = E + E_1 - mc^2 \sim E - mc^2. \tag{7.57}$$

In order to quantize the theory the fermion equation is used as described in UFT 226 to give the hamiltonian operator:

$$H\psi = (H_1 + H_2)\psi \tag{7.58}$$

7.3. LOW ENERGY NUCLEAR REACTIONS (LENR)

where:

$$H_1\psi = \frac{1}{2m}\left(\boldsymbol{\sigma}\cdot(-i\hbar\boldsymbol{\nabla}+\mathbf{p}_1)\,\boldsymbol{\sigma}\cdot(-i\hbar\boldsymbol{\nabla}+\mathbf{p}_1)\right)\psi \qquad (7.59)$$

and

$$H_2\psi = \left(-\boldsymbol{\sigma}\cdot(-i\hbar\boldsymbol{\nabla}+\mathbf{p}_1)\frac{E_1}{4m^2c^2}(-i\hbar\boldsymbol{\nabla}+\mathbf{p}_1)\right)\psi. \qquad (7.60)$$

In Eq. (7.58):

$$\boldsymbol{\sigma}\cdot(\mathbf{p}+\mathbf{p}_1)\,\boldsymbol{\sigma}\cdot(\mathbf{p}+\mathbf{p}_1) = p^2 + p_1^2 + \mathbf{p}_1\cdot\mathbf{p} + \mathbf{p}\cdot\mathbf{p}_1 + i\boldsymbol{\sigma}\cdot(\mathbf{p}_1\times\mathbf{p}+\mathbf{p}\times\mathbf{p}_1) \quad (7.61)$$

so the first type of hamiltonian becomes:

$$H_1 = -\frac{\hbar^2}{2m}\nabla^2 + \frac{p_1^2}{2m} + \frac{i\hbar}{2m}(\mathbf{p}_1\cdot\boldsymbol{\nabla}+\boldsymbol{\nabla}\cdot\mathbf{p}_1) + \frac{\hbar}{2m}\boldsymbol{\sigma}\cdot(\mathbf{p}_1\times\boldsymbol{\nabla}+\boldsymbol{\nabla}\times\mathbf{p}_1) \quad (7.62)$$

and operates on the wave function to give energy eigenvalues. As described in UFT 226 the hamiltonian operator may be simplified to give:

$$H_1 = -\frac{\hbar^2}{2m}\nabla^2 + \frac{p_1^2}{2m} + \frac{i\hbar}{2m}(\boldsymbol{\nabla}\cdot\mathbf{p}_1 + 2\mathbf{p}_1\cdot\boldsymbol{\nabla}) + \frac{\hbar}{2m}\boldsymbol{\sigma}\cdot\boldsymbol{\nabla}\times\mathbf{p}_1. \qquad (7.63)$$

In the generally covariant format of this theory the concept of mass is generalized to curvature R using the Hamilton Jacobi equation:

$$(p^\mu - \hbar\kappa^\mu)(p_\mu - \hbar\kappa_\mu) = m_0^2 c^2 \qquad (7.64)$$

as in UFT 182 on www.aias.us. Eq. (7.64) may be written as:

$$p^\mu p_\mu = \hbar^2 R_1 + m_0^2 c^2. \qquad (7.65)$$

Using this theory it is possible to consider the four momentum $p^\mu{}_1$ of particle 1 interacting with a matter wave 2 defined by the wave vector $\kappa^\mu{}_2$. Particle 1 is also a matter wave:

$$p^\mu{}_1 = \hbar\kappa^\mu{}_1. \qquad (7.66)$$

In UFT 182 it was shown that the interaction is described by:

$$\left(\Box + R_2 + \left(\frac{m_{10}c}{\hbar}\right)^2\right)\psi_1 = 0 \qquad (7.67)$$

where the R_2 parameter is:

$$R_2 = \left(\frac{m_2 c}{\hbar}\right)^2 \qquad (7.68)$$

and where the concept of interacting mass is defined as:

$$m_2 = \frac{\hbar}{c}\left(2\left(\frac{\omega_1\omega_2}{c^2} - \kappa_1\kappa_2\right) - \left(\frac{\omega_2^2}{c^2} - \kappa_2^2\right)\right)^{\frac{1}{2}}. \qquad (7.69)$$

Therefore in this general ECE theory it is possible to think of a quantum of space time energy being absorbed during a LENR reaction. This idea generalizes the Planck concept of photon energy to particle energy.

A low energy nuclear reaction can be exemplified as follows:

$$^{64}\text{Ni} + \text{p} = ^{63}\text{Cu} + 2\,\text{n}. \tag{7.70}$$

Here, ^{64}Ni has 36 neutrons and 28 protons, and ^{63}Cu has 34 neutrons and 29 protons. So ^{64}Ni is transmuted into ^{63}Cu with release of two neutrons. The theory must explain why this nuclear reaction occurs. Nickel is transmuted to copper with the release of usable heat and this reaction can be made to occur in simple apparatus in the laboratory. It does not need the vast amount of expenditure of conventional nuclear fusion research. Using the theory of this section the interacting mass is:

$$m = \frac{\hbar}{c}\left(\frac{\omega^2}{c^2} - \kappa^2\right)^{\frac{1}{2}} \tag{7.71}$$

and the total mass of the nickel atom during interaction increases to:

$$M = \left(m^2 + m_0^2\right)^{\frac{1}{2}} \tag{7.72}$$

with concomitant energy:

$$E_0 = Mc^2 \tag{7.73}$$

so that a nuclear reaction occurs, a LENR reaction.

This is a simple first theory, which is a plausible explanation of LENR. In UFT 227 a more general theory was considered to develop an expression for the mass M of a fused nucleus when reactants 1 and 2 produce products 3 and 4. Total energy momentum is conserved as follows:

$$p^\mu{}_1 + p^\mu{}_2 = p^\mu{}_3 + p^\mu{}_4. \tag{7.74}$$

As shown in UFT 227 this equation can be expressed as:

$$(E_1 + E_2)^2 - (\mathbf{p}_1 + \mathbf{p}_2)\cdot(\mathbf{p}_1 + \mathbf{p}_2) = M^2 c^4 \tag{7.75}$$

where:

$$M^2 = m_1{}^2 + m_2{}^2 + 2m_1 m_2\left(\gamma_1\gamma_2 - (\gamma_1{}^2 - 1)^{\frac{1}{2}}(\gamma_2{}^2 - 1)^{\frac{1}{2}}\cos\theta\right) \tag{7.76}$$

in which the angle θ is defined as

$$(\mathbf{p}_1 + \mathbf{p}_2)\cdot(\mathbf{p}_1 + \mathbf{p}_2) = p^2{}_1 + p^2{}_2 + 2p_1 p_2 \cos\theta. \tag{7.77}$$

In the non relativistic limit:

$$v_1 \ll c, v_2 \ll c \tag{7.78}$$

7.3. LOW ENERGY NUCLEAR REACTIONS (LENR)

Eq. (7.76) becomes:

$$M^2 = m_1{}^2 + m_2{}^2 + 2m_1 m_2 = (m_1 + m_2)^2 \tag{7.79}$$

so in this limit M is the sum of m_1 and m_2. Otherwise there is a mass discrepancy or difference:

$$\Delta m = \left(m_1{}^2 + m_2{}^2 - M^2\right)^{\frac{1}{2}} \tag{7.80}$$

which gives rise to the energy released in nuclear fusion as heat and light.

This classical relativistic theory was quantized in UFT 227 using the fermion equation for the fusion of two atoms 1 and 2. The attractive nuclear strong forces are denoted V_1 and V_2, their sum being:

$$V = V_1 + V_2. \tag{7.81}$$

The total relativistic energy of nuclei 1 and 2 is:

$$E = E_1 + E_2 \tag{7.82}$$

and their fused mass is M. The vector sum of their relativistic momenta is:

$$\mathbf{p} = \mathbf{p}_1 + \mathbf{p}_2. \tag{7.83}$$

The fermion equation for this nuclear fusion reaction is:

$$((E - V) + c\boldsymbol{\sigma} \cdot \mathbf{p})\phi^L = Mc^2 \phi^R \tag{7.84}$$

$$((E - V) + c\boldsymbol{\sigma} \cdot \mathbf{p})\phi^R = Mc^2 \phi^L \tag{7.85}$$

which can be developed as the Schroedinger type equation:

$$H\psi = E\psi \tag{7.86}$$

where the hamiltonian operator is:

$$H = H_1 + H_2 \tag{7.87}$$

where:

$$H_1 = Mc^2 + V - \frac{\hbar^2 \nabla^2}{2m} \tag{7.88}$$

and:

$$H_2 = \frac{1}{4M^2 c^2} \boldsymbol{\sigma} \cdot \mathbf{p}\, V\, \boldsymbol{\sigma} \cdot \mathbf{p} \tag{7.89}$$

giving the nuclear energy levels.

In UT 227 the well known Woods Saxon potential was used to model Eq. (7.86). It is described by:

$$V = -V_0 \left(1 + \exp\left(\frac{r-R}{a}\right)\right)^{-1} \tag{7.90}$$

where V_0 is the potential well depth, a is the surface thickness of the nucleus, and R is the nuclear radius. It can be approximated roughly by the harmonic oscillator potential:

$$V = \frac{1}{2}kr^2 - V_0 \qquad (7.91)$$

where k is the spring constant of Hooke's law, so Eq. (7.86) becomes:

$$H_1\psi = \left(-\frac{\hbar^2 \nabla^2}{2m} + \frac{1}{2}kr^2 + Mc^2 - V_0\right)\psi. \qquad (7.92)$$

The nuclear energy levels of the fused nucleus in this approximation are the well known energy levels of the harmonic oscillator:

$$E = \left(n + \frac{1}{2}\right)\hbar\omega \qquad (7.93)$$

where:

$$n = 0, 1, 2, \ldots \qquad (7.94)$$

and where:

$$\omega = \left(\frac{k}{M}\right)^{\frac{1}{2}}. \qquad (7.95)$$

In a rough approximation as described in UFT 227 the attractive nuclear strong force can be written as:

$$\mathbf{F}_N \sim \frac{1}{4a}\left(1 - \frac{r-R}{a}\right)\mathbf{e}_r \qquad (7.96)$$

and the spin orbit energy from the nuclear fermion equation (7.86) is:

$$H_{so}\psi = \frac{\hbar}{16M^2c^2a^2}\boldsymbol{\sigma}\cdot\mathbf{L}\psi. \qquad (7.97)$$

The spin orbit energy can be used to explain many features of nuclear physics and is its most important property.

The energy levels of the fused nucleus are in excited states, and the nucleus disintegrates to give products 3 and 4 accompanied by energy:

$$\Delta E_0 = (m_1 + m_2 - M)c^2 \qquad (7.98)$$

In UFT 228 quantum tunnelling theory was introduced by writing the Einstein energy equation:

$$E^2 = p^2c^2 + m^2c^4 \qquad (7.99)$$

as

$$E = \gamma mc^2 = \frac{1}{\gamma m}\left(p^2 + m^2c^2\right). \qquad (7.100)$$

7.3. LOW ENERGY NUCLEAR REACTIONS (LENR)

Eq. (7.100) becomes a Schroedinger equation:

$$H\psi = E\psi \tag{7.101}$$

with the hamiltonian:

$$H = \frac{1}{\gamma m}\left(p^2 + m^2 c^2\right) \tag{7.102}$$

and energy levels:

$$E = \gamma m c^2. \tag{7.103}$$

It follows that:

$$p^2 \psi = -\hbar^2 \nabla^2 \psi = m^2 c^2 \left(\gamma^2 - 1\right)\psi. \tag{7.104}$$

The four momentum is defined by:

$$p^\mu = i\hbar \partial^\mu = \hbar \kappa^\mu \tag{7.105}$$

where:

$$p^\mu = \left(\frac{E}{c}, \mathbf{p}\right), \tag{7.106}$$

$$\partial^\mu = \left(\frac{1}{c}\frac{\partial}{\partial t}, -\nabla\right), \tag{7.107}$$

$$\kappa^\mu = \left(\frac{\omega}{c}, \boldsymbol{\kappa}\right). \tag{7.108}$$

Here ω is the frequency of the matter wave, and κ the wave number. Therefore:

$$p^2 \psi = \hbar^2 \kappa^2 \psi = m^2 c^2 \left(\gamma^2 - 1\right)\psi = \left(\frac{E^2}{c^2} - m^2 c^2\right)\psi. \tag{7.109}$$

For a free wave / particle:

$$\kappa = \frac{mc}{\hbar}\left(\gamma^2 - 1\right)^{\frac{1}{2}}. \tag{7.110}$$

For the purposes of the development of quantum tunnelling theory denote:

$$k = \frac{mc}{\hbar}\left(\gamma^2 - 1\right)^{\frac{1}{2}} \tag{7.111}$$

In the presence of potential energy V the operator (7.102) becomes:

$$H = \frac{1}{\gamma m}\left(p^2 + m^2 c^2\right) + V \tag{7.112}$$

so:

$$p^2 \psi = \left(\gamma m \left(E - V\right) - m^2 c^2\right)\psi \tag{7.113}$$

and

$$\kappa^2 = \frac{1}{\hbar^2}\left(\gamma m\left(E-V\right) - m^2 c^2\right). \tag{7.114}$$

In quantum tunnelling theory E < V, so

$$E - V < 0. \tag{7.115}$$

Define:

$$\kappa = \frac{1}{\hbar}\left(\gamma m\left(V-E\right)\right)^{\frac{1}{2}}. \tag{7.116}$$

Denoting the rest wave number as:

$$\kappa_0 = \frac{mc}{\hbar} \tag{7.117}$$

we arrive at the definition:

$$\kappa^2 + \kappa_0^2 = \frac{\gamma m}{\hbar^2}\left(V-E\right). \tag{7.118}$$

Eq. (7.111) can be written as:

$$\kappa^2 + \kappa_0^2 = \gamma^2 \left(\frac{mc}{\hbar}\right)^2 \tag{7.119}$$

so:

$$\frac{p^2}{2m}\psi = \frac{mc^2}{2}\left(\gamma^2 - 1\right)\psi. \tag{7.120}$$

In the non-relativistic quantum limit, as shown in UFT 228:

$$\nabla^2 \psi = -\left(\frac{2mE}{\hbar^2}\right)\psi \tag{7.121}$$

giving the transmission coefficient:

$$T = 8\kappa^2 k^2 \left(\left(k^2 + \kappa^2\right)\cosh\left(4\kappa a\right) - \left(\kappa^4 + k^4 - 6\kappa k\right)\right)^{-1}, \tag{7.122}$$

for a potential of type:

$$\begin{aligned} V &= 0, \quad x < -a, \\ V &= V_0, \quad -a < x < a, \\ V &= 0, \quad x > a, \\ E &< V_0, \end{aligned} \tag{7.123}$$

in which:

$$k^2 = 2mE/\hbar^2, \quad E = mc^2 \frac{\gamma^2 - 1}{2}, \tag{7.124}$$

$$\kappa^2 = 2m\left(V_0 - E\right)/\hbar^2, \quad E = mc^2 \frac{\gamma^2 - 1}{2}.$$

7.3. LOW ENERGY NUCLEAR REACTIONS (LENR)

In a graphical analysis the transmission coefficient T of Eq. (7.122) has been calculated for the rectangular barrier (7.123). The coefficient depends on wave vectors k and κ and barrier half-width a. In Fig. 7.1 both a and k have been varied. It can be concluded that T is at maximum when ka as well as κ are minimal; this corresponds to quantum waves with lowest energy.

Since k and κ depend on the energy E and height of the potential well V_0 (see Eq. (7.124)), it is more conclusive to study the dependence on these parameters. For a special parameter combination, T is quite high in the "forbidden" region, showing the quantum mechanical tunneling behaviour. This is graphed in Fig. 7.2 in a 3D representation.

The tunnelling probability decreases drastically with slightly enhanced masses. Mass is a very sensitive parameter. This can be seen from Fig. 7.3 where we have graphed the mass dependence of T with relativistic velocity ratio v/c as a curve parameter. For $v \to c$ the transmission coefficient degenerates to a delta function at $m = 0$.

It is found using this analysis that the single most important factor is the mass of the incoming particle. The extra ingredient given by ECE theory is the possibility of augmenting the standard quantum tunnelling theory by resonant absorption of quanta of space time energy - energy from space time.

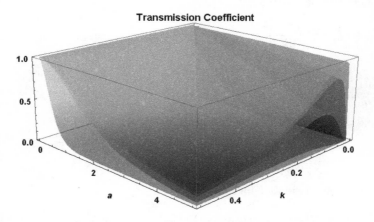

Figure 7.1: Transmission coefficient $T(k, a)$ for five values of κ.

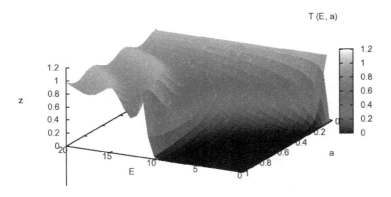

Figure 7.2: Transmission coefficient $T(E,a)$ for $m = \hbar = 1$, $V_0 = 10$.

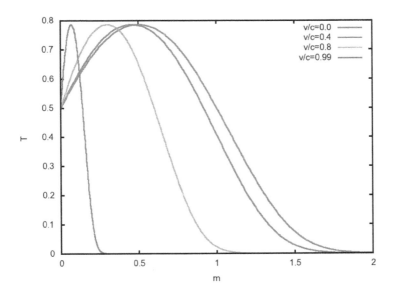

Figure 7.3: Mass dependence of the realtivistic transmission coefficient $T(m)$ for electron-electron tunneling, electrom mass is $m = 1$.

7.3. LOW ENERGY NUCLEAR REACTIONS (LENR)

Chapter 8

ECE Cosmology

8.1 Introduction

Astronomy is one of the oldest of the sciences and has become a precise subject area. Cosmology began to develop as a subject when the observations of the orbit of Mars by Tycho Brahe were analyzed by Johannes Kepler to give three planetary laws reduced by Newton to universal gravitation and the equivalence of gravitational and inertial mass. The famous Newtonian dynamics were developed to include rotational motions in non inertial frames by Euler, Bernoulli, Coriolis and others, and Laplace developed his elegant celestial mechanics. Lagrange developed the subject of dynamics from a different perspective, and using more general concepts which were taken up by Hamilton to produce the Hamilton equations and the idea of the Hamiltonian. The latter became the basis of quantum mechanics. Orbital theory can be developed elegantly with the idea of the Lagrangian and the Euler Lagrange equations. For example, conservation of angular momentum and the Euler Lagrange equations can be used to show that if the orbit of a mass m around a mass M is observed to be an ellipse, then the force between m and M is inversely proportional to the square of the distance r between m and M – the famous inverse square law as inferred by Newton. The same method also gives the three Kepler laws of planetary motion. However the Lagrangian method is more general than that of Newton because it can give the force law for any orbit.

In the eighteenth and early nineteenth centuries the orbits of all masses m around a mass M were thought to be ellipses to an excellent approximation, with M at one focus of the ellipse, so the subject was thought to be complete, and m travelled on the ellipse. The orbits of planets could be observed with precision, and objects such as galaxies were unknown. So the famous Newtonian concept of universal gravitation was thought to be as near to perfection as human intellect could devise. Newtonian dynamics worked for astronomy and also back on the ground. The apocryphal apple was governed by the acceleration due to gravity **g** of the earth. The apple and the moon were governed by the same law, universal gravitation.

8.1. INTRODUCTION

The gods however are offended by human pretence to perfection, the orbit of a planet precesses, a point of the ellipse such as its perihelion moves forward a little every orbit. In the Newtonian dynamics the elliptical orbit does not move forward if one considers only m and M and the force between them. From precise astronomical observations of orbits by ancient astronomers the precession of the perihelion had been known well before Newton's time. In Newton's time, the seventeenth century, it was thought to be caused by the gravitational pull of other planets. It is a very tiny effect so was not thought to be due to any flaw in Newton's universal gravitation. When the human intellect contrives something that it thinks to be perfect, no data are allowed to stand in the way, and it is human nature to hang on to a theory even though the data show that the theory is not quite right. Sometimes the theory is totally wrong and always gave an illusion of the truth. The precession of planetary orbits can indeed be explained to a large extent by Newtonian concepts, but there seems to be a tiny part of the precession that cannot be explained.

Following the Michelson Morley experiment the entire subject of dynamics was changed and the concept of special relativity introduced as described in chapter 1 of this book. The Newtonian and Lagrangian dynamics were recovered as limits of special relativity. However, special relativity is restricted to the Lorentz transform and a constant inter frame velocity. In order to consider acceleration and similar effects a new relativity was needed. Another profound change in thought occurred when Einstein and others decided to base dynamics on geometry. This was also Kepler's idea, and went back to the ancient Greeks, who thought of geometry as beauty itself, or perfect beauty. Effectively this means that the Lorentz transform becomes the general coordinate transform. It is not in any way clear to the human intuition that space should become part of time, that the familiar three dimensions should be abandoned, and that the familiar concepts of Euclid should be replaced by a different geometry. The very idea of a different geometry had been considered only by a few mathematicians up to about 1905.

Among the first to consider such as geometry was Riemann in the early nineteenth century, followed in the eighteen sixties by Christoffel. These two prominent mathematicians devised the concept of metric and connection. The metric is a symmetric object by definition, but the connection has no particular symmetry in the lower two of its three indices. About forty years later Ricci and Levi-Civita devised the concept of curvature of space of any dimension, including four dimensional spacetime, that of special relativity. In physics concomitant progress was being made by Noether, who linked the conservation laws of physics to symmetry laws. The subject of physics introduced the canonical energy momentum tensor, which is also symmetric in its indices. In mathematics, in about 1900, Levi-Civita defined the Christoffel connection as being symmetric. This was an axiom, or hypothesis, not a rigorous proof. In 1900 it was not known that there existed a fundamental property of any mathematical space in any dimension, the torsion.

In 1902 Bianchi inferred an identity in which a well defined cyclic sum of curvature tensors vanishes. This is known as the first Bianchi identity,

CHAPTER 8. ECE COSMOLOGY

from which the second Bianchi identity can be inferred. The two Bianchi identities were also inferred in ignorance of the existence of torsion, and using a symmetric connection. The ingredients available to Einstein from 1905 to 1915 were therefore the second Bianchi identity and the Noether Theorem, thought to be fundamental principles of geometry and physics. Proceeding on the ancient basis that geometry gives physics, Einstein attempted for a decade to arrive at a field equation linking the two concepts. This was finally published in 1915 and asserts that the second Bianchi identity is proportional to the covariant derivative of the canonical energy momentum tensor. With the benefit of hindsight this is an over complicated procedure. By Ockham's Razor a simpler theory is preferred, and that theory is ECE theory. In addition the Einstein field equation was arrived at in ignorance of torsion. So it was bound to fail qualitatively, and has indeed done so. The velocity curve of a whirlpool galaxy shows that the Einstein theory is incorrect qualitatively, or completely. The proof of this is given later in this chapter.

At first the field equation of Einstein seemed to be logical, but on closer inspection it contains an assumption made a priori, i. e. guesswork. This is the assumption of the symmetric connection made by Levi-Civita fifteen years before the field equation appeared. The second Bianchi identity used by Einstein relies on a symmetric connection, so is true if and only if the torsion is zero. This was of course unknown to Einstein and also unknown to Levi-Civita and Ricci. The procedure used in deriving the Einstein field equation is to reduce the second Bianchi identity to the covariant derivative of the Einstein tensor, which is symmetric in its lower two indices, and which is made up of a combination of the Ricci tensor and the Ricci scalar. Unknown to Einstein and all his contemporaries this procedure is true if and only if the torsion is zero. If the torsion is finite it fails completely as explained in UFT 88 on www.aias.us.

The field equation was criticized immediately and severely by Schwarzschild in a letter to Einstein of December 1915 as explained earlier in this book. Apart from the assumption of a symmetric connection, there are other flaws in the attempted first solution of the field equation by Einstein. Schwarzschild solved the equation using a metric which does not contain a singularity. So it was known as early as 1915 that there are no black holes and big bang, concepts which were ridiculed by Einstein and Hoyle independently. The cold truth is that these concepts are just mathematical flaws. Experimental data have shown many times over that there was no big bang, and black holes have never been discovered. They are simply asserted to exist by dogmatists. The confusion was greatly compounded by the introduction of a metric that was attributed falsely to Schwarzschild. This metric contains singularities or infinities, so by definition should be rejected as a valid solution of the Einstein field equation. The Schwarzschild metrics, true (1915), and false, fail completely in whirlpool galaxies. This fact has been known for sixty years. A plethora of such metrics have been inferred in a century of work on the Einstein field equation but all fail completely in view of the failure of the field equation in whirlpool galaxies and in view of the fact that they all neglect torsion (M. W. Evans, S. J. Crothers, H. Eckardt and K. Pendergast, "Criticisms of the Einstein Field Equation" referred to in chapter 1).

8.2. ECE THEORY OF LIGHT DEFLECTION DUE TO GRAVITATION

The existence of torsion is a fundamental building block of ECE theory, which set out in 2003 to rebuild general relativity using a rigorously correct geometry, one which does not contain guesswork. So it is essential to prove that torsion cannot be discarded in any valid geometry. In the Cartan geometry used in ECE theory the torsion is defined by the first Maurer Cartan structure equation, inferred in the twenties. This procedure has been explained earlier in this book and the basis of ECE cosmology and unified field theory is that torsion and curvature are identically non zero in any valid geometry. The reason is that they are both generated by the commutator of covariant derivatives acting on any tensor in any space of any dimension. They are always produced simultaneously, and the commutator always produces the two structure equations of Cartan simultaneously. The commutator always produces the torsion tensor as the difference of two anti symmetric connections, so the anti symmetry of the connection is the anti symmetry of the commutator.

A symmetric connection produces a symmetric commutator which vanishes, and a symmetric connection means that the torsion vanishes. This means that the curvature vanishes if the torsion vanishes because torsion and curvature are always produced simultaneously by the commutator. A null commutator means both a null torsion and null curvature, so a symmetric connection means a null torsion AND a null curvature.

The incorrect procedure used by the Einsteinian general relativity is to omit the torsion tensor, and to assume that the commutator produces only the curvature. This is mathematical nonsense that has become dogma. The fact that the torsion always exists means that the first and second Bianchi identities are changed completely in structure. The first Bianchi identity becomes the Cartan identity and the second Bianchi identity becomes the equation given in chapter 1. These mathematical flaws are obvious in retrospect, and were compounded greatly through the illusion of accuracy of the Einstein theory in the solar system. In chapter 8.2 the correct explanation for light deflection by gravitation is given in terms of the spin connection of ECE theory, which is also capable of giving a satisfactory explanation of the velocity curve of a whirlpool galaxy. Currently both the ECE and the Einsteinian theories are influential in science, but obvious and drastic flaws in geometry cannot remain indefinitely without being remedied. The fundamental aim of ECE theory is to improve on the ideas used by Einstein and his contemporaries, ideas which go back to Kepler and to ancient times.

8.2 ECE Theory of Light Deflection due to Gravitation

Consider as in UFT 215 the linear orbital velocity in cylindrical polar coordinates (r, θ):

$$\mathbf{v} = \dot{r}\mathbf{e}_r + r\dot{\theta}\mathbf{e}_\theta \tag{8.1}$$

where $\mathbf{e}_r$ and $\mathbf{e}_\theta$ are the unit vectors of the cylindrical polar system. The velocity squared is:

$$v^2 = \dot{r}^2 + r^2\dot{\theta}^2. \tag{8.2}$$

The precession of an elliptical orbit can be described by the equation:

$$r = \frac{\alpha}{1 + \epsilon\cos(x\theta)} \tag{8.3}$$

when x is near to unity. In this equation, α is the half right latitude and ϵ is the eccentricity. When x becomes large, some very interesting mathematical results are obtained, the subject area of precessing conical sections which show fractal behaviour as described and illustrated in the UFT papers on www.aias.us. However in astronomy the factor x is close to unity for all types of precessing orbits, in the solar system and in binary systems which exhibit the largest precessions. When x is exactly one, the subject of conical sections is recovered, for example static ellipse, the static hyperbola and so on.

Elementary kinematics of plane polar coordinates produce the acceleration:

$$\mathbf{a} = \left(\ddot{r} - r\dot{\theta}^2\right)\mathbf{e}_r + \left(r\ddot{\theta} + 2\dot{r}\dot{\theta}\right)\mathbf{e}_\theta. \tag{8.4}$$

This is a well known general result described in several UFT papers. From the equation (8.3) of precessing conical sections

$$\frac{dr}{d\theta} = \frac{x\epsilon}{\alpha}r^2\sin(x\theta). \tag{8.5}$$

From lagrangian dynamics the conserved orbital angular momentum is well known to be:

$$L = mr^2\dot{\theta} = mr^2\frac{d\theta}{dt}. \tag{8.6}$$

Therefore:

$$\dot{r} = \frac{dr}{dt} = \frac{dr}{d\theta}\frac{d\theta}{dt} = \frac{xL\epsilon}{m\alpha}\sin(x\theta) \tag{8.7}$$

and from Eq. (8.6):

$$\dot{\theta} = \frac{L}{mr^2}. \tag{8.8}$$

The second derivatives are:

$$\ddot{r} = \frac{x^2L^2\epsilon}{m^2\alpha r^2}\cos(x\theta) \tag{8.9}$$

and:

$$\ddot{\theta} = -\frac{2L^2x\epsilon}{m^2r^3\alpha}\sin(x\theta) \tag{8.10}$$

8.2. ECE THEORY OF LIGHT DEFLECTION DUE TO GRAVITATION

and the angular dependent part of the acceleration vanishes:

$$r\ddot{\theta} + 2\dot{r}\dot{\theta} = 0. \tag{8.11}$$

The radial part is given by:

$$\ddot{r} - r\dot{\theta}^2 = \frac{x^2 L^2 \epsilon}{m^2 \alpha r^2} \cos(x\theta) - \frac{L^2}{m^2 r^3}. \tag{8.12}$$

From Eq. (8.3):

$$\cos(x\theta) = \frac{1}{\epsilon}\left(\frac{\alpha}{r} - 1\right) \tag{8.13}$$

and the acceleration of an object in orbit is:

$$\mathbf{a} = \left(\frac{L}{m}\right)^2 \left(\frac{(x^2 - 1)}{r^3} - \frac{x^2}{\alpha r^2}\right) \mathbf{e}_r. \tag{8.14}$$

The force is defined conventionally as:

$$\mathbf{F} = m\mathbf{a}. \tag{8.15}$$

If there is no precession then:

$$x = 1 \tag{8.16}$$

and the force law reduces to the inverse square law:

$$\mathbf{F} = -\frac{L^2}{m\alpha r^2}\mathbf{e}_r. \tag{8.17}$$

This is the Newtonian inverse square law if:

$$\alpha = \frac{L^2}{m^2 MG}. \tag{8.18}$$

The same force law is obtained elegantly from Lagrangian dynamics, which gives the following equation for any orbit:

$$\frac{d^2}{d\theta^2}\left(\frac{1}{r}\right) + \frac{1}{r} = -\frac{mr^2}{L}F(r). \tag{8.19}$$

From Eqs. (8.3) and (8.19):

$$F(r) = \frac{L^2}{m}\left(\frac{(x^2-1)}{r^3} - \frac{x^2}{\alpha r^2}\right) \tag{8.20}$$

which is the same as Eq. (8.14).

The square of the orbital velocity can therefore be expressed as:

$$v^2 = \left(\frac{L}{m\alpha}\right)^2 \left[\frac{2x^2 \alpha}{r} - x^2(1-\epsilon^2) + \frac{\alpha^2}{r^2}(1-x^2)\right] \tag{8.21}$$

and when
$$x = 1 \tag{8.22}$$

the Keplerian equation for orbital linear velocity is obtained:

$$v^2 \xrightarrow[x=1]{} \left(\frac{L}{m\alpha}\right)^2 \left[\frac{2\alpha}{r} - (1-\epsilon^2)\right] \tag{8.23}$$

thus checking that the theory is correct and self consistent. At the distance R_0 of closest approach of m to M in an orbit:

$$R_0 = \frac{\alpha}{1+\epsilon} \tag{8.24}$$

so Eq. (8.21) becomes:

$$v^2 = \frac{L^2}{m^2 R_0}\left[\frac{x^2}{\alpha}(1+\epsilon) - \frac{(x^2-1)}{R_0}\right] \tag{8.25}$$

and solving for the eccentricity ϵ gives:

$$\epsilon = \frac{m^2 \alpha R_0}{x^2 L^2}\left(v^2 - \frac{L^2}{m^2}\left(\frac{x^2-1}{R_0}\right)\right) - 1. \tag{8.26}$$

This equation can be used in the problem of determining the angle of deflection of a hyperbolic orbit of m around M.

The total deflection for a hyperbola, as in UFT 216, is 2ψ:

$$\Delta\psi = 2\psi = 2\sin^{-1}\frac{1}{\epsilon} \tag{8.27}$$

where

$$\psi = \tan^{-1}\frac{a}{b} \tag{8.28}$$

where a and b are the major and minor semi axes. Therefore:

$$\Delta\psi = 2\sin^{-1}\frac{1}{\epsilon} = 2\tan^{-1}\frac{a}{b} \tag{8.29}$$

where the eccentricity is defined by:

$$\epsilon = \left(1 + \frac{b^2}{a^2}\right)^{1/2}. \tag{8.30}$$

The half right latitude is defined by:

$$\alpha = \frac{b^2}{a}. \tag{8.31}$$

At the distance of closest approach of m to M in a hyperbolic orbit:

$$R_0 = \frac{\alpha}{1+\epsilon} \tag{8.32}$$

8.2. ECE THEORY OF LIGHT DEFLECTION DUE TO GRAVITATION

so:

$$\cos(x\theta) = 1 \tag{8.33}$$

as in Eq. (8.24).

For very small angles of deflection such as that observed in the deflection of light from a distant source by the sun:

$$\sin\psi \sim \psi = \frac{1}{\epsilon} = \left[\frac{m^2 \alpha R_0}{x^2 L^2}\left(v^2 - \frac{L^2}{m^2}\left(\frac{x^2-1}{R_0^2}\right)\right) - 1\right]^{-1}. \tag{8.34}$$

If v could be measured experimentally, m can be found. For light v is very close to c and m is the mass of the photon. Theoretically, photon mass can be obtained in this way. In the Newtonian limit:

$$x = 1 \tag{8.35}$$

and

$$\sin\psi \sim \psi = \frac{1}{\epsilon} = \left[\frac{m^2 \alpha R_0 v^2}{L^2} - 1\right]^{-1} \tag{8.36}$$

in which the Newtonian half right latitude is:

$$\alpha = \frac{L^2}{m^2 MG}. \tag{8.37}$$

So the well known Newtonian theory of the orbital deflection is recovered:

$$\sin\psi \sim \psi = \frac{1}{\epsilon} = \left(\frac{R_0 v^2}{MG} - 1\right)^{-1}. \tag{8.38}$$

Note that m cancels out of the calculation in the Newtonian limit, but does not cancel in the rigorous equation (8.34). If the photon velocity is assumed to be c for all practical purposes, i.e. to be very close to c, then

$$\Delta\psi = 2\psi = \frac{2MG}{R_0 c^2} \tag{8.39}$$

to an excellent approximation. This is the famous Newtonian value for light deflection by gravitation.

The experimentally observed value is always:

$$\Delta\psi = 2\psi = \frac{4MG}{R_0 c^2} \tag{8.40}$$

to high precision, for electromagnetic radiation grazing any object of mass M. This is twice the Newtonian value.

The reason for this famous result cannot be found in the deeply flawed Einsteinian theory, but a straightforward explanation can be found using the principles of this book.

CHAPTER 8. ECE COSMOLOGY

Consider the vector format of the first Maurer Cartan structure equation given here in the notation of chapter one:

$$\mathbf{T}^a(\text{orb}) = -\boldsymbol{\nabla} q^a{}_0 - \frac{\partial \mathbf{q}^a}{\partial t} - \omega^a{}_{0b} \mathbf{q}^b + q^b{}_0 \boldsymbol{\omega}^a{}_b \tag{8.41}$$

and

$$\mathbf{T}^a(\text{spin}) = \boldsymbol{\nabla} \times \mathbf{q}^a - \boldsymbol{\omega}^a{}_b \times \mathbf{q}^b. \tag{8.42}$$

The fundamental ECE hypothesis was devised for electromagnetism and defines the electromagnetic potential in terms of the tetrad:

$$A^a{}_\mu = A^{(0)} q^a{}_\mu. \tag{8.43}$$

Now define the linear momentum tetrad:

$$p^a{}_\mu = p^{(0)} q^a{}_\mu \tag{8.44}$$

in an analogous manner, using the minimal prescription:

$$p^a{}_\mu \to p^a{}_\mu + eA^a{}_\mu. \tag{8.45}$$

It follows from Eqs. (8.41 and 8.44) that the orbital force of ECE theory is:

$$\mathbf{F}^a(\text{orb}) = -\boldsymbol{\nabla} \phi^a{}_0 - \frac{\partial \mathbf{p}^a}{\partial t} - \omega^a{}_{0b} \mathbf{p}^b + \phi^b{}_0 \boldsymbol{\omega}^a{}_b \tag{8.46}$$

and that the spin force is:

$$\mathbf{F}^a(\text{spin}) = \boldsymbol{\nabla} \times \mathbf{p}^a - \boldsymbol{\omega}^a{}_b \times \mathbf{p}^b. \tag{8.47}$$

In the simplified single polarization theory:

$$\mathbf{F}(\text{orb}) = -\boldsymbol{\nabla}\phi - \frac{\partial \mathbf{p}}{\partial t} - \omega_0 \mathbf{p} + \phi \boldsymbol{\omega} \tag{8.48}$$

and:

$$\mathbf{F}(\text{spin}) = \boldsymbol{\nabla} \times \mathbf{p} - \boldsymbol{\omega} \times \mathbf{p}. \tag{8.49}$$

In the non relativistic limit the spin connection vanishes and:

$$\mathbf{F}(\text{orb}) = -\boldsymbol{\nabla}\phi - \frac{\partial \mathbf{p}}{\partial t}. \tag{8.50}$$

The famous equivalence of inertial and gravitational mass is recovered from Eq. (8.50) using the anti symmetry law of ECE theory described earlier in this book. So:

$$-\frac{\partial \mathbf{p}}{\partial t} = -\boldsymbol{\nabla}\phi \tag{8.51}$$

and:

$$\phi = -\frac{mMG}{r} \tag{8.52}$$

8.2. ECE THEORY OF LIGHT DEFLECTION DUE TO GRAVITATION

where ϕ is the gravitational potential. This is defined in direct analogy to the electromagnetic scalar potential ϕ_e as follows:

$$p^a{}_\mu = \left(\frac{\phi^a}{c}, -\mathbf{p}^a\right) \tag{8.53}$$

and

$$A^a{}_\mu = \left(\frac{\phi^a_e}{c}, -\mathbf{A}^a\right). \tag{8.54}$$

In Newtonian dynamics:

$$\phi = -\frac{mMG}{r} \tag{8.55}$$

so the force is:

$$F = -\frac{mMG}{r^2} \tag{8.56}$$

and the acceleration due to gravity is:

$$g = -\frac{MG}{r^2}. \tag{8.57}$$

This powerful and precise result of ECE theory was first inferred in UFT 141. The ECE theory is therefore precise to one part in ten to the power seventeen, the precision of the experimental proof of the equivalence of gravitational and inertial mass. The equivalence is due to Cartan geometry.

The calculation of light deflection due to gravitation proceeds by applying the ECE anti symmetry law to Eq. (8.48) to find that:

$$-\boldsymbol{\nabla}\phi + \boldsymbol{\omega}\phi = -\frac{d\mathbf{p}}{dt} - \omega_0 \mathbf{p} \tag{8.58}$$

in which it has been assumed that:

$$\frac{d\mathbf{p}}{dt} = \frac{\partial \mathbf{p}}{\partial t}. \tag{8.59}$$

So the force is:

$$\mathbf{F} = 2\left(-\frac{d\mathbf{p}}{dt} - \omega_0 \mathbf{p}\right) = -2\left(\boldsymbol{\nabla}\phi - \boldsymbol{\omega}\phi\right). \tag{8.60}$$

The factor two in Eq. (8.60) can be eliminated without affecting the physics by assuming that:

$$p^a{}_\mu = \frac{p^{(0)}}{2} q^a{}_\mu \tag{8.61}$$

so the orbital force becomes:

$$\mathbf{F} = -\frac{d\mathbf{p}}{dt} - \omega_0 \mathbf{p} = -\boldsymbol{\nabla}\phi - \boldsymbol{\omega}\phi \tag{8.62}$$

an equation which gives the equivalence principle (8.51) for vanishing spin connection. Now define:

$$\mathbf{p} = p_r \mathbf{e}_r, \tag{8.63}$$

$$\boldsymbol{\omega} = \omega_r \mathbf{e}_r \tag{8.64}$$

and compare Eqs. (8.20) and (8.62) to find that:

$$\mathbf{F} = -\frac{\partial \phi}{\partial r} + \phi \omega_r = -\frac{kx^2}{r^2} - \frac{k(1-x^2)\alpha}{r^3}. \tag{8.65}$$

For small deviations from a Newtonian orbit as in planetary precession or any observable precession in astronomy:

$$-\frac{\partial \phi}{\partial r} = -\frac{kx^2}{r^2} \tag{8.66}$$

i. e.:

$$x \sim 1 \tag{8.67}$$

to an excellent approximation. From Eqs. (8.63) and (8.64):

$$\phi \omega_r = -\frac{k\alpha}{r^3}(1-x^2) \tag{8.68}$$

in an almost Newtonian approximation. In this approximation the gravitational potential is well known to be:

$$\phi = -\frac{k}{r} \tag{8.69}$$

so the spin connection can be expressed in terms of x as follows:

$$\omega_r = (1-x^2)\frac{\alpha}{r} = (1-x^2)\frac{b^2}{ar^2}. \tag{8.70}$$

Using Eq. (8.70), the correction needed to produce Eq. (8.40) from Eq. (8.39) is:

$$\frac{R_0 c^2}{MG} \to \frac{R_0 c^2}{MG} + \frac{\alpha}{R_0}\left(\frac{1-x^2}{x^2}\right). \tag{8.71}$$

Using Eq. (8.32) it is found that:

$$2\psi = 2\frac{R_0 c^2}{MG} + 2(1+\epsilon)\left(\frac{1-x^2}{x^2}\right). \tag{8.72}$$

Experimentally:

$$(1+\epsilon)\left(\frac{1-x^2}{x^2}\right) = \frac{R_0 c^2}{MG} \tag{8.73}$$

8.2. ECE THEORY OF LIGHT DEFLECTION DUE TO GRAVITATION

and using Eq. (8.27):

$$\frac{1}{\epsilon} = \sin\left(\frac{\Delta\psi}{2}\right). \tag{8.74}$$

For small deflections:

$$\frac{1}{\epsilon} \sim \frac{\Delta\psi}{2} \tag{8.75}$$

so to an excellent approximation:

$$\left(1 + \frac{2}{\Delta\psi}\right)\left(\frac{1-x^2}{x^2}\right) = \frac{R_0 c^2}{MG}. \tag{8.76}$$

However by experiment:

$$\Delta\psi = \frac{4R_0 c^2}{MG}, \tag{8.77}$$
$$x \sim 1,$$

so using Eq. (8.70):

$$\omega_r = \frac{\Delta\psi}{4}\left(1 + \frac{2}{\Delta\psi}\right)^{-1}\frac{\alpha}{r^2}. \tag{8.78}$$

From Eq. (8.32):

$$\alpha = R_0(1+\epsilon) = R_0\left(1 + \frac{2}{\Delta\psi}\right) \tag{8.79}$$

and from Eqs. (8.78) and (8.79):

$$\omega_r = \frac{\Delta\psi}{4}\frac{R_0}{r^2}. \tag{8.80}$$

This is a universal spin connection that describes all electromagnetic deflections from any relevant object M in the universe. This spin connection also describes planetary precession through its relation to x, Eq. (8.70). An example on a cosmic scale is the precession of the Hulse-Taylor pulsar, Fig. 8.1. The procedure used to derive this result also gives the equivalence principle. Finally at distance of closest approach:

$$\omega_r = \frac{\Delta\psi}{4R_0} \tag{8.81}$$

a very simple result that can be tabulated in astronomy for any relevant object of mass M.

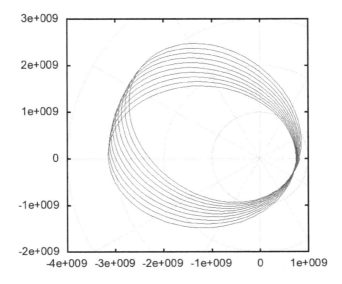

Figure 8.1: Calculated orbit of the Hulse-Taylor pulsar ($\alpha = 1.207718 \cdot 10^9$ m, $\epsilon = 0.617131$, $x = 1.0117$).

8.3 The Velocity Curve of a Whirlpool Galaxy

Whirlpool galaxies are familiar objects in cosmology and are very complex in structure. However there is one feature that makes them useful for the study of the fundamental theories of cosmology such as those of Newton and Einstein, and ECE, and that is the velocity curve, the plot of the velocity of a star orbiting the centre of a galaxy versus the distance between the star and the centre. It was discovered experimentally in the late fifties that the velocity becomes constant as r goes to infinity. The first part of this section will give the basic kinematics of the orbit and will show that both the Newton and Einstein theories fail completely to describe the velocity curve. The second part will describe how ECE theory gives a plausible explanation of the velocity curve without the use of random empiricism such as dark matter. It appears that the theory of dark matter has been refuted experimentally, leaving ECE cosmology as the only explanation.

Consider the radial vector in the plane of any orbit:

$$\mathbf{r} = r\mathbf{e}_r \qquad (8.82)$$

where $\mathbf{e}_r$ is the radial unit vector. The velocity of an object of mass m in orbit is defined as:

$$\mathbf{v} = \frac{d\mathbf{r}}{dt} = \frac{dr}{dt}\mathbf{e}_r + r\frac{d\mathbf{e}_r}{dt} \qquad (8.83)$$

because in plane polar coordinates the unit vector $\mathbf{e}_r$ is a function of time so the Leibnitz theorem applies. In the Cartesian system the unit vectors $\mathbf{i}$ and

8.3. THE VELOCITY CURVE OF A WHIRLPOOL GALAXY

j are not functions of time. The unit vectors of the plane polar system are defined by:

$$\mathbf{e}_r = \cos\theta\, \mathbf{i} + \sin\theta\, \mathbf{j} \tag{8.84}$$

3

$$\mathbf{e}_\theta = -\sin\theta\, \mathbf{i} + \cos\theta\, \mathbf{j} \tag{8.85}$$

and it follows that:

$$\frac{d\mathbf{e}_r}{dt} = \frac{d\theta}{dt}\mathbf{e}_\theta = \omega \mathbf{e}_\theta \tag{8.86}$$

as described in UFT 236. The velocity in a plane is therefore:

$$\begin{aligned}\mathbf{v} &= \frac{dr}{dt}\mathbf{e}_r + \omega r \mathbf{e}_\theta \\ &= \frac{dr}{dt}\mathbf{e}_r + \boldsymbol{\omega} \times \mathbf{r}\end{aligned} \tag{8.87}$$

in which the angular velocity vector:

$$\boldsymbol{\omega} = \frac{d\theta}{dt}\mathbf{k} \tag{8.88}$$

is the Cartan spin connection as proven in UFT 235 on www.aias.us. Therefore this spin connection is related to the universal spin connection inferred in Section 8.2 giving a coherent cosmology for the solar system and whirlpool galaxies. As we shall prove, the Newton and Einstein theories fail completely to do so.

Using the chain rule:

$$\frac{dr}{dt} = \frac{dr}{d\theta}\frac{d\theta}{dt} \tag{8.89}$$

it is found that the velocity is defined for any orbit by:

$$v^2 = \omega^2 \left(\left(\frac{dr}{d\theta}\right)^2 + r^2\right) \tag{8.90}$$

and is therefore defined by the angular velocity or spin connection magnitude:

$$\omega = \frac{d\theta}{dt}. \tag{8.91}$$

The orbit itself is defined by $dr/d\theta$, because any planar orbit is defined by r as a function of θ. The angular momentum of any planar orbit is defined by:

$$\mathbf{L} = \mathbf{r} \times \mathbf{p} = m\mathbf{r} \times \mathbf{v} \tag{8.92}$$

and its magnitude is:

$$L = mr^2\omega. \tag{8.93}$$

Therefore for any planar orbit:

$$v^2 = \left(\frac{L}{mr}\right)^2 + \left(\frac{L}{mr^2}\left(\frac{dr}{d\theta}\right)\right)^2 \tag{8.94}$$

and as r becomes infinite:

$$r \to \infty \tag{8.95}$$

the velocity reaches the limit:

$$\frac{dr}{d\theta} = \left(\frac{mv_\infty}{L}\right)r^2 \tag{8.96}$$

where v_∞ is the velocity for infinite r. In whirlpool galaxies v_∞ is a constant by experimental observation. Therefore:

$$\frac{d\theta}{dr} = \left(\frac{L}{mv_\infty}\right)\frac{1}{r^2} \tag{8.97}$$

and

$$\theta = \frac{L}{mv_\infty}\int\frac{dr}{r^2} = -\left(\frac{L}{mv_\infty}\right)\frac{1}{r} \tag{8.98}$$

which is the equation of a hyperbolic spiral orbit. In UFT 76 on www.aias.us this hyperbolic spiral orbit was compared with the observed M101 whirlpool galaxy (see Fig. 8.2). So the essentials of galactic dynamics can be understood from the simple first principles of kinematics, defining the angular velocity as the spin connection of ECE theory.

Newtonian dynamics fails completely to describe this result because it produces a static conical section:

$$r = \frac{\alpha}{1 + \epsilon\cos\theta} \tag{8.99}$$

with an inverse square law of attraction. From Eq. (8.99):

$$\frac{dr}{d\theta} = \frac{\epsilon r^2}{\alpha}\sin\theta \tag{8.100}$$

and using this result in Eq. (8.90):

$$v^2 = \omega^2 r^2\left(1 + \left(\frac{\epsilon r}{\alpha}\right)^2\sin^2\theta\right) \tag{8.101}$$

where:

$$\sin^2\theta = 1 - \cos^2\theta = 1 - \frac{1}{\epsilon^2}\left(\frac{\alpha}{r} - 1\right)^2. \tag{8.102}$$

So the Newtonian velocity is:

$$v^2 = \omega^2 r^2\left(\frac{2\alpha}{r} - \left(\frac{r}{\alpha}\right)^2(1 - \epsilon^2)\right). \tag{8.103}$$

8.3. THE VELOCITY CURVE OF A WHIRLPOOL GALAXY

Figure 8.2: Sprial galaxy M101 with hyperbolic spirals fittet to galaxy arms.

The semi major axis of an elliptical orbit is defined by:

$$a = \frac{\alpha}{1 - \epsilon^2} \tag{8.104}$$

so Newtonian dynamics produces:

$$v^2 = \frac{1}{\alpha}\left(\frac{L}{m}\right)^2\left(\frac{2}{r} - \frac{1}{a}\right). \tag{8.105}$$

Using the Newtonian half right latitude:

$$\alpha = \frac{L^2}{m^2 MG} \tag{8.106}$$

gives:

$$v^2 = MG\left(\frac{2}{r} - \frac{1}{a}\right). \tag{8.107}$$

Note that:

$$\frac{1}{a} = \frac{1 - \epsilon^2}{\alpha} = \frac{1}{r}(1 + \epsilon\cos\theta)(1 - \epsilon^2) \tag{8.108}$$

so the Newtonian velocity is:

$$v^2(\text{Newton}) = \frac{MG}{r}\left(2 - (1 - \epsilon^2)(1 + \epsilon\cos\theta)\right). \tag{8.109}$$

It follows that:

$$v(\text{Newton}) \xrightarrow[r \to \infty]{} 0 \tag{8.110}$$

so the theory fails completely to describe the velocity curve of a whirlpool galaxy.

The Einstein theory does no better because it produces a precessing ellipse, Eq. (8.3), from which:

$$\frac{dr}{d\theta} = \frac{x\epsilon r^2}{\alpha} \sin(x\theta). \tag{8.111}$$

Using Eq. (8.111) in Eq. (8.90) gives:

$$v^2 = \left(\frac{L}{mr}\right)^2 \left(1 + \left(\frac{x\epsilon \sin(x\theta)}{1 + \epsilon \cos(x\theta)}\right)^2\right) \tag{8.112}$$

and again it is found that:

$$v(\text{Einstein}) \xrightarrow[r \to \infty]{} 0 \tag{8.113}$$

and the Einstein theory fails completely to describe the dynamics of a whirlpool galaxy. This leaves ECE theory as the only correct and general theory of cosmology. The latter can be developed by considering again the acceleration in plane polar coordinates

$$\mathbf{a} = \frac{d\mathbf{v}}{dt} = \left(\ddot{r} - r\dot{\theta}^2\right)\mathbf{e}_r + \left(r\ddot{\theta} + 2\dot{r}\dot{\theta}\right)\mathbf{e}_\theta. \tag{8.114}$$

As shown in UFT 235 this can be expressed as:

$$\left(\ddot{r} - r\dot{\theta}^2\right)\mathbf{e}_r = \frac{d^2r}{dt^2}\mathbf{e}_r + \boldsymbol{\omega} \times (\boldsymbol{\omega} \times \mathbf{r}) \tag{8.115}$$

and

$$\left(r\ddot{\theta} + 2\dot{r}\dot{\theta}\right)\mathbf{e}_\theta = \frac{d\boldsymbol{\omega}}{dt} \times \mathbf{r} + 2\boldsymbol{\omega} \times \dot{\mathbf{r}}. \tag{8.116}$$

Eq. (8.116) is the Coriolis acceleration and $\boldsymbol{\omega} \times (\boldsymbol{\omega} \times \mathbf{r})$ is the centrifugal acceleration. In the UFT papers it is shown that the Coriolis acceleration vanishes for all planar orbits (see Eq. (8.11)). Using the chain rule it can be shown as in the UFT papers that:

$$\frac{d^2r}{dt^2} = \left(\frac{L}{mr}\right)^2 \left(\frac{dr}{d\theta}\right) \frac{d}{dr}\left(\frac{1}{r^2}\frac{dr}{d\theta}\right). \tag{8.117}$$

The centrifugal acceleration is defined by:

$$\boldsymbol{\omega} \times (\boldsymbol{\omega} \times \mathbf{r}) = -\omega^2 r \mathbf{e}_r = -\frac{L^2}{m^2 r^3}\mathbf{e}_r \tag{8.118}$$

so the total acceleration is defined by:

$$\mathbf{a} = \left(\frac{L}{mr}\right)^2 \left[\left(\frac{dr}{d\theta}\right)\frac{d}{dr}\left(\frac{1}{r^2}\frac{dr}{d\theta}\right) - \frac{1}{r}\right]\mathbf{e}_r \tag{8.119}$$

8.3. THE VELOCITY CURVE OF A WHIRLPOOL GALAXY

for all planar orbits.
In this equation:

$$\frac{d}{dr}\left(\frac{1}{r^2}\frac{dr}{d\theta}\right) = \frac{d\theta}{dr}\frac{d}{d\theta}\left(\frac{1}{r^2}\frac{dr}{d\theta}\right) \tag{8.120}$$

so:

$$\mathbf{a} = \left(\frac{L}{mr}\right)^2\left[\frac{d}{d\theta}\left(\frac{1}{r^2}\frac{dr}{d\theta}\right) - \frac{1}{r}\right]\mathbf{e}_r. \tag{8.121}$$

Now note that:

$$\frac{d}{d\theta}\left(\frac{1}{r}\right) = \frac{d}{dr}\left(\frac{1}{r}\right)\frac{dr}{d\theta} \tag{8.122}$$

so:

$$\frac{d}{d\theta}\left(\frac{1}{r^2}\frac{dr}{d\theta}\right) = \frac{1}{r^2}\frac{d}{d\theta}\left(\frac{dr}{d\theta}\right) = -\frac{d^2}{d\theta^2}\left(\frac{1}{r}\right). \tag{8.123}$$

Therefore the acceleration is:

$$\mathbf{a} = -\left(\frac{L}{mr}\right)^2\left(\frac{d^2}{d\theta^2}\left(\frac{1}{r}\right) + \frac{1}{r}\right)\mathbf{e}_r \tag{8.124}$$

and using the definition of force:

$$\mathbf{F} = m\mathbf{a} \tag{8.125}$$

which is Eq (8.19) derived from lagrangian dynamics. This analysis of any planar orbit is therefore rigorously self consistent.

The Lagrangian method of deriving Eq. (8.125) sets up the Lagrangian:

$$\mathscr{L} = \frac{1}{2}mv^2 - U \tag{8.126}$$

in which the velocity is defined by:

$$v^2 = \left(\frac{dr}{dt}\right)^2 + r^2\left(\frac{d\theta}{dt}\right)^2. \tag{8.127}$$

The force is derived from the potential energy as follows:

$$F = -\frac{\partial U}{\partial r}. \tag{8.128}$$

The two Euler Lagrange equations are:

$$\frac{\partial \mathscr{L}}{\partial \theta} = \frac{d}{dt}\left(\frac{\partial \mathscr{L}}{\partial \dot{\theta}}\right), \quad \frac{\partial \mathscr{L}}{\partial r} = \frac{d}{dt}\left(\frac{\partial \mathscr{L}}{\partial \dot{r}}\right) \tag{8.129}$$

and the angular momentum is defined by the lagrangian to be a constant of motion:

$$L = \frac{\partial \mathscr{L}}{\partial \dot{\theta}} = mr^2\frac{d\theta}{dt} = \text{constant}. \tag{8.130}$$

Eq. (8.124) is the result of pure kinematics in a plane, and is also an equation of Cartan geometry. It is the result of the fundamental expression for acceleration in a plane. Eq. (8.124) is also an equation of Cartan geometry because the spin connection is the angular velocity.

The covariant derivative of Cartan may be defined for use in classical kinematics in three dimensional space. For any vector $\mathbf{V}$ the covariant derivative is:

$$\frac{D\mathbf{V}}{dt} = \left(\frac{d\mathbf{V}}{dt}\right)_{\text{axes fixed}} + \boldsymbol{\omega} \times \mathbf{V} \tag{8.131}$$

where the spin connection vector is the angular velocity $\boldsymbol{\omega}$. In plane polar coordinates define:

$$\mathbf{V} = V\mathbf{e}_r \tag{8.132}$$

for simplicity of development. The velocity is then defined by:

$$\mathbf{v} = \frac{D\mathbf{r}}{dt} = \frac{d\mathbf{r}}{dt} + \boldsymbol{\omega} \times \mathbf{r} \tag{8.133}$$

where:

$$\frac{d\mathbf{r}}{dt} = \left(\frac{d\mathbf{r}}{dt}\right)_{\text{axes fixed}}. \tag{8.134}$$

By definition:

$$\frac{D\mathbf{r}}{dt} = \frac{D}{dt}(r\mathbf{e}_r) = \frac{dr}{dt}\mathbf{e}_r + r\frac{d\mathbf{e}_r}{dt} \tag{8.135}$$

so:

$$\left(\frac{d\mathbf{r}}{dt}\right)_{\text{axes fixed}} = \left(\frac{dr}{dt}\right)\mathbf{e}_r \tag{8.136}$$

and

$$\boldsymbol{\omega} \times \mathbf{r} = r\frac{d\mathbf{e}_r}{dt}. \tag{8.137}$$

The acceleration is defined by:

$$\mathbf{a} = \frac{D\mathbf{v}}{dt} = \frac{d\mathbf{v}}{dt} + \boldsymbol{\omega} \times \mathbf{v} \tag{8.138}$$

where:

$$\frac{d\mathbf{v}}{dt} = \left(\frac{d\mathbf{v}}{dt}\right)_{\text{axes fixed}}. \tag{8.139}$$

From fundamental kinematics as described above:

$$\mathbf{a} = \frac{d\mathbf{v}}{dt} + \boldsymbol{\omega} \times \mathbf{v} = (\ddot{r} - \omega^2 r)\mathbf{e}_r + \left(r\frac{d\omega}{dt} + 2\frac{dr}{dt}\omega\right)\mathbf{e}_\theta \tag{8.140}$$

8.3. THE VELOCITY CURVE OF A WHIRLPOOL GALAXY

where the unit vectors of the plane polar coordinates system are defined by:

$$\mathbf{e}_r \times \mathbf{e}_\theta = \mathbf{k}, \tag{8.141}$$

$$\mathbf{k} \times \mathbf{e}_r = \mathbf{e}_\theta, \tag{8.142}$$

$$\mathbf{e}_\theta \times \mathbf{k} = \mathbf{e}_r. \tag{8.143}$$

Therefore:

$$\frac{d\mathbf{v}}{dt} + \boldsymbol{\omega} \times \mathbf{v} = \frac{d^2 r}{dt^2}\mathbf{e}_r + \boldsymbol{\omega} \times (\boldsymbol{\omega} \times \mathbf{r}) + \frac{d\boldsymbol{\omega}}{dr} \times \mathbf{r} + 2\boldsymbol{\omega} \times \left(\frac{dr}{dt}\mathbf{e}_r\right). \tag{8.144}$$

From Eq. (8.133)

$$\mathbf{v} = \frac{d\mathbf{r}}{dt} + \boldsymbol{\omega} \times \mathbf{r} \tag{8.145}$$

so in Eq. (8.138):

$$\mathbf{a} = \frac{d^2 r}{dt^2}\mathbf{e}_r + \frac{d\boldsymbol{\omega}}{dt} \times \mathbf{r} + \boldsymbol{\omega} \times \left(\frac{d\mathbf{r}}{dt}\right)_{\text{axes fixed}} + \boldsymbol{\omega} \times \frac{dr}{dt}\mathbf{e}_r + \boldsymbol{\omega} \times (\boldsymbol{\omega} \times \mathbf{r}). \tag{8.146}$$

In this equation:

$$\boldsymbol{\omega} \times \left(\frac{d\mathbf{r}}{dt}\right)_{\text{axes fixed}} = \boldsymbol{\omega} \times \frac{dr}{dt}\mathbf{e}_r \tag{8.147}$$

so:

$$\mathbf{a} = \frac{d^2 r}{dt^2}\mathbf{e}_r + \boldsymbol{\omega} \times (\boldsymbol{\omega} \times \mathbf{r}) + \frac{d\boldsymbol{\omega}}{dt} \times \mathbf{r} + 2\boldsymbol{\omega} \times \left(\frac{dr}{dt}\mathbf{e}_r\right) \tag{8.148}$$

which is Eq. (8.144), Q. E. D.

The covariant derivatives used in these calculations are examples of the Cartan covariant derivative:

$$D_\mu V^a = \partial_\mu V^a + \omega^a{}_{\mu b} V^b. \tag{8.149}$$

The well known centripetal acceleration:

$$\mathbf{a} = \boldsymbol{\omega} \times (\boldsymbol{\omega} \times \mathbf{r}) \tag{8.150}$$

and the Coriolis acceleration:

$$\mathbf{a} = \frac{d\boldsymbol{\omega}}{dt} \times \mathbf{r} + 2\boldsymbol{\omega} \times \left(\frac{dr}{dt}\mathbf{e}_r\right) \tag{8.151}$$

are produced by the plane polar system of coordinates. These accelerations do not exist in the Cartesian system and depend entirely on the existence of the spin connection of Cartan. As shown already the Coriolis acceleration vanishes for all closed planar orbits and the acceleration simplifies to:

$$\mathbf{a} = (\ddot{r} - \omega^2 r)\mathbf{e}_r = \frac{d^2 r}{dt^2}\mathbf{e}_r + \boldsymbol{\omega} \times (\boldsymbol{\omega} \times \mathbf{r}). \tag{8.152}$$

For example the acceleration due to gravity is:

$$\mathbf{g} = \frac{d^2 r}{dt^2}\mathbf{e}_r + \boldsymbol{\omega} \times (\boldsymbol{\omega} \times \mathbf{r}) \tag{8.153}$$

and includes the centripetal acceleration:

$$\boldsymbol{\omega} \times (\boldsymbol{\omega} \times \mathbf{r}) = -\omega^2 r \mathbf{e}_r. \tag{8.154}$$

The acceleration due to gravity in the plane polar system is the sum of $\mathbf{g}$ in the Cartesian system:

$$\mathbf{g}(\text{Cartesian}) = \frac{d^2 r}{dt^2}\mathbf{e}_r \tag{8.155}$$

and the centripetal acceleration. To make this point clearer consider the acceleration of an elliptical orbit or closed elliptical trajectory in the plane polar system. It is:

$$\mathbf{a} = -\frac{L^2}{m^2 r^2 \alpha}\mathbf{e}_r \tag{8.156}$$

where the angular momentum is a constant of motion and defined by:

$$L = |\mathbf{L}| = |\mathbf{r} \times \mathbf{p}| = m r^2 \omega. \tag{8.157}$$

The acceleration due to gravity of the elliptical motion of a mass m is:

$$\mathbf{g} = -\frac{L^2}{m^2 r^2 \alpha}\mathbf{e}_r \tag{8.158}$$

in plane polar coordinates. The Newtonian result is recovered using the half right latitude:

$$\alpha = \frac{L^2}{m^2 MG} \tag{8.159}$$

so:

$$\mathbf{g} = -\frac{MG}{r^2}\mathbf{e}_r. \tag{8.160}$$

The only force present in the plane polar system of coordinates is:

$$\mathbf{F} = m\mathbf{g} = -\frac{mMG}{r^2}\mathbf{e}_r \tag{8.161}$$

which is the equivalence principle, Q. E. D.

The acceleration in the Cartesian system of coordinates from Eq. (8.153) is:

$$\mathbf{a}(\text{Cartesian}) = \mathbf{g} - \boldsymbol{\omega} \times (\boldsymbol{\omega} \times \mathbf{r}) \tag{8.162}$$

in which the centrifugal acceleration is:

$$-\boldsymbol{\omega} \times (\boldsymbol{\omega} \times \mathbf{r}) = \omega^2 r \mathbf{e}_r. \tag{8.163}$$

Therefore in the Cartesian system the acceleration produced by the same elliptical trajectory is:

$$\left(\frac{d^2r}{dt^2}\right)_{\text{Cartesian}} \mathbf{e}_r = \left(-\frac{L^2}{m^2 r^2 \alpha} + \omega^2 r\right) \mathbf{e}_r. \tag{8.164}$$

It generalizes the Newtonian theory to give:

$$\left(\frac{d^2r}{dt^2}\right)_{\text{Cartesian}} \mathbf{e}_r = \left(-\frac{MG}{r^2} + \frac{L^2}{m^2 r^3}\right) \mathbf{e}_r \tag{8.165}$$

and the familiar force:

$$\mathbf{F} = m\left(\frac{d^2r}{dt^2}\right)_{\text{Cartesian}} \mathbf{e}_r = \left(-\frac{mMG}{r^2} + \frac{L^2}{mr^3}\right) \mathbf{e}_r \tag{8.166}$$

of the textbooks. From a comparison of Eqs. (8.161) and (8.166) the forces in the plane polar and Cartesian systems are different. If the frame of reference is static with respect to the observer the force is Eq (8.166). If the frame of reference is rotating with respect to the observer the force is defined by Eq. (8.161).

The easiest way to approach this analysis is always to calculate the acceleration firstly in plane polar coordinates and to realize that one term of the resultant expression is the acceleration in the Cartesian system. For an observer on earth orbiting the sun, the relevant expression is that in the Cartesian frame, because the latter is also fixed on the earth and does not move with respect to the observer. In other words the observer is in his own frame of reference. For an observer on the sun the relevant expression is that in the plane polar system of coordinates, because the earth rotates with respect to the observer fixed on the sun.

The observer on the earth experiences the centrifugal acceleration:

$$-\boldsymbol{\omega} \times (\boldsymbol{\omega} \times \mathbf{r}) = \omega^2 r \mathbf{e}_r \tag{8.167}$$

directed outwards from the earth. This is the origin of the everyday centrifugal force. The observer on the sun experiences the centripetal acceleration:

$$\boldsymbol{\omega} \times (\boldsymbol{\omega} \times \mathbf{r}) = -\omega^2 r \mathbf{e}_r \tag{8.168}$$

directed towards the sun and towards the observer. The entire analysis rests on the spin connection and on the fact that in the plane polar system the frame itself is rotating and thus generates the spin connection by definition.

8.4 Description of Orbits with the Minkowski Force Equation

In UFT 238 on www.aias.us an entirely new approach to orbital theory was taken using the Minkowski force equation. This is a course that relativity theory could have taken, but cosmology followed the use of Einstein's flawed geometry, a subject that became known as general relativity. The Minkowski force equation is the Newton force equation with proper time τ replacing time t. This equation was inferred by Minkowski shortly after Einstein's introduction of the idea of relativistic momentum. A completely general kinematic theory of orbits can be developed in this way. It reduces to the Newtonian theory but never to the Einsteinian theory. Newtonian dynamics does not give any of the forces that are generated as discussed in Section 8.3 using plane polar coordinates and a system of rotating coordinates. It turns out that the space part of the Minkowski 4-force produces new and unexpected orbital properties that can be tested experimentally.

The relativistic force law and relativistic orbits of the Minkowski equation can be derived by considering the relativistic velocity in plane polar coordinates:

$$\mathbf{v} = \frac{d\mathbf{r}}{d\tau} = \gamma \frac{d\mathbf{r}}{dt} \tag{8.169}$$

where τ is the proper time and γ the Lorentz factor:

$$\gamma = \left(1 - \frac{v^2}{c^2}\right)^{-1/2}. \tag{8.170}$$

The relativistic acceleration is:

$$\mathbf{a} = \frac{d}{d\tau}\left(\frac{d\mathbf{r}}{d\tau}\right) = \frac{d}{d\tau}\left(\gamma \frac{d\mathbf{r}}{dt}\right) = \gamma \frac{d}{dt}\left(\gamma \frac{d\mathbf{r}}{dt}\right). \tag{8.171}$$

Using the Leibnitz Thoerem:

$$\mathbf{a} = \gamma \left(\frac{d\gamma}{dt}\frac{d\mathbf{r}}{dt} + \gamma \frac{d}{dt}\left(\frac{d\mathbf{r}}{dt}\right)\right). \tag{8.172}$$

The velocity v appearing in the Lorentz factor is defined by the infinitesimal line element:

$$ds^2 = c^2 d\tau^2 = c^2 dt^2 - d\mathbf{r} \cdot d\mathbf{r} \tag{8.173}$$

where:

$$d\mathbf{r} \cdot d\mathbf{r} = v^2 dt^2. \tag{8.174}$$

Therefore

$$c^2 d\tau^2 = (c^2 - v^2) dt^2 \tag{8.175}$$

and the Lorentz factor is:

$$\gamma = \frac{dt}{d\tau} = \left(1 - \frac{v^2}{c^2}\right)^{-1/2}. \tag{8.176}$$

In plane polar coordinates:

$$d\mathbf{r} \cdot d\mathbf{r} = dr^2 + r^2 d\theta^2. \tag{8.177}$$

8.4. DESCRIPTION OF ORBITS WITH THE MINKOWSKI FORCE...

Therefore:
$$v^2 = \left(\frac{dr}{dt}\right)^2 + r^2 \left(\frac{d\theta}{dt}\right)^2. \tag{8.178}$$

The radial vector in plane polar coordinates is:
$$\mathbf{r} = r\mathbf{e}_r \tag{8.179}$$

therefore the non relativistic velocity is:
$$\begin{aligned}\mathbf{v} &= \frac{d}{dt}(r\mathbf{e}_r) = \frac{dr}{dt}\mathbf{e}_r + r\frac{d\mathbf{e}_r}{dt} = \frac{dr}{dt}\mathbf{e}_r + \omega r\mathbf{e}_\theta \\ &= \frac{dr}{dt}\mathbf{e}_r + \boldsymbol{\omega} \times \mathbf{r} = \left(\frac{L_0}{m}\right)\left(\frac{1}{r}\mathbf{e}_\theta - \frac{d}{d\theta}\left(\frac{1}{r}\right)\mathbf{e}_r\right).\end{aligned} \tag{8.180}$$

For a particle of mass m in an orbit, its relativistic momentum is:
$$\mathbf{p} = \gamma m \frac{d\mathbf{r}}{dt} = m\frac{d\mathbf{r}}{d\tau}, \tag{8.181}$$

an equation which can be rearranged as follows:
$$p^2 c^2 = \gamma^2 m^2 c^4 \left(\frac{v}{c}\right)^2 = \gamma^2 m^2 c^4 \left(1 - \frac{1}{\gamma^2}\right) = \gamma^2 m^2 c^4 - m^2 c^4 \tag{8.182}$$

giving the Einstein energy equation:
$$E^2 = c^2 p^2 + m^2 c^4 \tag{8.183}$$

in which
$$E = \gamma mc^2 \tag{8.184}$$

is the total energy and
$$E_0 = mc^2 \tag{8.185}$$

is the rest energy. The relativistic total angular momentum is:
$$L = mr^2 \frac{d\theta}{d\tau} = \gamma L_0. \tag{8.186}$$

The concept of Minkowski force equation uses acceleration, so this is a plausible new approach to all orbits. The Einstein energy equation can be derived from the infinitesimal line element (8.173) and developed as:
$$\begin{aligned}mc^2 &= mc^2 \left(\frac{dt}{d\tau}\right)^2 - \left(\frac{dr}{d\tau}\right)^2 - r^2 \left(\frac{d\theta}{d\tau}\right)^2 \\ &= \gamma^2 mc^2 - \left(\left(\frac{dr}{d\tau}\right)^2 + r^2 \left(\frac{d\theta}{d\tau}\right)^2\right) = \frac{E^2}{mc^2} - \frac{p^2}{c^2}.\end{aligned} \tag{8.187}$$

So
$$E^2 = c^2 p^2 + m^2 c^4 \tag{8.188}$$

Q. E. D. The relativistic linear momentum in Eq. (8.187) is:

$$p^2 = m^2 \left(\left(\frac{dr}{d\tau}\right)^2 + r^2 \left(\frac{d\theta}{d\tau}\right)^2 \right) \tag{8.189}$$

which is Eq. (8.181), Q. E. D. The definition of relativistic acceleration is

$$\mathbf{a} = \frac{d}{d\tau}\left(\frac{d\mathbf{r}}{d\tau}\right) = \gamma \left(\frac{d\gamma}{dt}\frac{d\mathbf{r}}{dt} + \gamma \frac{d}{dt}\left(\frac{d\mathbf{r}}{dt}\right) \right) \tag{8.190}$$

in which:

$$\frac{d\mathbf{r}}{dt} = \frac{dr}{dt}\mathbf{e}_r + \boldsymbol{\omega} \times \mathbf{r} \tag{8.191}$$

and:

$$\frac{d}{dt}\left(\frac{d\mathbf{r}}{dt}\right) = \frac{d^2 r}{dt^2}\mathbf{e}_r + \frac{d\boldsymbol{\omega}}{dt} \times \mathbf{r} + 2\boldsymbol{\omega} \times \frac{dr}{dt}\mathbf{e}_r + \boldsymbol{\omega} \times (\boldsymbol{\omega} \times \mathbf{r}). \tag{8.192}$$

Using the chain rule:

$$\frac{d\gamma}{dt} = \frac{d\gamma}{dv}\frac{dv}{dt} \tag{8.193}$$

where v is the velocity of the Lorentz factor defined in Eq. (8.176). Therefore:

$$\frac{d\gamma}{dv} = \frac{d}{dv}\left(1 - \frac{v^2}{c^2}\right)^{-1/2} = \gamma^3 \frac{v}{c^2} \tag{8.194}$$

and in plane polar coordinates:

$$\begin{aligned}\mathbf{a} &= \gamma^4 \frac{v}{c^2}\frac{dv}{dt}\frac{d\mathbf{r}}{dt} + \gamma^2 \frac{d}{dt}\left(\frac{d\mathbf{r}}{dt}\right) \\ &= \left(\frac{d\gamma}{d\tau}\frac{dr}{dt} + \gamma^2 \frac{d^2 r}{dt^2}\right)\mathbf{e}_r + \gamma^2 \boldsymbol{\omega} \times (\boldsymbol{\omega} \times \mathbf{r}) + \frac{d\gamma}{d\tau}\boldsymbol{\omega} \times \mathbf{r} \\ &\quad + \gamma^2 \left(\frac{d\boldsymbol{\omega}}{dt} \times \mathbf{r} + 2\boldsymbol{\omega} \times \frac{dr}{dt}\mathbf{e}_r\right). \end{aligned} \tag{8.195}$$

In static Cartesian coordinates on the other hand;

$$\mathbf{a} = \frac{d}{d\tau}\left(\gamma \frac{d\mathbf{r}}{dt}\right) = \gamma \frac{d\gamma}{dt}\frac{d\mathbf{r}}{dt} + \gamma^2 \frac{d}{dt}\left(\frac{d\mathbf{r}}{dt}\right) \tag{8.196}$$

so:

$$\mathbf{a}(\text{Cartesian}) = \left(\gamma \frac{d\gamma}{dt}\frac{dr}{dt} + \gamma^2 \frac{d^2 r}{dt^2}\right)\mathbf{e}_r \tag{8.197}$$

8.4. DESCRIPTION OF ORBITS WITH THE MINKOWSKI FORCE...

in which:

$$v = \frac{dr}{dt}, \quad \frac{d^2r}{dt^2} = \frac{dv}{dt}, \quad \frac{d\gamma}{dv} = \gamma^3 \frac{v}{c^2} \tag{8.198}$$

and

$$\frac{d\gamma}{dt} = \frac{d\gamma}{dv}\frac{dv}{dt} = \frac{\gamma^3 v}{c^2}\frac{dv}{dt}. \tag{8.199}$$

Therefore:

$$\mathbf{a}(\text{Cartesian}) = \left(\gamma^4 \frac{v^2}{c^2} + \gamma^2\right)\frac{dv}{dt}\mathbf{e}_r \tag{8.200}$$

in which:

$$\frac{v^2}{c^2} = 1 - \frac{1}{\gamma^2}. \tag{8.201}$$

Therefore the Cartesian acceleration is:

$$\mathbf{a}(\text{Cartesian}) = \gamma^4 \frac{d^2r}{dt^2}\mathbf{e}_r. \tag{8.202}$$

Using Eq. (8.202) in Eq. (8.195):

$$\mathbf{a}(\text{plane polar}) = \gamma^4 \frac{d^2r}{dt^2}\mathbf{e}_r + \gamma^2 \boldsymbol{\omega} \times (\boldsymbol{\omega} \times \mathbf{r})$$
$$+ \frac{d\gamma}{d\tau}\boldsymbol{\omega} \times \mathbf{r} + \gamma^2 \left(\frac{d\boldsymbol{\omega}}{dt} \times \mathbf{r} + 2\boldsymbol{\omega} \times \frac{dr}{dt}\mathbf{e}_r\right) \tag{8.203}$$

which is the expression for relativistic acceleration in plane polar coordinates.

It can be proven as follows that the relativistic Coriolis acceleration vanishes for all planar orbits. The general expression for relativistic Coriolis acceleration is:

$$\mathbf{a}(\text{Coriolis}) = \gamma^2 \left(r\frac{d}{dt}\frac{d\theta}{dt} + 2\frac{dr}{dt}\frac{d\theta}{dt}\right)\mathbf{e}_\theta \tag{8.204}$$

in which the total non relativistic angular momentum is:

$$L_0 = mr^2 \frac{d\theta}{dt}. \tag{8.205}$$

It follows that:

$$\frac{d}{dt}\left(\frac{d\theta}{dt}\right) = \frac{d}{dt}\left(\frac{L_0}{mr^2}\right) = \frac{d}{dr}\left(\frac{L_0}{mr^2}\right)\frac{dr}{dt} = -\frac{2L_0}{mr^3}\frac{dr}{dt} \tag{8.206}$$

so:

$$\mathbf{a}(\text{Coriolis}) = \left(-\frac{2L_0}{mr^2}\frac{dr}{dt} + \frac{2L_0}{mr^2}\frac{dr}{dt}\right)\mathbf{e}_\theta = \mathbf{0} \tag{8.207}$$

CHAPTER 8. ECE COSMOLOGY

Q. E. D.

Therefore the relativistic acceleration for all planar orbits is:

$$\mathbf{a} = \gamma^4 \frac{d^2r}{dt^2}\mathbf{e}_r + \gamma^2 \boldsymbol{\omega} \times (\boldsymbol{\omega} \times \mathbf{r}) + \frac{d\gamma}{d\tau}\boldsymbol{\omega} \times \mathbf{r}. \tag{8.208}$$

The relativistic centripetal component of this orbit is:

$$\mathbf{a}(\text{centripetal}) = \gamma^2 \boldsymbol{\omega} \times (\boldsymbol{\omega} \times \mathbf{r}) = -\frac{L^2}{m^2 r^3}\mathbf{e}_r. \tag{8.209}$$

In Eq. (8.208):

$$\frac{d\gamma}{d\tau} = \gamma \frac{d\gamma}{dv}\frac{dv}{dt} = \frac{\gamma^4}{c^2} v \frac{dv}{dt} = \frac{\gamma^4}{c^2}\frac{dr}{dt}\frac{d^2r}{dt^2} \tag{8.210}$$

and therefore the acceleration becomes:

$$\mathbf{a} = \gamma^4 \frac{d^2r}{dt^2}\mathbf{e}_r - \frac{L^2}{m^2 r^3}\mathbf{e}_r + \frac{\gamma^4}{c^2}\frac{dr}{dt}\frac{d^2r}{dt^2}\omega r \mathbf{e}_\theta \tag{8.211}$$

in which the relativistic total angular momentum is

$$L = \gamma L_0 = mr^2 \frac{d\theta}{d\tau} = \gamma m r^2 \omega. \tag{8.212}$$

The relativistic force law is therefore the mass m multiplied by the relativistic acceleration:

$$\mathbf{a} = \left(\gamma^4 \frac{d^2r}{dt^2} - \frac{L^2}{m^2 r^3}\right)\mathbf{e}_r + \frac{\gamma^4}{c^2}\frac{dr}{dt}\frac{d^2r}{dt^2}\boldsymbol{\omega} \times \mathbf{r} \tag{8.213}$$

in which:

$$\boldsymbol{\omega} \times \mathbf{r} = \omega r \mathbf{e}_\theta. \tag{8.214}$$

This equation can be transformed into a format where the relativistic force can be calculated from the observation of any planar orbit. The result is the relativistic generalization of Eq. (8.124).

Consider the relativistic acceleration:

$$\mathbf{a} = \gamma^4 \frac{d^2r}{dt^2}\mathbf{e}_r + \gamma^2 \boldsymbol{\omega} \times (\boldsymbol{\omega} \times \mathbf{r}) + \frac{d\gamma}{d\tau}\boldsymbol{\omega} \times \mathbf{r} \tag{8.215}$$

in which the relativistic momentum is:

$$\mathbf{p} = m\frac{d\mathbf{r}}{d\tau}. \tag{8.216}$$

It follows that:

$$\frac{d^2r}{dt^2} = -\left(\frac{L}{\gamma m r}\right)^2 \frac{d^2}{d\theta^2}\left(\frac{1}{r}\right) \tag{8.217}$$

8.4. DESCRIPTION OF ORBITS WITH THE MINKOWSKI FORCE...

and that:

$$\mathbf{a} = -\left(\left(\frac{\gamma L}{mr}\right)^2 \frac{d^2}{d\theta^2}\left(\frac{1}{r}\right) + \frac{L^2}{m^2 r^3}\right)\mathbf{e}_r + \frac{\gamma^4}{c^2}\frac{dr}{dt}\frac{d^2 r}{dt^2}\omega r \mathbf{e}_\theta. \qquad (8.218)$$

It also follows as in UFT 238 that:

$$\frac{dr}{dt} = -\frac{L}{m\gamma}\frac{d}{d\theta}\left(\frac{1}{r}\right) \qquad (8.219)$$

so the required relativistic generalization of Eq. (8.124) is:

$$\mathbf{a} = -\left(\frac{L}{mr}\right)^2\left(\gamma^2 \frac{d^2}{d\theta^2}\left(\frac{1}{r}\right) + \frac{1}{r}\right)\mathbf{e}_r + \frac{L^4}{m^4 c^2 r^3}\frac{d}{d\theta}\left(\frac{1}{r}\right)\frac{d^2}{d\theta^2}\left(\frac{1}{r}\right)\mathbf{e}_\theta. \qquad (8.220)$$

For the purposes of graphics and animation it is convenient to express the Lorentz factor in terms of the angle θ. The result as derived in UFT 238 is:

$$v^2 = \left(\frac{L_0}{m\alpha}\right)^2 \left(1 + \epsilon^2 + 2\epsilon \cos\theta\right). \qquad (8.221)$$

In summary, the relativistic force for any planar orbit is defined by:

$$\mathbf{F} = -\frac{L^2}{mr^2}\left(\gamma^2 \frac{d^2}{d\theta^2}\left(\frac{1}{r}\right) + \frac{1}{r}\right)\mathbf{e}_r + \frac{L^4}{m^4 c^2 r^3}\frac{d}{d\theta}\left(\frac{1}{r}\right)\frac{d^2}{d\theta^2}\left(\frac{1}{r}\right)\mathbf{e}_\theta \qquad (8.222)$$

in which the Lorentz factor is:

$$\gamma = 1 - \left(\frac{L_0}{mc}\right)^2 \left(\frac{1}{r^2} + \left(\frac{d}{d\theta}\left(\frac{1}{r}\right)\right)^2\right)^{-1/2} \qquad (8.223)$$

and in which the relativistic total angular momentum is:

$$L = \gamma L_0 = \gamma m r^2 \frac{d\theta}{d\tau} = \gamma m r^2 \omega. \qquad (8.224)$$

In Figs. 8.3 and 8.4 the radial and angular force component of a precessing ellipse are graphed for two values of angular momentum L. The precession factor was $x = 1.1$. The angular force component, non existing in the Newtonian case, is much smaller than the radial component. The angular component takes both signs, leading to zero crossings and a different angular dependency than the radial component. The asymmetry of the radial component increases with L.

CHAPTER 8. ECE COSMOLOGY

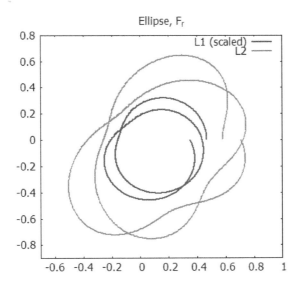

Figure 8.3: Radial force component of a precessing ellipse, polar plot.

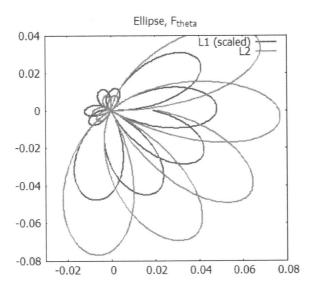

Figure 8.4: Angular force component of a precessing ellipse, polar plot.

Chapter 9

Relativistic Cosmology and Einstein's "Gravitational Waves"

9.1 Introduction

The LIGO Scientific Collaboration and Virgo Collaboration have announced [38] that on the 14th of September 2015, at 09:50:45 UTC, they detected a transient Einstein gravitational wave, designated GW150914, produced by two merging black holes forming a single black hole. Not so long ago similar media excitement surrounded the announcement by the BICEP2 Team of detection of primordial gravitational waves imprinted in B-mode polarisations of a Cosmic Microwave Background, which proved to be naught. The two black holes that merged are reported to have been at a distance of some 1.3 billion light years from Earth, of ≈29 solar masses and ≈36 solar masses respectively, the newly formed black hole at ≈62 solar masses, radiating away ≈3 solar masses as Einstein gravitational waves. The insurmountable problem for the credibility of the LIGO-Virgo Collaboration claims is the falsity of the theoretical assumptions upon which they are based.

The reported detection was obtained, not during an observation run of LIGO, but during an engineering test run, prior to the first scheduled observation run on 18 September 2015.

> "In the early morning hours of September 14, 2015 - during an engineering run just days before official data-taking started - a strong signal, consistent with merging black holes, appeared simultaneously in LIGO's two observatories, located in Hanford, Washington and Livingston, Louisiana." Conover [39]

> "The eighth engineering run (ER8) began on 2015 August 16 at 15:00 and critical software was frozen by August 30. The rest of

ER8 was to be used to calibrate the detectors, to carry out diagnostic studies, to practice maintaining a high coincident duty cycle, and to train and tune the data analysis pipelines. Calibration was complete by September 12 and O1 was scheduled to begin on September 18. On 2015 September 14, cWB reported a burst candidate to have occurred at 09:50:45 with a network signal-to-noise ratio (S/N) of 23.45 ..." Abbott et al. [40]

Magnitudes with error margins have been presented by the LIGO-Virgo Collaborations for the masses of the black holes, along with other related source quantities, in their TABLE I [38], reproduced herein as Figure 9.1.

TABLE I. Source parameters for GW150914. We report median values with 90% credible intervals that include statistical errors, and systematic errors from averaging the results of different waveform models. Masses are given in the source frame; to convert to the detector frame multiply by $(1+z)$ [90]. The source redshift assumes standard cosmology [91].

Primary black hole mass	$36^{+5}_{-4} M_\odot$
Secondary black hole mass	$29^{+4}_{-4} M_\odot$
Final black hole mass	$62^{+4}_{-4} M_\odot$
Final black hole spin	$0.67^{+0.05}_{-0.07}$
Luminosity distance	410^{+160}_{-180} Mpc
Source redshift z	$0.09^{+0.03}_{-0.04}$

Figure 9.1: Reproduced from Abbott, B.P. et al., Observation of Gravitational Waves from a Binary Black Hole Merger, PRL 116, 061102 (2016), DOI: 10.1103/PhysRevLett.116.061102

There are two ways by which the LIGO report can be analysed: (a) examination of the LIGO interferometer operation and data acquisition, and (b) consideration of the theories of black holes and gravitational waves upon which all else relies. Only theoretical considerations are considered herein, as there is no need to analyse the LIGO apparatus and its signal data to understand that the claim for detection of Einstein gravitational waves and black holes is built upon theoretical fallacies and conformational bias.

9.2 Gravitational waves, black holes and big bangs combined

Presumably the gravitational waves reported by LIGO-Virgo are present inside some big bang expanding universe as there has been no report that big bang cosmology has been abandoned. Of the gravitational wave detection the LIGO-Virgo Collaborations have stated,

> "*It matches the waveform predicted by general relativity for the inspiral and merger of a pair of black holes and the ringdown of the resulting single black hole.*" Abbott et al. [38]

All purported black hole equations are solutions to corresponding specific forms of Einstein's nonlinear field equations. Gravitational waves on the other hand are obtained from a linearisation of Einstein's nonlinear field equations, combined with a deliberate selection of coordinates that produce the assumed aforehand propagation at the speed of light in vacuo. Because General Relativity is a nonlinear theory, the Principle of Superposition does not hold.

> "*In a gravitational field, the distribution and motion of the matter producing it cannot at all be assigned arbitrarily - on the contrary it must be determined (by solving the field equations for given initial conditions) simultaneously with the field produced by the same matter.*" Landau & Lifschitz [41]

> "*An important characteristic of gravity within the framework of general relativity is that the theory is nonlinear. Mathematically, this means that if g_{ab} and γ_{ab} are two solutions of the field equations, then $ag_{ab} + b\gamma_{ab}$ (where a, b are scalars) may not be a solution. This fact manifests itself physically in two ways. First, since a linear combination may not be a solution, we cannot take the overall gravitational field of the two bodies to be the summation of the individual gravitational fields of each body.*" McMahon [42]

Let **X** be some black hole universe and **Y** be some big bang universe. Then the linear combination (i.e. superposition) **X** + **Y** is not a universe. Indeed, **X** and **Y** pertain to completely different sets of Einstein field equations and so they have absolutely nothing to do with one another. The same argument holds if **X** and **Y** are both black hole universes, be they the same or not, and if **X** and **Y** are big bang universes, be they the same or not. Consequently, a black hole universe cannot co-exist with any other black hole universe or with any big bang universe.

All black hole universes:

1. are spatially infinite

2. are eternal

3. contain only one mass

4. are not expanding (i.e. are not non-static)

5. are asymptotically flat (or, even more exotically, asymptotically curved).

All big bang universes:

1. are either spatially finite (1 case; $k = 1$) or spatially infinite (2 different cases; $k = -1, k = 0$)

2. are of finite age (≈ 13.8 billion years)

9.2. GRAVITATIONAL WAVES, BLACK HOLES AND BIG BANGS...

3. contain radiation and many masses

4. are expanding (i.e. are non-static)

5. are not asymptotically anything.

Note also that no black hole universe even possesses a big bang universe k-curvature. It is clearly evident that black holes and big bang universes cannot co-exist by their very definitions.

All the different black hole 'solutions' are also applicable to stars, planets and the like. Thus, these equations do not permit the presence of more than one star or planet in the universe. In the case of a body such as a star, the only significant difference is that its spacetime does not bend to infinite curvature at the star because there is no singularity and no event horizon in this case. Newton's theory on the other hand, places no restriction on the number of masses that can be present.

Since a black hole universe is a solution to a specific set of Einstein's nonlinear field equations it is not possible to extract from it any gravitational waves that are produced from linearised field equations. No gravitational waves can in fact be extracted from Einstein's nonlinear field equations [43]. Superposing solutions obtained from the nonlinear system with those from the linearised system violates the mathematical structure of General Relativity. Accordingly, contrary to the report of the LIGO-Virgo Collaborations, gravitational waves cannot exist in any black hole universe. Neither can they exist in any big bang universe because all big bang models are in fact single mass universes by mathematical construction [43, 44]. Nonetheless the LIGO-Virgo Collaborations superpose everything [38], depicted in Figure 9.2 herein.

Superposition where superposition does not hold is a standard method of cosmologists.

> "*From what I have said, collapse of the kind I have described must be of frequent occurrence in the Galaxy; and black-holes must be present in numbers comparable to, if not exceeding, those of the pulsars. While the black-holes will not be visible to external observers, they can nevertheless interact with one another and with the outside world through their external fields.*
>
> "*In considering the energy that could be released by interactions with black holes, a theorem of Hawking is useful. Hawking's theorem states that **in the interactions involving black holes, the total surface area of the boundaries of the black holes can never decrease**; it can at best remain unchanged (if the conditions are stationary).*
>
> "*Another example illustrating Hawking's theorem (and considered by him) is the following. Imagine two spherical (Schwarzschild) black holes, each of mass $\frac{1}{2}M$, coalescing to form a single black hole; and let the black hole that is eventually left be, again, spherical and have a mass M. Then Hawking's theorem requires that*
>
> $$16\pi \overline{M}^2 \geq 16\pi \left[2\left(\frac{1}{2}M\right)^2 \right]^2$$

CHAPTER 9. RELATIVISTIC COSMOLOGY AND EINSTEIN'S...

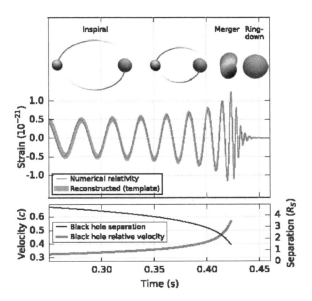

Figure 9.2: *"Top: Estimated gravitational-wave strain amplitude from GW150914 projected onto H1. This shows the full bandwidth of the waveforms, without the filtering used for Fig. 1. The inset images show numerical relativity models of the black hole horizons as the black holes coalesce. Bottom: The Keplerian effective black hole separation in units of Schwarzschild radii ($R_S = 2GM/c^2$) and the effective relative velocity given by the post-Newtonian parameter $v/c = (GM\pi f/c^3)^{1/3}$, where f is the gravitational-wave frequency calculated with numerical relativity and M is the total mass (value from Table I)."* Reproduced from Abbott, B.P. et al., Observation of Gravitational Waves from a Binary Black Hole Merger, PRL 116, 061102 (2016), DOI: 10.1103/PhysRevLett.116.061102

or

$$\overline{M} \geq \frac{M}{\sqrt{2}}$$

"Hence the maximum amount of energy that can be released in such a coalescence is

$$M\left(1 - \frac{1}{\sqrt{2}}\right)c^2 = 0.293Mc^2\text{"}$$

Chandrasekhar [45]

"Also, suppose two black holes collided and merged together to form a single black hole. Then the area of the event horizon of the final black hole would be greater than the sum of the areas of the event horizons of the original black holes." Hawking [46]

9.2. GRAVITATIONAL WAVES, BLACK HOLES AND BIG BANGS...

"*Hawking's area theorem: in any physical process involving a horizon, the area of the horizon cannot decrease in time. ... This fundamental theorem has the result that, while two black holes can collide and coalesce, a single black hole can never bifurcate spontaneously into two smaller ones.*

"*Black holes produced by supernovae would be much harder to observe unless they were part of a binary system which survived the explosion and in which the other star was not so highly evolved.*" Schutz [47]

"*The extreme RN in isotropic coordinates is*

$$ds^2 = V^{-2}dt^2 + V^2\left(d\rho^2 + \rho^2 d\Omega^2\right)$$

where

$$V = 1 + \frac{M}{\rho}$$

This is a special case of the multi black hole solution

$$ds^2 = V^{-2}dt^2 + V^2 d\vec{x} \cdot d\vec{x}$$

where $d\vec{x} \cdot d\vec{x}$ is the Euclidean 3-metric and V is any solution of $\nabla^2 V = 0$. In particular

$$V = 1 + \Sigma_{i=1}^{N}\frac{M_i}{|\vec{x} - \vec{\bar{x}}|}$$

yields the metric for N extreme black holes of masses M_i at positions $\bar{x}_i$." Townsend [48]

"*We not only accept the existence of black holes, we also understand how they can actually form under various circumstances. Theory allows us to calculate the behavior of material particles, fields or other substances near or inside a black hole. What is more, astronomers have now identified numerous objects in the heavens that completely match the detailed descriptions theoreticians have derived. These objects cannot be interpreted as anything else but black holes. The 'astronomical black holes' exhibit no clash whatsoever with other physical laws. Indeed, they have become rich sources of knowledge about physical phenomena under extreme conditions. General Relativity itself can also now be examined up to great accuracies.*" 't Hooft [49]

"*Black holes can be in the vicinity of other black holes.*" 't Hooft [43]

Much of the justification for the notion of irresistible gravitational collapse into an infinitely dense point-mass 'physical' singularity where spacetime is infinitely curved, and hence the formation of a black hole, is due to Oppenheimer and Snyder [50].

> "*In an idealized but illustrative calculation, Oppenheimer and Snyder ... showed in 1939 that a black hole in fact does form in the collapse of ordinary matter. They considered a 'star' constructed out of a pressureless 'dustball'. By Birkhof's Theorem, the entire exterior of this dustball is given by the Schwarzschild metric .... Due to the self-gravity of this 'star', it immediately begins to collapse. Each mass element of the pressureless star follows a geodesic trajectory toward the star's center; as the collapse proceeds, the star's density increases and more of the spacetime is described by the Schwarzschild metric. Eventually, the surface passes through $r = 2M$. At this point, the Schwarzschild exterior includes an event horizon: A black hole has formed. Meanwhile, the matter which formerly constituted the star continues collapsing to ever smaller radii. In short order, all of the original matter reaches $r = 0$ and is compressed (classically!) into a singularity[4]."*
> Hughes [51]

> "[4] *Since all of the matter is squashed into a point of zero size, this classical singularity must be modified in a complete, quantum description. However, since all the singular nastiness is hidden behind an event horizon where it is causally disconnected from us, we need not worry about it (at least for astrophysical black holes)."*

Note that the 'Principle of Superposition' has again been incorrectly applied by Oppenheimer and Snyder, from the outset. They first assume a relativistic universe in which there are multiple mass elements present *a priori*, where the 'Principle of Superposition' however, does not apply. Then mass elements "*collapse*" into a central point (zero volume, finite mass, infinite density), due to 'self-gravity'. But the 'collapse' cannot be due to Newtonian gravitation, because gravity is not a force in General Relativity, and with the resulting black hole, which does not occur in Newton's theory of gravitation. A Newtonian universe cannot collapse into a non-Newtonian universe. Neither can a non-Newtonian universe collapse into a Newtonian universe. Furthermore, the black hole so formed is in an empty universe, since the 'Schwarzschild black hole' relates to $R_{\mu\nu} = 0$, a spacetime that by construction contains no matter. Nonetheless, Oppenheimer and Snyder permit, within the context of General Relativity, the presence of and the gravitational interaction of many mass elements, which coalesce and collapse into a point of zero volume to form an infinitely dense point-mass singularity, when there is no demonstrated general relativistic or Newtonian mechanism by which any of this can occur. Moreover, nobody has ever observed a body, celestial or otherwise, undergo irresistible gravitational collapse, and there is no laboratory evidence whatsoever for the existence of such a phenomenon.

In the 'self-gravity' of a star the cosmologists invoke Newtonian gravitational forces.

> "*Assume that it obeys an equation of state. If, according to this equation of state, the pressure stays sufficiently low, one can*

calculate that this ball of matter will contract under its own weight." 't Hooft (see [43])

"One must ask what happens when larger quantities of mass are concentrated in a small enough volume. If no stable soution (sic) exists, this must mean that the system collapses under its own weight." 't Hooft (see [43])

Weight is a force due to gravity, but in General Relativity gravity is not a force. Contrary to the practice of cosmologists, Newton's gravitational forces cannot be invoked in General Relativity, because there are none.

"In Einstein's new theory, gravitation is of a much more fundamental nature: it becomes almost a property of space. ... Gravitation is thus, properly speaking, not a 'force' in the new theory." de Sitter [52]

9.3 Gravitational wave propagation speed and the linearisation game

The LIGO-Virgo Collaborations opened the *Introduction* to their paper with the following:

"In 1916, the year after the final formulation of the field equations of general relativity, Albert Einstein predicted the existence of gravitational waves. He found that the linearized weak-field equations had wave solutions: transverse waves of spatial strain that travel at the speed of light, generated by time variations of the mass quadrupole moment of the source." Abbott et al. [38]

The impression given here that the speed of propagation of Einstein's gravitational waves is uniquely that of light in vacuo is false. Propagation speed of Einstein's gravitational waves is arbitrary, because it is coordinate dependent. That Einstein's waves seem to travel uniquely at the speed of light *in vacuo* is simply because this speed was deliberately selected in order to conform to the presupposition. The method employed to determine the wave equation for Einstein's gravitational waves is the weak-field 'linearisation' of his field equations and concomitant selection of a specific set of coordinates.

Maxwell's electromagnetic theory predicts sinusoidal electromagnetic-wave propagation *in vacuo* at the unique speed v, given by,

$$v = \frac{1}{\sqrt{\epsilon_0 \mu_0}} = c \qquad (9.1)$$

The speed of light changes according to the permittivity and permeability of the dielectric medium in which it travels,

$$v = \frac{1}{\sqrt{\epsilon \mu}} = \frac{c}{\sqrt{\kappa \kappa_m}} \qquad (9.2)$$

CHAPTER 9. RELATIVISTIC COSMOLOGY AND EINSTEIN'S...

wherein κ and κ_m are the dielectric constant and relative permeability respectively of the medium. Note that the speed of electromagnetic wave propagation *in vacuo* is not arbitrary. Since the speed of light 'in vacuo' plays a central role in Einstein's Relativity Theory, the motive for choosing coordinates in order to make gravitational waves, emerging from the linearisation game[1], travel at the speed of light 'in vacuo', is abundantly clear.

Einstein's gravitational waves are extracted from his linearisation of his nonlinear field equations. Accordingly the metric tensor is written as,

$$g_{\mu\nu} = \eta_{\mu\nu} + h_{\mu\nu} \qquad (9.3)$$

where the $h_{\mu\nu} \ll 1$ and $\eta_{\mu\nu}$ represents the Galilean values $(1, -1, -1, -1)$. In the linearisation game the $h_{\mu\nu}$ slightly perturb the flat Minkowski spacetime $g_{\mu\nu} = \eta_{\mu\nu}$ from its flatness, and so suffixes are raised and lowered on the $h_{\mu\nu}$ by the $\eta_{\mu\nu}$. Here the $h_{\mu\nu}$ and their first derivatives $\partial h_{\mu\nu}/\partial x^\sigma \equiv h_{\mu\nu,\sigma}$, and higher derivatives, are small, and all products of them are neglected. Since the $\eta_{\mu\nu}$ are constants, the derivatives of Eq.(9.3) are $g_{\mu\nu,\sigma} = h_{\mu\nu,\sigma}$. The validity of the linearisation game is merely taken on trust because it leads to the desired result.

The selection of a specific coordinate system in order to ensure that gravitational waves propagate at the presupposed speed of light $c = 2.998 \times 10^8 \text{m/s}$ is exposed by the approximation of the Ricci tensor $R_{\mu\nu}$, in accordance with Eq.(9.3). The Ricci tensor can be first written in the following form by a contraction on the Riemann-Christoffel curvature tensor $R_{\rho\mu\nu\sigma}$, as follows,

$$R_{\mu\nu} = g^{\sigma\rho} \left[\frac{1}{2} (g_{\sigma\rho,\mu\nu} - g_{\nu\rho,\sigma\mu} - g_{\mu\sigma,\nu\rho} + g_{\mu\nu,\sigma\rho}) + \Gamma_{\beta\rho\sigma}\Gamma^\beta_{\mu\nu} - \Gamma_{\beta\rho\nu}\Gamma^\beta_{\mu\sigma} \right] \qquad (9.4)$$

Since the last two terms of Eq.(9.4) are products of the components of the metric tensor $g_{\mu\nu}$ and their first derivatives, by the linearisation game they are neglected, so that the Ricci tensor reduces to,

$$R_{\mu\nu} = g^{\sigma\rho} \left[\frac{1}{2} (g_{\sigma\rho,\mu\nu} - g_{\nu\rho,\sigma\mu} - g_{\mu\sigma,\nu\rho} + g_{\mu\nu,\sigma\rho}) \right] \qquad (9.5)$$

which can be broken into two parts,

$$R_{\mu\nu} = \frac{1}{2} g^{\sigma\rho} g_{\mu\nu,\sigma\rho} + \frac{1}{2} g^{\sigma\rho} \left[(g_{\sigma\rho,\mu\nu} - g_{\nu\rho,\sigma\mu} - g_{\mu\sigma,\nu\rho}) \right] \qquad (9.6)$$

If the coordinates x^α are chosen so that the second part of Eq.(9.6) vanishes, the Ricci tensor reduces further as follows,

$$R_{\mu\nu} = \frac{1}{2} g^{\sigma\rho} g_{\mu\nu,\sigma\rho} = \frac{1}{2} g^{\sigma\rho} \frac{\partial^2 g_{\mu\nu}}{\partial x^\rho \partial x^\sigma} \qquad (9.7)$$

$$g^{\sigma\rho} (g_{\sigma\rho,\mu\nu} - g_{\nu\rho,\sigma\mu} - g_{\mu\sigma,\nu\rho}) = 0 \qquad (9.8)$$

---

[1] "*The rules of the 'linearization game' are as follows: (a) $h_{\mu\nu}$ together with its first derivatives $h_{\mu\nu,\rho}$ and higher derivatives are small, and all products of these are ignored; (b) suffixes are raised and lowered using $\eta^{\mu\nu}$ and $\eta_{\mu\nu}$, rather than $g^{\mu\nu}$ and $g_{\mu\nu}$.*" Foster & Nightingale [53]

9.3. GRAVITATIONAL WAVE PROPAGATION SPEED AND THE ...

According to Eq.(9.3), $g_{\mu\nu,\beta} = h_{\mu\nu,\beta}$ and so on for higher derivatives. Hence,

$$R_{\mu\nu} = \frac{1}{2}\eta^{\sigma\rho} h_{\mu\nu,\sigma\rho} = \frac{1}{2}\eta^{\sigma\rho}\frac{\partial^2 h_{\mu\nu}}{\partial x^\rho \partial x^\sigma} \tag{9.9}$$

$$\eta^{\sigma\rho}\left(h_{\sigma\rho,\mu\nu} - h_{\nu\rho,\sigma\mu} - h_{\mu\sigma,\nu\rho}\right) = 0 \tag{9.10}$$

(remembering that suffixes on the kernel h are raised and lowered by $\eta^{\mu\nu}$ according to tensor type). Contracting Eq.(9.9) yields the Ricci scalar,

$$R = \eta^{\nu\mu}R_{\mu\nu} = \frac{1}{2}\eta^{\nu\mu}\eta^{\sigma\rho}\frac{\partial^2 h_{\mu\nu}}{\partial x^\rho \partial x^\sigma} = \frac{1}{2}\eta^{\rho\sigma}\frac{\partial^2 h}{\partial x^\sigma \partial x^\rho} \tag{9.11}$$

Einstein's field equations (without cosmological constant) are,

$$R_{\mu\nu} - \frac{1}{2}Rg_{\mu\nu} = -\kappa T_{\mu\nu} \tag{9.12}$$

In terms of $h_{\mu\nu}$ these become, using Eq.(9.9) and Eq.(9.11),

$$\eta^{\sigma\rho}\frac{\partial^2 h_{\mu\nu}}{\partial x^\rho \partial x^\sigma} - \frac{1}{2}\eta^{\sigma\rho}\frac{\partial^2 h}{\partial x^\sigma \partial x^\rho}\eta_{\mu\nu} = -2\kappa T_{\mu\nu} \tag{9.13}$$

The d'Almbertian operator is defined by,

$$\Box \equiv -\eta^{\mu\nu}\frac{\partial}{\partial x^\mu}\frac{\partial}{\partial x^\nu} \tag{9.14}$$

Recalling that $\eta^{\mu\nu}$ represents the Galilean values and that hence $\eta^{\mu\nu} = 0$ when $\mu \neq \nu$, Eq.(9.14) gives,

$$\Box = \frac{\partial^2}{\partial x^2} + \frac{\partial^2}{\partial y^2} + \frac{\partial^2}{\partial z^2} - \frac{1}{c^2}\frac{\partial^2}{\partial t^2} = \nabla^2 - \frac{1}{c^2}\frac{\partial^2}{\partial t^2} \tag{9.15}$$

where ∇ is the differential operator *del* (or *nabla*), defined as,

$$\nabla \equiv \left\langle \frac{\partial}{\partial x}, \frac{\partial}{\partial y}, \frac{\partial}{\partial z} \right\rangle \tag{9.16}$$

Taking the dot product of *del* with itself gives the Laplacian operator ∇^2,

$$\nabla^2 = \frac{\partial^2}{\partial x^2} + \frac{\partial^2}{\partial y^2} + \frac{\partial^2}{\partial z^2} \tag{9.17}$$

Setting $x_0 = ct, x_1 = x, x_2 = y, x_3 = z$, Eq.(9.13) can be written as,

$$\Box\left(h_{\mu\nu} - \frac{1}{2}\delta_{\mu\nu}h\right) = -2\kappa T_{\mu\nu} \tag{9.18}$$

These are the linearised field equations. They are subject to the condition Eq.(9.8), which can be condensed to the following condition [54],

$$\frac{\partial}{\partial x^\alpha}\left(h^\alpha_\mu - \frac{1}{2}\delta^\alpha_\mu h\right) = 0 \tag{9.19}$$

Using Eq.(9.14), Eq.(9.9) can be written as,

$$\Box h_{\mu\nu} = 2R_{\mu\nu} \qquad (9.20)$$

For empty space this becomes,

$$\Box h_{\mu\nu} = 0 \qquad (9.21)$$

which by Eq.(9.15) describes a wave propagating at the speed of light *in vacuo*.

However, the crucial point of the foregoing mathematical development is that Einstein's gravitational waves do not have a unique speed of propagation. The speed of the waves is coordinate dependent, as the condition at Eq.(9.8) attests. It is the constraint at Eq.(9.8) that selects a set of coordinates to produce the propagation speed c. A different set of coordinates yields a different speed of propagation, as Eq.(9.5) does not have to be constrained by Eq.(9.8). Einstein deliberately chose a set of coordinates that yields the desired speed of propagation at that of light in vacuum (i.e. $c = 2.998 \times 10^8$m/s) in order to satisfy the presupposition that propagation is at speed c. There is no *a priori* reason why this particular set of coordinates is better than any other. The sole purpose for the choice is to obtain the desired and presupposed result.

> "*All the coordinate-systems differ from Galilean coordinates by small quantities of the first order. The potentials $g_{\mu\nu}$ pertain not only to the gravitational influence which has objective reality, but also to the coordinate-system which we select arbitrarily. We can 'propagate' coordinate-changes with* **the speed of thought***, and these may be mixed up at will with the more dilatory propagation discussed above. There does not seem to be any way of distinguishing a physical and a conventional part in the changes of $g_{\mu\nu}$.*
>
> "*The statement that in the relativity theory gravitational waves are propagated with the speed of light has, I believe, been based entirely upon the foregoing investigation; but it will be seen that it is only true in a very conventional sense. If coordinates are chosen so as to satisfy a certain condition which has no very clear geometrical importance, the speed is that of light; if the coordinates are slightly different the speed is altogether different from that of light. The result stands or falls by the choice of coordinates and, so far as can be judged, the coordinates here used were purposely introduced in order to obtain the simplification which results from representing the propagation as occurring with the speed of light. The argument thus follows a vicious circle.*" Eddington [54]

9.4 A black hole is a universe

Each and every black hole is an independent universe by its very definition, no less than the big bang universes are independent universes, because the black hole universe is not contained within its event horizon. Its spacetime extends indefinitely far from its singularity. All types of black hole universes are spatially infinite and eternal, and are either asymptotically flat or, in more esoteric cases, asymptotically curved. There is no bound on asymptotic, for otherwise it would not be asymptotic. Thus every type of black hole constitutes an independent infinite and eternal universe; bearing in mind also that each different type of black hole universe pertains to a different set of Einstein field equations and therefore have nothing to do with one another. Without the asymptotic condition the black hole equations do not result, and one can then write as many non-asymptotic solutions to the corresponding Einstein field equations for the supposed different types of black holes as one pleases [43, 55], none of which produces a black hole.

> "*Black holes were first discovered as purely mathematical solutions of Einstein's field equations. This solution, the Schwarzschild black hole, is a nonlinear solution of the Einstein equations of General Relativity. It contains no matter, and exists forever in an asymptotically flat space-time.*" Dictionary of Geophysics, Astrophysics and Astronomy [56]

> "*The Kerr-Newman solutions ... are explicit asymptotically flat stationary solutions of the Einstein-Maxwell equation ($\lambda = 0$) involving just three free parameters **m**, **a** and **e**. ...the mass, as measured asymptotically, is the parameter **m** (in gravitational units). The solution also possesses angular momentum, of magnitude **am**. Finally, the total charge is given by **e**. When **a**=**e**=0 we get the Schwarzschild solution.*" Penrose [57]

> "*The charged Kerr metrics are all stationary and axisymmetric ... They are asymptotically flat ...*" Wald [58]

All the different black hole equations are also applicable to stars and planets. Thus, these equations do not permit the presence of more than one star or planet in the universe. In the case of a body such as a star, the only significant difference is that the spacetime does not go to infinite curvature at the star, because there is no singularity and no event horizon in the case of a star (or planet).

9.5 Black hole gravity

Cosmology maintains that the finite mass of a black hole is concentrated at its 'singularity', where volume is zero, density is infinite, and spacetime curvature is infinite. There are two types of black hole singularity proposed by cosmologists, according to whether or not the black hole is rotating. In the case of no rotation the singularity is a point, adorned with mass: a 'point-mass'. In the case of rotation the singularity is the circumference of a circle, adorned with mass: a circumference-mass. Cosmologists call them 'physical singularities'. These and other mathematical singularities of black hole equations are reified so as to contain the masses of black holes and to locate their event horizons[2].

---

[2] An event horizon is also called 'a trapped surface' or 'a Schwarzschild sphere'.

CHAPTER 9. RELATIVISTIC COSMOLOGY AND EINSTEIN'S...

Singularities are actually only places in an equation where the equation is undefined, owing for example, to a division by zero. Although they have been construed as such by cosmology, singularities are not in fact physical entities. A singularity also occurs in Newton's theory; it is called the *centre of gravity* or the *centre of mass*. The centre of gravity of a body is not a physical object, rather a mathematical artifice.

> "*Let me be more precise as to what one means by a black hole. One says that a black hole is formed when the gravitational forces on the surface become so strong that light cannot escape from it.*
> "*...A trapped surface is one from which light cannot escape to infinity.*" Chandrasekhar [45]

There are forces in General Relativity but gravity is not one of them, because it is spacetime curvature. It immediately follows that according to cosmologists, gravity is infinite at a black hole singularity. Infinities of density, spacetime curvature, and gravity are claimed to be physically real.

> "*The work that Roger Penrose and I did between 1965 and 1960 showed that, according to general relativity, there must be a singularity of infinite density and space-time curvature, within the black hole. This is rather like the big bang at the beginning of time ...*" Hawking [97]

> "*Once a body of matter, of any mass m, lies inside its Schwarzschild radius 2m it undergoes gravitational collapse ... and the singularity becomes physical, not a limiting fiction.*" Dodson and Poston [59]

> "*A nonrotating black hole has a particularly simple structure. At the center is the singularity, a point of zero volume and infinite density where all of the black hole's mass is located. Spacetime is infinitely curved at the singularity. ... The black hole's singularity is a real physical entity. It is not a mathematical artifact ...*" Carroll and Ostlie [60]

> "*As r decreases, the space-time curvature mounts (in proportion to r^{-3}), becoming theoretically infinite at $r = 0$.*" Penrose [57]

> "*One says that '$r = 0$ is a physical singularity of spacetime.'*" Misner, Thorne & Wheeler [61]

> "*Black holes, the most remarkable consequences of Einstein's theory, are not just theoretical constructs. There are huge numbers of them in our Galaxy and in every other galaxy, each being the remnant of a star and weighing several times as much as the Sun. There are much larger ones, too, in the centers of galaxies.*" Rees [62]

> "*We not only accept the existence of black holes, we also understand how they can actually form under various circumstances. Theory allows us to calculate the behavior of material particles, fields or other substances near or inside a black hole. What is more, astronomers have now identified numerous objects in the heavens that completely match the detailed descriptions theoreticians have derived.*" 't Hooft [63]

> "*We've got the black holes cornered.*" Stern [64]

However, no finite mass possesses zero volume, infinite density, or infinite gravity, anywhere.

9.6 The mathematical theory of black holes

The LIGO-Virgo Collaborations have invoked a binary black hole system, merging to cause emission of their reported gravitational waves.

> "*The basic features of GW150914 point to it being produced by the coalescence of two black holes-i.e., their orbital inspiral and merger, and subsequent final black hole ringdown. Over 0.2s, the signal increases in frequency and amplitude in about 8 cycles from 35 to 150 Hz, where the amplitude reaches a maximum. The most plausible explanation for this evolution is the inspiral of two orbiting masses, m_1 and m_2, due to gravitational-wave emission.*" Abbott et al. [38]

In the *Introduction* of their paper the LIGO-Virgo Collaborations incorrectly attribute the black hole to K. Schwarzschild.

> "*Also in 1916, Schwarzschild published a solution for the field equations [41] that was later understood to describe a black hole [42, 43], and in 1963 Kerr generalized the solution to rotating black holes [43].*" Abbott et al. [38]

The resultant black hole type is identified in [38].

> "*A pair of neutron stars, while compact, would not have the required mass, while a black hole neutron star binary with the deduced chirp mass would have a very large total mass, and would thus merge at much lower frequency. This leaves black holes as the only known objects compact enough to reach an orbital frequency of 65 Hz without contact. Furthermore, the decay of the waveform after it peaks is consistent with the damped oscillations of a black hole relaxing to a final stationary Kerr configuration.*" Abbott et al. [38]

All the black holes are identified in [65].

CHAPTER 9. RELATIVISTIC COSMOLOGY AND EINSTEIN'S...

> "*Here we perform several studies of GW150914, aimed at detecting deviations from the predictions of GR. Within the limits set by LIGO's sensitivity and by the nature of GW150914, we find no statistically significant evidence against the hypothesis that, indeed, GW150914 was emitted by a binary system composed of two black holes (i.e., by the Schwarzschild [53] or Kerr [55] GR solutions), that the binary evolved dynamically toward merger, and that it formed a merged rotating black hole consistent with the GR solution.*" Abbott et al. [65]

Note the invalid superposition of the two 'Schwarzschild' or 'Kerr' black holes, due to violation of their asymptotic flatness (each encounters infinite spacetime curvature i.e. infinite gravity, at the singularity of the other). The Kerr configuration subsumes the Schwarzschild configuration and so depends upon the existence of the latter. The Schwarzschild solution has no physical meaning because it is the solution to a set of physically meaningless equations (see §6 and §7 below). Furthermore, all black hole equations are obtained by violations of the rules of pure mathematics, which will now be proven.

Satisfaction of the Einstein field equations is a necessary but insufficient condition for determination of Einstein's gravitational field. Einstein's field equations "*in the absence of matter*" [66] are,

$$R_{\mu\nu} = 0 \qquad (9.22)$$

To determine his gravitational field *in the absence of matter*, Einstein prescribed the following conditions:

1. the solution must be static

2. it must be spherically symmetric

3. it must satisfy the field equations

4. it must be asymptotically flat.

Consider Schwarzschild's [67] actual solution to Eq.(9.22), which is not the solution that has been assigned to him by cosmologists:

$$ds^2 = \left(1 - \frac{\alpha}{R}\right) dt^2 - \left(1 - \frac{\alpha}{R}\right)^{-1} dR^2 - R^2 \left(d\theta^2 + \sin^2\theta \, d\varphi^2\right)$$

$$R = \left(r^3 + \alpha^3\right)^{\frac{1}{3}}, \quad 0 \leq r \qquad (9.23)$$

Here α is a positive real-valued constant and $r = \sqrt{x^2 + y^2 + z^2}$. The speed of light *in vacuo* is set to unity, i.e. $c = 1$. Eq.(9.23) is singular (i.e. undefined) only at $r = 0$ (i.e. when $x = y = z = 0$). Contrast this with the so-called 'Schwarzschild solution' attributed to Schwarzschild but actually due to the German mathematician D. Hilbert [68]-[70],

$$ds^2 = \left(1 - \frac{2M}{r}\right) dt^2 - \left(1 - \frac{2M}{r}\right)^{-1} dr^2 - r^2 \left(d\theta^2 + \sin^2\theta \, d\varphi^2\right) \qquad (9.24)$$

$$0 \leq r$$

9.6. THE MATHEMATICAL THEORY OF BLACK HOLES

Here $c = 1$, Newton's gravitational constant $G = 1$, and M is claimed to be the mass that produces the gravitational field. Note that *prima facie* Eq.(9.24) is singular (i.e. undefined) at $r = 2M$ and at $r = 0$. Eq.(9.24) is not equivalent to Eq.(9.23) owing to $0 \leq r$ in Eq.(9.24). If they are equivalent then in Eq.(9.23) it must be that $-\alpha \leq r = \sqrt{x^2 + y^2 + z^2}$.

Eq.(9.24) is somewhat deceptive. Rewriting it explicitly with c and G is much more informative,

$$ds^2 = c^2 \left(1 - \frac{2GM}{c^2 r}\right) dt^2 - \left(1 - \frac{2GM}{c^2 r}\right)^{-1} dr^2 - r^2 \left(d\theta^2 + \sin^2\theta \, d\varphi^2\right)$$
$$0 \leq r$$
(9.25)

In Eq.(9.24) the so-called 'Schwarzschild radius' is $r_s = 2M$. From Eq.(9.25) the Schwarzschild radius r_s is easily identified in full,

> "*This value is known as the Schwarzschild radius. In terms of the mass of the object that is the source of the gravitational field, it is given by*
> $$r_s = \frac{2GM}{c^2}$$
> *For ordinary stars, the Schwarzschild radius lies buried deep in the stellar interior.*" McMahon [42]

> "*Remarkably, as we shall see this is exactly the modern formula for the radius of a black hole in general relativity ...*" Schutz [47]

According to cosmologists, the Schwarzschild radius (or 'gravitational radius') is the radius of the event horizon of a black hole. That r is incorrectly treated as the radius by cosmologists is most clearly evident by the very 'Schwarzschild radius', which for stars, "*lies buried deep in the stellar interior*" [42].

> "*The Schwarzschild radius for the Earth is about 1.0 cm and that of the Sun is 3.0 km.*" d'Inverno [71]

> "*For example, a Schwarzschild black hole of mass equal to that of the Earth, $M_E = 6 \times 10^{26} g$, has $r_s = 2GM_E/c^2 \approx 1$ cm. ... A black hole of one solar mass has a Schwarzschild radius of only 3km.*" Wald [58]

> "*... 'ordinary' stars and planets contain matter ($T_{\mu\nu} = 0$) within a certain radius $r > 2M$, so that for them the validity of the Schwarzschild solution stops there.*" 't Hooft [49]

In relation to Hilbert's solution, the cosmologists Celotti, Miller and Sciama [72], make the following assertion:

> "*The 'mean density' $\overline{\rho}$ of a black hole (its mass M divided by $\frac{4}{3}\pi r_s^3$) is proportional to $1/M^2$*" [72]

CHAPTER 9. RELATIVISTIC COSMOLOGY AND EINSTEIN'S...

where r_s is the 'Schwarzschild radius'. However, the expression $\frac{4}{3}\pi r^3$ gives the volume of a Euclidean sphere where r is the radius of the sphere. It does not give the volume of the non-Euclidean sphere within Hilbert's solution, where the volume is in fact given by [73]- [85],

$$V = \int_0^\pi \sin\theta\, d\theta \int_0^{2\pi} d\varphi \int_{2M}^r \frac{r^2\, dr}{\sqrt{1-\frac{2M}{r}}} = 4\pi \int_{2M}^r \frac{r^2\, dr}{\sqrt{1-\frac{2m}{r}}} \qquad (9.26)$$

which is a particular case of the general expression [73]- [74],

$$V = \int_0^\pi \sin\theta\, d\theta \int_0^{2\pi} d\varphi \int_{r_o}^r \frac{R_c^2(r)}{\sqrt{1-\frac{\alpha}{R_c(r)}}} \frac{dR_c}{dr} dr = 4\pi \int_{r_o}^r \frac{R_c^2(r)}{\sqrt{1-\frac{\alpha}{R_c(r)}}} \frac{dR_c}{dr} dr \qquad (9.27)$$

wherein the value of the real number r_o, although arbitrary, affects the form of $R_c(r)$.

Cosmology confounds the quantity r as radial distance, which ultimately gives rise to the 'Schwarzschild radius' r_s. It is variously and vaguely called the 'areal radius', the '*Schwarzschild r-coordinate*', the '*distance*', '*the radius*', the '*radius of a 2-sphere*', the '*radial coordinate*', the '*reduced circumference*', the '*radial space coordinate*', the '*coordinate radius*', and even "*a gauge choice: it determines the coordinate r*" 't Hooft [43]. None of these mere labels correctly identifies the geometric significance of the quantity r in Hilbert's solution.

Cosmologists maintain that the Schwarzschild radius $r = r_s$ is a 'coordinate' or 'apparent' or 'removable' singularity, and that $r = 0$ is a 'physical singularity' (because it is endowed with the fantastic physical properties in §5 above).

The quantity R in Schwarzschild's solution and the quantity r in Hilbert's solution can be replaced by any analytic real-valued function $R_c(r)$ of the real variable r without violating $R_{\mu\nu} = 0$ or spherical symmetry. However, not simply any analytic function of r is permissible. Satisfaction of the field equations is a necessary but insufficient condition for determination of Einstein's 'gravitational field'. For example, replace Hilbert's r with $R_c(r) = e^r$. The resulting metric is singular only at $r = \ln(2M)$. At $r = 0$ nothing special happens; on the unproven assumption that $0 \leq r \leq \ln(2M)$ is permissible. But $R_c(r) = e^r$ is forbidden because the resulting metric is not asymptotically flat. The infinite equivalence class of permissible analytic functions $R_c(r)$ must be ascertained.

Let r' be the radius of a Euclidean sphere. It is routinely claimed for Eq.(9.24) and Eq.(9.25) that $r = r' = \sqrt{x^2+y^2+z^2}$ (Einstein [85]), from which the black hole was constructed. This is incorrect [43, 44, 55], [73]- [85] because here,

$$r = \sqrt{x_o^2+y_o^2+z_o^2} + \sqrt{(x-x_o)^2+(y-y_o)^2+(z-z_o)^2} = r_o + r' \qquad (9.28)$$

where

$$r_o = \sqrt{x_o^2+y_o^2+z_o^2} = \frac{2GM}{c^2} \qquad (9.29)$$

9.6. THE MATHEMATICAL THEORY OF BLACK HOLES

Notwithstanding, r is neither the radius nor even a distance in any black hole equation [43, 44, 55], [73]- [85]; a mathematical fact which subverts the entire theory of black holes. The reader is referred to [43, 44, 55], [73]- [85] for all the mathematical details, which I only summarise herein.

Geometrically speaking Eq.(9.25) means that the black hole is the result of unwittingly moving a sphere, initially centred at the origin of coordinates, to some other place, but leaving its centre behind. Analytically this is revealed by,

$$ds^2 = \left(1 - \frac{\alpha}{R_c}\right) dt^2 - \left(1 - \frac{\alpha}{R_c}\right)^{-1} dR_c^2 - R_c^2 \left(d\theta^2 + \sin^2\theta \, d\varphi^2\right) \tag{9.30}$$

$$R_c = \left(|r - r_o|^n + \alpha^n\right)^{\frac{1}{n}}, \quad r, r_o \in \Re, \quad n \in \Re^+$$

Eqs.(9.30) satisfy Einstein's prescription, and constitute an infinite equivalence class because every element of the class describes the very same metric space.

The radius R_p for Eq.(9.30) is,

$$R_p = \int \frac{dR_c}{\sqrt{1 - \frac{\alpha}{R_c}}} = \sqrt{R_c(R_c - \alpha)} + \alpha \ln\left(\frac{\sqrt{R_c} + \sqrt{R_c - \alpha}}{\sqrt{\alpha}}\right) \tag{9.31}$$

Note that $R_c(r_o) = \alpha \; \forall r_o \; \forall n$ and $R_p(r_o) = 0 \; \forall r_o \; \forall n$. The constants r_o and n are arbitrary. Setting $r_o = 0, n = 3, r_o \leq r$ yields Schwarzschild's actual solution [53]. Setting $r_o = 0, n = 1, r_o \leq r$ yields Brillouin's solution [73]. Setting $r_o = \alpha, n = 1, r_o \leq r$ yields Droste's solution [73]. Hilbert's solution is not an element of the infinite equivalence class. Note that Hilbert's solution is an alleged 'extension' of Droste's solution to $0 \leq r$, for an 'event horizon' at 'the radius' $r = \alpha$ and a 'physical singularity' at 'the origin' $r = 0$. Although $r = 0$ denotes the origin of a coordinate system, it does not denote the centre of spherical symmetry of Eq.(9.24) and Eq.(9.25), as Eq.(9.30) reveals. The centre of spherical symmetry is at $r = r_o$. When a sphere initially centred at the origin of coordinates is moved, it takes its centre with it, and the position of the sphere is specified by the coordinates of its centre (x_0, y_0, z_0) so that whereupon the radius r' of the sphere is no longer given by $r = r' = \sqrt{x^2 + y^2 + z^2}$, but by $r' = \sqrt{(x - x_0)^2 + (y - y_0)^2 + (z - z_0)^2}$. The intrinsic geometry of a sphere is not altered by changing its position and so its radius does not change with a change of position. When Hilbert set r^2 as the coefficient of $(d\theta^2 + \sin^2\theta \, d\varphi)$ in the derivation of his solution, he unwittingly shifted the centre of Schwarzschild's Euclidean sphere from $r = r_o = 0$ to the coordinates (x_0, y_0, z_0) at the distance $r = 2m$ from the origin of coordinates, mistakenly thinking the centre of that sphere still at $r = 0$. Hilbert shifted Schwarzschild's Euclidean sphere but left its centre behind. The result was fantastic. David Hilbert had separated the Euclidean sphere from its centre and even placed its centre outside the sphere!

Owing to equivalence, if any element of the infinite equivalence class determined by Eq.(9.30) cannot be extended then none can be extended, owing to equivalence. It is immediately apparent that none can be extended because

$|r-r_o|^n \geq 0$. This is amplified by the case $r_o = 0, n = 2$, in which case Eq.(9.30) is defined for all real values of r except $r = r_o = 0$. In this case the black hole requires that,

$$-\alpha^2 \leq r^2 = (x^2 + y^2 + z^2) \tag{9.32}$$

Thus, the 'Schwarzschild' black hole is invalid because it violates the rules of pure mathematics - it requires the square of a real number to be less than zero. In general, the mathematical theory of black holes requires that the positive power of the absolute value of a real number must take on values less that zero. The same violation of the rules of pure mathematics produces all the black hole universes [43,44,55] [73]- [84]. All purported means of extending Droste's solution to Hilbert's are consequently invalid [43,83,84].

Schwarzschild spacetime can be written in the 'isotropic coordinates'. The infinite equivalence class in this case is given by [43,44,86], [41,43,71][3],

$$ds^2 = \left(\frac{4R_c - \alpha}{4R_c + \alpha}\right)^2 dt^2 - \left(1 + \frac{\alpha}{4R_c}\right)^4 [dR_c^2 + R_c^2 (d\theta^2 + \sin^2\theta \, d\varphi^2)]$$

$$R_c = \left[|r - r_o|^n + \left(\frac{\alpha}{4}\right)^n\right]^{\frac{1}{n}}, \quad r, r_o \in \Re, \; n \in \Re^+ \tag{9.33}$$

and the radius is,

$$R_p = \int \left(1 + \frac{\alpha}{4R_c}\right)^2 dR_c = R_c + \frac{\alpha}{2} \ln\left(\frac{4R_c}{\alpha}\right) - \frac{\alpha^2}{8R_c} + \frac{\alpha}{4} \tag{9.34}$$

Note that $R_c(r_o) = \alpha/4 \; \forall r_o \; \forall n$ and $R_p(r_o) = 0 \; \forall r_o \; \forall n$. Once again it is evident that no black hole is possible without a violation of the rules of pure mathematics, as the case $r_o = 0, n = 2$ again amplifies.

The Kerr-Newman solution adds charge q and angular momentum a to the 'Schwarzschild solution'[4]. The infinite equivalence class for Kerr-Newman spacetime is given by [43,55,77],

$$ds^2 = -\frac{\Delta}{\rho^2}(dt - a\sin^2\theta \, d\varphi)^2 + \frac{\sin^2\theta}{\rho^2}[(R_c^2 + a^2)d\varphi - adt]^2 + \frac{\rho^2}{\Delta}dR_c^2 + \rho^2 d\theta^2$$

$$\Delta = R_c^2 - \alpha R_c + a^2 + q^2, \quad \rho^2 = R_c^2 + a^2\cos^2\theta$$

$$R_c = (|r - r_o|^n + \zeta^n)^{\frac{1}{n}}, \quad r, r_o \in \Re, \; n \in \Re^+$$

$$\zeta = \frac{\alpha + \sqrt{\alpha^2 - 4q^2 - 4a^2\cos^2\theta}}{2}, \quad a^2 + q^2 < \frac{\alpha^2}{4} \tag{9.35}$$

The infinite equivalence class for Kerr spacetime is obtained from Eqs.(9.35) by setting $q = 0$. Similarly, the infinite equivalence class for Reissner-Nordström

---

[3] Here $c = 1$.
[4] The pronumeral a is called 'the angular momentum parameter': $a = J/M$ where J is angular momentum and M is the mass of the source of a 'gravitational field' (i.e. the mass of a star or a black hole).

9.6. THE MATHEMATICAL THEORY OF BLACK HOLES

spacetime is obtained from Eqs.(9.35) by setting $a = 0$. Setting $a = 0$ and $q = 0$ in Eqs.(9.35) yields the infinite equivalence class for Schwarzschild spacetime. No black hole is possible without a violation of the rules of pure mathematics, as the case $r_o = 0, n = 2$ yet again amplifies.

Black hole 'escape velocity'

On the one hand, cosmologists assign an escape speed to the black hole. At the event horizon it is the speed of light. Rearranging the 'Schwarzschild radius' for c gives,

$$c = \sqrt{\frac{2GM}{r_s}} \qquad (9.36)$$

which is immediately recognised as the Newtonian expression for escape speed. Although only one term for mass appears in this expression (i.e. M), it is an implicit two-body relation: one body 'escapes' from another body. Consequently, the Newtonian expression for escape speed cannot rightly appear in a solution for a one-body problem. The Schwarzschild solution is supposedly for a one-body problem. It is by this incorrect insinuation of the Newtonian expression for escape speed that cosmologists assign the black hole an escape speed, especially at its 'event horizon'.

> "**black hole** *A region of spacetime from which the escape velocity exceeds the velocity of light. In Newtonian gravity the escape velocity from the gravitational pull of a spherical star of mass M and radius R is*
>
> $$v_{esc} = \sqrt{\frac{2GM}{R}},$$
>
> *where G is Newton's constant. Adding mass to the star (increasing M), or compressing the star (reducing R) increases v_{esc}. When the escape velocity exceeds the speed of light c, even light cannot escape, and the star becomes a black hole. The required radius R_{BH} follows from setting v_{esc} equal to c:*
>
> $$R_{BH} = \frac{2GM}{c^2}.$$
>
> *"... In General Relativity for spherical black holes (Schwarzschild black holes), exactly the same expression R_{BH} holds for the surface of a black hole. The surface of a black hole at R_{BH} is a null surface, consisting of those photon trajectories (null rays) which just do not escape to infinity. This surface is also called the black hole horizon."* Dictionary of Geophysics, Astrophysics and Astronomy [56]

> "**black hole** *A massive object so dense that no light or any other radiation can escape from it; its escape velocity exceeds the speed of light."* Collins Encyclopaedia of the Universe [87]

"*A black hole is characterized by the presence of a region in space-time from which no trajectories can be found that escape to infinity while keeping a velocity smaller than that of light.*" 't Hooft [63]

On the other hand, nothing can even leave the event horizon of a black hole, not even light.

"*The problem we now consider is that of the gravitational collapse of a body to a volume so small that a trapped surface forms around it; as we have stated, from such a surface no light can emerge.*" Chandrasekhar [45]

"*It is clear from this picture that the surface $r = 2m$ is a one-way membrane, letting future-directed timelike and null curves cross only from the outside (region I) to the inside (region II).*" d'Inverno [71]

"*Things can go into the horizon (from $r > 2M$ to $r < 2M$), but they cannot get out; once inside, all causal trajectories (timelike or null) take us inexorably into the classical singularity at $r = 0$. ... The defining property of black holes is their event horizon. Rather than a true surface, black holes have a 'one-way membrane' through which stuff can go in but cannot come out.*" Hughes [51]

"*Einstein predicts that nothing, not even light, can be successfully launched outward from the horizon ... and that light launched outward EXACTLY at the horizon will never increase its radial position by so much as a millimeter.*" Taylor and Wheeler [88]

"*In the exceptional case of a ∂_v photon parametrizing the positive v axis, $r = 2M$, though it is racing 'outward' at the speed of light the pull of the black hole holds it hovering at rest. ... No particle, whether material or lightlike, can escape from the black hole.*" O'Neill [89]

"*Thus we cannot have direct observational knowledge of the region $r < 2m$. Such a region is called a black hole, because things can fall into it (taking an infinite time, by our clocks, to do so) but nothing can come out.*" Dirac [90]

"*The most obvious asymmetry is that the surface $r = 2m$ acts as a one-way membrane, letting future-directed timelike and null curves cross only from the outside ($r > 2m$) to the inside ($r < 2m$).*" Hawking and Ellis [91]

"*It turned out that, at least in principle, a space traveller could go all the way in such a 'thing' but never return. Not even light could emerge out of the central region of these solutions. It was John Archibald Wheeler who dubbed these strange objects 'black holes'.*" 't Hooft [63]

9.6. THE MATHEMATICAL THEORY OF BLACK HOLES

Escape speed however means that things can either leave or escape from some other body, depending upon initial speed at the place of departure. It does not mean that nothing can leave. To escape from some body, the escapee must achieve the escape speed. If it fails to do so it can leave, but not escape, unless its initial speed is precisely 0 m/s, in which case it neither leaves nor escapes, because its does not move. If it achieves the escape speed it can leave and escape. Escape speed does not mean that nothing can leave. The black hole event horizon has an escape speed, the speed of light c, yet nothing, not even light, can leave (light hovers forever at the event horizon as it tries to 'escape'). As the foregoing citations attest, cosmologists assert that the black hole event horizon has the unique property of having and not having an escape speed simultaneously at the same place. However, no material body can have and not have an escape speed simultaneously, anywhere.

> "A black hole is, ah, a massive object, and it's something which is so massive that light can't even escape. ...some objects are so massive that the escape speed is basically the speed of light and therefore not even light escapes. ...so black holes themselves are, are basically inert, massive and nothing escapes." Bland-Hawthorn [92]

If the escape speed at the event horizon of a black hole is the speed of light, and light travels at the speed of light, then, by the very definition of escape speed, light must escape. Cosmologists however assert the opposite; that the escape speed at the event horizon is the speed of light, so light cannot escape! In fact, light cannot even leave the event horizon, hovering there instead, forever. In other words, the speed of light at the event horizon along a radially outward direction is $c = 0$ m/s and thereby light cannot either leave or escape, because light is not moving. On the other hand, the speed of light at the event horizon, the 'escape' speed, according to the cosmologists, is $c = 2.998 \times 10^8 \text{m/s} = \sqrt{2GM/r_s}$, Einstein's 'speed of light *in vacuo*'. Thus, the speed of light at the black hole event horizon has a split personality; two different values at the same place, *in vacuo*. Furthermore, if the escape speed is zero, any speed greater than zero must ensure leaving and escape. Presumably, no physical object can even achieve the escape speed $c = 0$, because, according to the cosmologists, nothing at all can even leave the event horizon, let alone escape from it. In Relativity Theory the speed of any material body is always restricted to values less than that of $c = 2.998 \times 10^8 \text{m/s}$, not to $c = 0$ m/s. If the escape speed at the event horizon is 0 m/s, this contradicts the escape speed obtained from the 'Schwarzschild radius': $v_{esc} = \sqrt{\frac{2GM}{r_s}} = c = 2.998 \times 10^8 \text{m/s}$ which is > 0. In fact, the 'Schwarzschild radius' is itself obtained by setting $v_{esc} = c = 2.998 \times 10^8 \text{m/s}$ in the Newtonian expression for escape speed. Thus, on the one hand, according to the cosmologists, the escape speed at the event horizon of a black hole is the speed of light $c = 2.998 \times 10^8 \text{m/s}$.

By various mathematical approaches which amount to the same thing, the cosmologists on the other hand claim that the escape speed at the event horizon (the speed of light) is 0m/s. One of their means is to set $\theta = const.$ and $\varphi = const.$ in Hilbert's solution to yield for 'radial motion'. For light

CHAPTER 9. RELATIVISTIC COSMOLOGY AND EINSTEIN'S...

$ds = c\,d\tau = 0$, because the so-called 'proper time' $\tau = 0$[5]. Hence,

$$0 = c^2\left(1 - \frac{2GM}{c^2 r}\right) dt^2 - \left(1 - \frac{2GM}{c^2 r}\right)^{-1} dr^2 \qquad (9.37)$$

Rearrangement for what cosmologists call 'the radial velocity' [93, 94] gives,

$$v = \frac{dr}{dt} = \pm c\left(1 - \frac{2GM}{c^2 r}\right) \qquad (9.38)$$

" *The $+$ sign is for a light ray heading outwards i.e. r increasing with time, and the $-$ is for a light ray heading inwards, i.e. r decreasing with time.*" Rennie [94]

At the event horizon $r = r_s = 2GM/c^2$ (the 'Schwarzschild radius'). Putting this value into Eq.(9.38) yields,

$$v = v_{esc} = \frac{dr}{dt} = \pm c\left(1 - \frac{2GM}{c^2}\frac{c^2}{2GM}\right) = 0 \qquad (9.39)$$

Thus, according to the cosmologists, the speed of light at the event horizon is zero for light travelling either outward or inward.

" *We find that the velocity of light at the event horizon is zero.*" Rennie [94]

This is the other cosmologist 'escape speed' (here the outward radial speed for the $+$ sign) at the black hole event horizon. Consequently, light cannot leave or escape because it is unable to even move. Contrast this with the 'escape speed' at the black hole event horizon obtained from the 'Schwarzschild radius': $v_{esc} = \sqrt{\frac{2GM}{r_s}} = c = 2.998 \times 10^8$ m/s. A body freely falling from rest 'at infinity' along a radial line acquires a speed equal to that of the escape speed, according to the 'Schwarzschild radius' r_s, because r_s is obtained from the Newtonian relation for escape speed. Note that in Eq.(9.39) the cosmologists give the speed of light two different values: the escape speed $c = 2.998 \times 10^8$ m/s by $r_s = 2GM/c^2$ and the escape speed 0 m/s by means of $dr/dt = 0$. The speed of light (the 'escape speed') cannot have two different values in the one equation. This logical absurdity however does not stop cosmologists.

The proof that the Newtonian relation for escape speed is a two-body relation is elementary. According to Newton's theory,

$$F_g = -\frac{GMm}{r^2} = ma = m\frac{dv}{dt} = mv\frac{dv}{dr} \qquad (9.40)$$

where G is the gravitational constant and r is the distance between the centre of mass of m and the centre of mass of M. A centre of mass is not a physical object; merely a mathematical artifice. Although Newton's F_g is singular at

---
[5]The motion of light is 'light-like', or 'null': hence $ds = 0$.

9.6. THE MATHEMATICAL THEORY OF BLACK HOLES

$r = 0$, this does not produce a 'physical singularity'. Separating variables and integrating gives,

$$\int_v^0 mv\, dv = \lim_{r_f \to \infty} \int_R^{r_f} -GMm\frac{dr}{r^2} \tag{9.41}$$

whence,

$$v = \sqrt{\frac{2GM}{R}} \tag{9.42}$$

where R is the radius of the mass M. Thus, although Eq.(9.42) contains only one mass term (M), escape speed necessarily involves two bodies: m and M.

In any event, contrary to cosmology, the Newtonian implicit two-body escape speed relation cannot be involved because the black hole pertains to a universe that contains only one mass (that of the black hole itself) by hypothesis. The impossible duality of cosmology's black hole 'escape velocity' is now clear. The black hole event horizon has an escape speed and no escape speed simultaneously at the same place. But, contrary to cosmology, nothing can have and not have an escape speed simultaneously at the same place. Furthermore, according to Einstein, no material body can move with a speed that is equal to or greater than the speed of light *in vacuo*, i.e. $c = 2.998 \times 10^8$m/s, but can certainly move with a speed v such that $0 < v < c = 2.998 \times 10^8$m/s. Cosmologists, with an escape speed $v_{esc} = c = 0$ do not permit any material object to have a speed greater than zero at their event horizon, contrary to Einstein's fundamental tenet, because, they say, no material body can move at or greater than the speed of light. In other circles this is called 'an each-way bet'.

Since Hilbert's solution is utilised by cosmologists, $0 \leq r$. Therefore, if $0 < r < 2M$ the escape speed from Eq.(9.38) becomes negative and hence is no longer an escape speed. Beneath the event horizon, say cosmologists, the 'escape velocity' is greater than that of light. Cosmologists have an additional and equally bizarre interpretation for this: time-convergence "*inexorably into the classical singularity at $r = 0$*" [51], into the black hole's 'physical singularity', because Hilbert's metric changes its signature and becomes time-dependent (i.e. non-static). Eqs.(9.30) maintain a fixed signature, $(+,-,-,-)$. It is not possible for the signature to change to $(-,+,-,-)$, for instance. Cosmologists admit that when $0 < r < 2m$ in Hilbert's Eq.(9.24), the roles of t and r are interchanged. This produces a non-static solution to a static problem, i.e. a solution that is time-dependent for a problem that is time-independent. To further illustrate this violation, when $2m < r$ the signature of Eq.(9.24) is $(+,-,-,-)$; but if $0 < r < 2m$ in Eq.(9.24), then

$$g_{oo} = \left(1 - \frac{2M}{r}\right) \text{ is negative, and } g_{11} = -\left(1 - \frac{2M}{r}\right)^{-1} \text{ is positive.} \tag{9.43}$$

So the signature of Eq.(9.24) changes from $(+,-,-,-)$ to $(-,+,-,-)$. Thus the roles of t and r are exchanged. According to Misner, Thorne and Wheeler, who use the spacetime signature $(-,+,+,+)$ instead of $(+,-,-,-)$,

CHAPTER 9. RELATIVISTIC COSMOLOGY AND EINSTEIN'S...

> "The most obvious pathology at $r = 2M$ is the reversal there of the roles of t and r as timelike and spacelike coordinates. In the region $r > 2M$, the t direction, $\partial/\partial t$, is timelike ($g_{tt} < 0$) and the r direction, $\partial/\partial r$, is spacelike ($g_{rr} > 0$); but in the region $r < 2M$, $\partial/\partial t$, is spacelike ($g_{tt} > 0$) and $\partial/\partial r$, is timelike ($g_{rr} < 0$).
>
> "What does it mean for r to 'change in character from a spacelike coordinate to a timelike one'? The explorer in his jet-powered spaceship prior to arrival at $r = 2M$ always has the option to turn on his jets and change his motion from decreasing r (infall) to increasing r (escape). Quite the contrary in the situation when he has once allowed himself to fall inside $r = 2M$. Then the further decrease of r represents the passage of time. No command that the traveler can give to his jet engine will turn back time. That unseen power of the world which drags everyone forward willy-nilly from age twenty to forty and from forty to eighty also drags the rocket in from time coordinate $r = 2M$ to the later time coordinate $r = 0$. No human act of will, no engine, no rocket, no force (see exercise 31.3) can make time stand still. As surely as cells die, as surely as the traveler's watch ticks away 'the unforgiving minutes', with equal certainty, and with never one halt along the way, r drops from $2M$ to 0.
>
> "At $r = 2M$, where r and t exchange roles as space and time coordinates, g_{tt} vanishes while g_{rr} is infinite." Misner, Thorne and Wheeler [61]

Note that at $r = 2M$, $g_{rr} = (1 - 2M/r)^{-1}$ is not in fact infinite. At $r = 2M$, $g_{rr} = 1/0$, which is undefined. Similarly, if $r = 0$, $2M/r = 2M/0$ which is undefined. Contrary to the cosmologists, division by zero does not produce 'infinity', it is actually undefined, and infinity is not even a number[6].

> "There is no alternative to the matter collapsing to an infinite density at a singularity once a point of no-return is passed. The reason is that once the event horizon is passed, all time-like trajectories must necessarily get to the singularity: 'all the King's horses and all the King's men' cannot prevent it." Chandrasekhar [45]

> "This is worth stressing; not only can you not escape back to region I, you cannot even stop yourself from moving in the direction of decreasing r, since this is simply the timelike direction. (This could have been seen in our original coordinate system; for $r < 2GM$, t becomes spacelike and r becomes timelike.) Thus you can no more stop moving toward the singularity than you can stop getting older." Carroll [60]

> "For $r < 2GM/c^2$, however, the component g_{oo} becomes negative, and g_{rr}, positive, so that in this domain, the role of time-like coordinate is played by r, whereas that of space-like coordinate by t.

---

[6]Cantor's theory of 'transfinite numbers' has no relevance here either.

9.6. THE MATHEMATICAL THEORY OF BLACK HOLES

Thus in this domain, the gravitational field depends significantly on time (r) and does not depend on the coordinate t." Vladmimirov, Mitskiévich and Horský [95]

To amplify this, set $t = r^*$ and $r = t^*$. Then for $0 < r < 2M$, Eq.(9.24) becomes,

$$ds^2 = \left(1 - \frac{2M}{t^*}\right) dr^{*2} - \left(1 - \frac{2M}{t^*}\right)^{-1} dt^{*2} - t^2 \left(d\theta^2 + \sin^2\theta \, d\varphi^2\right) \quad (9.44)$$

$$0 \leq t^* < 2M$$

It now becomes quite clear that this is a time-dependent metric since all the components of the metric tensor are functions of the timelike t^*. Therefore this metric bears no relationship to the original time-independent problem to be solved. In other words, this metric is a non-static solution to a static problem (see also Brillouin [96]).

Infinite densities

The 'infinite density' of the black hole's 'physical singularity' produced by irresistible 'gravitational collapse' violates Special Relativity. The singularity of big bang cosmology is also infinitely dense. Yet according to Special Relativity, infinite densities are forbidden because their existence implies that a material object can acquire the speed of light c in vacuo i.e. 2.998×10^8m/s (or equivalently, the existence of infinite kinetic energy), thereby violating the very basis of Special Relativity.

"Eventually when a star has shrunk to a certain critical radius, the gravitational field at the surface becomes so strong that the light cones are bent inward so much that the light can no longer escape. According to the theory of relativity, nothing can travel faster than light. Thus, if light cannot escape, neither can anything else. Everything is dragged back by the gravitational field. So one has a set of events, a region of space-time from which it is not possible to escape to reach a distant observer. This region is what we now call a black hole. Its boundary is called the event horizon. It coincides with the paths of the light rays that just fail to escape from the black hole." Hawking [97]

Since General Relativity cannot violate Special Relativity, General Relativity must therefore also forbid infinite densities. Therefore, point-mass singularities are forbidden by the Theory of Relativity. Let a cuboid rest-mass m_o have sides of length L_o. Let m_o have a relative speed $v < c$ in the direction of one of three mutually orthogonal Cartesian axes attached to an observer of rest-mass M_o. According to Einstein [98] the observer M_o reckons the moving mass m is,

$$m = \frac{m_o}{\sqrt{1 - \frac{v^2}{c^2}}} \quad (9.45)$$

and the volume is,
$$V = L_o^2 \sqrt{1 - \frac{v^2}{c^2}}. \tag{9.46}$$
The density of m according to M_o is therefore,
$$D = \frac{m}{V} = \frac{m_o}{L_o^2 \left(1 - \frac{v^2}{c^2}\right)}. \tag{9.47}$$
Hence, $v \to c \Rightarrow D \to \infty$. Since, according to Special Relativity, no material object can acquire the speed c, infinite densities are forbidden by Special Relativity, and so point-mass singularities and circumference-mass singularities are forbidden. Since General Relativity cannot repudiate Special Relativity, it too must thereby forbid infinite densities and hence forbid point-mass singularities and circumference-mass singularities. It does not matter how it is alleged that a 'physical singularity' is generated by General Relativity because the infinitely dense physical singularity cannot be reconciled with Special Relativity. Point-charges and circumference-charges too are therefore forbidden by the Theory of Relativity since there can be no charge without mass.

Curvature invariants

The squared differential element of arc of a curve in a surface is given by the First Fundamental Quadratic Form for a surface,
$$ds^2 = E\,du^2 + 2F\,du\,dv + G\,dv^2 \tag{9.48}$$
wherein u and v are curvilinear coordinates. If either u or v is constant the resulting line-elements are called parametric curves in the surface. The differential element of surface area is given by,
$$dA = \sqrt{EG - F^2}\,du\,dv \tag{9.49}$$
An expression which depends only on E, F, G, and their first and second derivatives, is called *a bending invariant*. It is an intrinsic (or absolute) property of a surface. The Gaussian (or Total) curvature of a surface is an important intrinsic property of a surface.

The 'Theorema Egregium' of Gauss: *The Gaussian curvature K at any point P of a surface depends only on the values at P of the coefficients in the First Fundamental Quadratic Form and their first and second derivatives.*

Hence, the Gaussian curvature of a surface is a bending invariant.

The Euclidean plane has a constant Gaussian curvature of $K = 0$. A surface of positive constant Gaussian curvature is called *a spherical surface*. A surface of constant negative curvature is called *a pseudo-spherical surface*.

Being an intrinsic geometric property of a surface, Gaussian curvature is independent of any embedding space.

> *"And in any case, if the metric form of a surface is known for a certain system of intrinsic coordinates, then all the results concerning the intrinsic geometry of this surface can be obtained without appealing to the embedding space."* Efimov [99]

9.6. THE MATHEMATICAL THEORY OF BLACK HOLES

All black hole spacetime metrics contain a surface from which various invariants and geometric identities can be deduced by purely mathematical means. Such identities are independent of the area of the surface and of the length of any curve in the surface. The Kerr-Newman form subsumes the Kerr, Reissner-Nordström, and Schwarzschild forms. The Gaussian curvature of the surface in the Kerr-Newman metric therefore subsumes the Gaussian curvatures of the surfaces in the subordinate forms to which it can be reduced. Gaussian curvature reveals the type of surface and uniquely identifies the terms that appear in its general form. Gaussian curvature reveals that no purported black hole metric can in fact be extended to produce the black hole it is said to contain.

The Gaussian curvature K of a surface can be calculated by means of the following relation,

$$K = \frac{R_{1212}}{g} \qquad (9.50)$$

where R_{1212} is a component of the Riemann-Christoffel curvature tensor of the first kind and g is the determinant of the metric tensor. Note that neither the area of the surface nor the length of any curve in the surface is involved.

If $r = const. \neq 0$ and $t = const.$, Eq.(9.35) reduces to the surface [43],

$$ds^2 = \rho^2 d\theta^2 + \frac{(R_c^2 + a^2) - a^2 \Delta \sin^2 \theta}{\rho^2} \sin^2 \theta \, d\varphi^2 \qquad (9.51)$$

where

$$\rho^2 = R_c^2 + a^2 \cos^2 \theta, \quad \Delta = R_c^2 - \alpha R_c + a^2 + q^2, \quad R_c = (|r - r_o|^n + \zeta^n)^{\frac{1}{n}},$$

$$r, r_o \in \Re, \quad n \in \Re^+$$

$$\zeta = \frac{\alpha + \sqrt{\alpha^2 - 4q^2 - 4a^2 \cos^2 \theta}}{2}, \quad a^2 + q^2 < \frac{\alpha^2}{4}$$

$$(9.52)$$

The Gaussian curvature K of this surface is given by [41],

$$K = \frac{1}{2hf} \frac{\partial \beta}{\partial \theta} \frac{\partial h}{\partial \theta} - \frac{a^2 \cos^2 \theta}{h^2} - \frac{1}{2f} \frac{\partial^2 \beta}{\partial \theta^2} + \frac{1}{h} + \frac{a^2 \sin \theta \cos \theta}{2hf} \frac{\partial \beta}{\partial \theta} + \frac{h}{4f^2} \left(\frac{\partial \beta}{\partial \theta}\right)^2$$

$$+ \frac{2a^2 (f - \Delta h) \cos^2 \theta}{h^2 f} \qquad (9.53)$$

where

$$f = (R_c^2 + a^2)^2 - a^2 \Delta \sin^2 \theta, \quad h = R_c^2 + a^2 \cos^2 \theta, \quad \beta = \frac{f}{h}$$

$$\Delta = R_c^2 - \alpha R_c + a^2 + q^2, \quad R_c = (|r - r_o|^n + \zeta^n)^{\frac{1}{n}}, \quad r, r_o \in \Re, \quad n \in \Re^+$$

$$\zeta = \frac{\alpha + \sqrt{\alpha^2 - 4q^2 - 4a^2 \cos^2 \theta}}{2}, \quad a^2 + q^2 < \frac{\alpha^2}{4}$$

$$(9.54)$$

It is clearly evident from this that the Gaussian curvature is not a positive constant and so the surface is not a spherical surface. Thus, the Kerr-Newman metric is not spherically symmetric. Furthermore, by virtue of this result, the quantity R_c is neither the radius nor even a distance because it is defined by the intrinsic geometry of the surface. Since the intrinsic geometry of a surface is independent of any embedding space the quantity R_c retains its identity when the surface is embedded in Kerr-Newman spacetime. If $r_o = \zeta$ and $n = 1$, then $R_c = r$. Hence, r is not the radius of anything nor even a distance in Kerr-Newman spacetime. This result is independent of the area of the surface or the length of any curve in the surface.

The Gaussian curvature for the Kerr-Newman surface is dependent on θ, because it is axially-symmetric. When $\theta = 0$ and $\theta = \pi$, it becomes [43],

$$K = \frac{R_c^2}{(R_c^2 + a^2)^2} - \frac{a^2(\alpha R_c - q^2)}{(R_c^2 + a^2)^3} \qquad (9.55)$$

Since $R_c(r_o) = \zeta \, \forall r_o \, \forall n$, for $\theta = 0$ and $\theta = \pi$ the Guassian curvature becomes [41]

$$K_{r_o} = \frac{\zeta^2}{(\zeta^2 + a^2)^2} - \frac{a^2(\alpha\zeta - q^2)}{(\zeta^2 + a^2)^3} \qquad (9.56)$$

Similarly, when $\theta = \pi/2$, the Gaussian curvature becomes [41],

$$K = \frac{1}{R_c^2} + \frac{a^2(R_c^2 + a^2)(\alpha R_c - q^2)}{R_c^4[R_c^2(R_c^2 + a^2) + a^2(\alpha R_c - q^2)]} \qquad (9.57)$$

Since $R_c(r_o) = \zeta \, \forall r_o \, \forall n$, for $\theta = \pi/2$ the Guassian curvature becomes [41],

$$K_{r_o} = \frac{1}{\zeta^2} + \frac{a^2(\zeta^2 + a^2)(\alpha\zeta - q^2)}{\zeta^4[\zeta^2(\zeta^2 + a^2) + a^2(\alpha\zeta - q^2)]} \qquad (9.58)$$

If $a = 0$ then the Gaussian curvature is independent of θ and reduces to the spherically-symmeytric Reissner-Nordström form [41],

$$K = \frac{1}{R_c^2} \qquad (9.59)$$

and hence, when $r = r_o$ [41],

$$K_{r_o} = \frac{1}{\zeta^2} = \frac{1}{\left[\frac{\alpha}{2} + \sqrt{\frac{\alpha^2}{4} - q^2}\right]^2} \qquad (9.60)$$

where ζ is reduced accordingly. This is an invariant for the Reissner-Nordström form.

If both $a = 0$ and $q = 0$ then the Gaussian curvature reduces to the spherically-symmetric Schwarzschild form [43],

$$K = \frac{1}{R_c^2} \qquad (9.61)$$

9.6. THE MATHEMATICAL THEORY OF BLACK HOLES

so that when $r = r_o$,

$$K_{r_o} = \frac{1}{\alpha^2} \tag{9.62}$$

which is an invariant for the Schwarzschild form.

The minimum value for Δ is,

$$\Delta_{min} = a^2 \sin^2 \theta \tag{9.63}$$

which occurs when $r = r_o$, irrespective of the values of r_o and n. $\Delta_{min} = 0$ only when $\theta = 0$ and when $\theta = \pi$, in which cases the metric is undefined.

Similarly, the minimum value of R_c^2 is,

$$R_c^2 = \zeta^2 + a^2 \cos^2 \theta \tag{9.64}$$

which occurs when $r = r_o$, irrespective of the values of r_o and n. Since ζ^2 is always greater than zero, R_c^2 can never be less than or equal to zero.

Note that if $a = 0$ and $q = 0$, the Gaussian curvature for the surface embedded in Kerr-Newman spacetime reduces to that for the surface in the Schwarzschild metric ground-form [43],

$$K = \frac{1}{R_c^2} \tag{9.65}$$

Because $R_c(r_o) = \alpha \, \forall r_o \, \forall n$,

$$K_{r_o} = \frac{1}{R_c^2(r_o)} = \frac{1}{\alpha^2} \tag{9.66}$$

If, further, $r_o = \alpha$ and $n = 1$, then $R_c = r$ and,

$$K = \frac{1}{r^2} \tag{9.67}$$

Hence, r in Hilbert's solution is not the radius of anything, or even a distance therein. Once again, this result is independent of the area of the surface or the length of any curve in the surface. Indeed, the length of a curve in the surface and the area of the surface are determined by the metric and r. The length L of a closed geodesic (a closed parametric curve where $r = const. \neq 0$, $\theta = \pi/2$) in the surface embedded in Hilbert's metric space is given by,

$$L = \int_0^{2\pi} r \, d\varphi = 2\pi r \tag{9.68}$$

Applying the relation for the area A of a surface, the area of the surface embedded in Hilbert's spacetime is,

$$A = r^2 \int_0^\pi \sin \theta \, d\theta \int_0^{2\pi} d\varphi = 4\pi r^2 \tag{9.69}$$

Since this is a surface, r is not the radius of anything, nor is it even a distance in the surface. The geometric identity of r is not lost when the surface is embedded in any other space because the Gaussian curvature of a surface is intrinsic. It is now clear why the cosmologist notions of 'areal radius' ($r = \sqrt{A/4\pi}$) and 'reduced circumference' ($r = L/2\pi$) are vacuous. Neither the length of any curve in a surface nor the area of the surface or part thereof determines the geometric identity of r in Hilbert's metric.

The impossibility of a black hole is reaffirmed by Riemannian curvature. Riemannian curvature is a generalisation of Gaussian curvature to dimensions greater than two. The Riemannian curvature K_S at any point in a metric space of dimensions $n > 2$ depends upon the Riemann-Christoffel curvature tensor of the first kind, R_{ijkl}, the components of the metric tensor g_{ik}, and two arbitrary n-dimensional linearly independent contravariant direction vectors U^i and V^i, as follows:

$$K_S = \frac{R_{ijkl} U^i V^j U^k V^l}{G_{pqrs} U^p V^q U^r V^s}, \quad G_{pqrs} = g_{pr} g_{qs} - g_{ps} g_{qr} \tag{9.70}$$

Definition 1: *If the Riemannian curvature at any point is independent of direction vectors at that point then the point is called an isotropic point.*

The Riemannian curvature K_S for Schwarzschild spacetime is given by [41],

$$K_S = \frac{A}{B}$$

$A = 2\alpha \left(R_c - \alpha\right) W_{0101} - \alpha R_c \left(R_c - \alpha\right)^2 W_{0202} - \alpha R_c \left(R_c - \alpha\right)^2 W_{0303} \sin^2 \theta$
$+ \alpha R_c^3 W_{1212} + \alpha R_c^3 W_{1313} \sin^2 \theta - 2\alpha R_c^4 \left(R_c - \alpha\right) W_{2323} \sin^2 \theta$

$B = -2R_c^3 \left(R_c - \alpha\right) W_{0101} - 2R_c^4 \left(R_c - \alpha\right)^2 W_{0202} - 2R_c^4 \left(R_c - \alpha\right)^2 W_{0303} \sin^2 \theta$
$+ 2R_c^6 W_{1212} + 2R_c^6 W_{1313} \sin^2 \theta + 2R_c^6 \left(R_c - \alpha\right) W_{2323} \sin^2 \theta$

$$W_{ijki} = \begin{vmatrix} U^i & U^j \\ V^i & V^j \end{vmatrix} \begin{vmatrix} U^k & U^l \\ V^k & V^l \end{vmatrix}, \quad R_c = \left(|r - r_o|^n + \alpha^n\right)^{\frac{1}{n}},$$

$$r, r_o \in \Re, \quad n \in \Re^+, \tag{9.71}$$

$$r = \sqrt{x_o^2 + y_o^2 + z_o^2} + \sqrt{(x - x_o)^2 + (y - y_o)^2 + (z - z_o)^2}$$

Since $R_c(r_o) = \alpha$ irrespective of the values of r_o and n, at $r = r_o$ the Riemannian curvature is,

$$K_S(r_o) = \frac{1}{2\alpha^2} = \frac{K_{r_o}}{2}$$

which is entirely independent of the direction vectors U^i and V^j, and of θ. Thus, $r = r_o$ produces an isotropic point (the only isotropic point), which reaffirms that Schwarzschild spacetime cannot be extended to produce a black hole. Note that $K_S(r_o) = K_{r_o}/2$, i.e. at $r = r_o$ the Riemannian curvature invariant of Schwarzschild 4-dimensional spacetime is half the Gaussian curvature invariant of the embedded spherical surface.

9.6. THE MATHEMATICAL THEORY OF BLACK HOLES

Similarly, the Riemannian curvature of Schwarzschild spacetime in isotropic coordinates is [43],

$$K_S = \frac{A+B}{C+D}$$

$$A = \frac{16\alpha(4R_c-\alpha)^2}{R_c(4R_c+\alpha)^4} W_{0101} - \frac{8\alpha R_c(4R_c-\alpha)^2}{(4R_c+\alpha)^4} W_{0202} - \frac{8\alpha R_c(4R_c-\alpha)^2 \sin^2\theta}{(4R_c+\alpha)^4} W_{0303}$$

$$B = \frac{\alpha(8R_c-\alpha)(4R_c+\alpha)^2}{2 \cdot 4^3 R_c^4} W_{1212} + \frac{\alpha(4R_c+\alpha)^2 \sin^2\theta}{2 \cdot 4^2 R_c^3} W_{1313} - \frac{\alpha(4R_c+\alpha)^2 \sin^2\theta}{4^2 R_c} W_{2323}$$

$$C = -\frac{\alpha(4R_c-\alpha)^2(4R_c+\alpha)^2}{2 \cdot 4^4 R_c^4} W_{0101} - \frac{\alpha(4R_c-\alpha)^2(4R_c+\alpha)^2}{2 \cdot 4^4 R_c^2} W_{0202}$$
$$- \frac{\alpha(4R_c-\alpha)^2(4R_c+\alpha)^2 \sin^2\theta}{2 \cdot 4^4 R_c^2} W_{0303}$$

$$D = \frac{(4R_c+\alpha)^8}{4^8 R_c^6} W_{1212} + \frac{(4R_c+\alpha)^8 \sin^2\theta}{4^8 R_c^6} W_{1313} + \frac{(4R_c+\alpha)^8 \sin^2\theta}{4^8 R_c^4} W_{2323}$$

$$W_{ijkl} = \begin{vmatrix} U^i & U^j \\ V^i & V^j \end{vmatrix} \begin{vmatrix} U^k & U^l \\ V^k & V^l \end{vmatrix}, \quad R_c = \left[|r-r_o|^n + \left(\frac{\alpha}{4}\right)^n\right]^{\frac{1}{n}},$$

$$r, r_o \in \Re, \quad n \in \Re^+,$$

$$r = \sqrt{x_o^2 + y_o^2 + z_o^2} + \sqrt{(x-x_o)^2 + (y-y_o)^2 + (z-z_o)^2}$$
(9.72)

This isotropic Riemannian curvature depends upon θ.

When $R_c(r_o) = \alpha/4 \,\forall r_o \,\forall n$, so the Riemannian curvature becomes,

$$K_S(r_o) = \frac{8(W_{1212} + W_{1313}\sin^2\theta) - \alpha^2 W_{2323}\sin^2\theta}{16\alpha^2(W_{1212} + W_{13123}\sin^2\theta) + \alpha^2 W_{2323}\sin^2\theta}$$
(9.73)

Note that this differs from that for the ordinary Schwarzschild equivalence class only by the terms in W_{2323} (i.e. if not for the W_{2323} terms the Riemannian curvature $K_S(r_o)$ would be $1/2\alpha^2$ as for the ordinary Schwarzschild form). For $\theta = 0$ and $\theta = \pi$ it reduces to the Riemannian curvature invariant for the Schwarzschild form:

$$K_S(r_o) = \frac{1}{2\alpha^2}$$
(9.74)

Hence, for $\theta = 0$ and $\theta = \pi$, r_o is an isotropic point (the only isotropic point). The W_{2323} terms appear due to the conformal mapping of ordinary Schwarzschild equivalence class into isotropic Schwarzschild equivalence class [43].

When $\theta = \pi/2$ and $r = r_o$ the Riemannian curvature is,

$$K_S(r_o) = \frac{8(W_{1212} + W_{1313}) - \alpha^2 W_{2323}}{16\alpha^2(W_{1212} + W_{13123}) + \alpha^2 W_{2323}}$$
(9.75)

CHAPTER 9. RELATIVISTIC COSMOLOGY AND EINSTEIN'S...

The Riemannian curvature for the Reissner-Nordström equivalence class is [41],

$$K_S = \frac{A+B+C}{D+E+F}$$

$$A = 2\left(R_c^2 - \alpha R_c + q^2\right)\left(\alpha R_c - 3q^2\right) W_{0101} - \left(R_c^2 - \alpha R_c + q^2\right)^2 \left(\alpha R_c - 2q^2\right) W_{0202}$$

$$B = -\left(R_c^2 - \alpha R_c + q^2\right)^2 \left(\alpha R_c - 2q^2\right) \sin^2\theta\, W_{0303} + R_c^4 \left(\alpha R_c - 2q^2\right) W_{1212}$$

$$C = R_c^4 \left(\alpha R_c - 2q^2\right) \sin^2\theta\, W_{1313} - 2R_c^4 \left(\alpha R_c - 2q^2\right)\left(R_c^2 - \alpha R_c + q^2\right) \sin^2\theta\, W_{2323}$$

$$D = -2R_c^4 \left(R_c^2 - \alpha R_c + q^2\right) W_{0101} - 2R_c^4 \left(R_c^2 - \alpha R_c + q^2\right)^2 W_{0202}$$

$$E = -2R_c^4 \left(R_c^2 - \alpha R_c + q^2\right)^2 \sin^2\theta\, W_{0303} + 2R_c^8 W_{1212}$$

$$F = 2R_c^8 \sin^2\theta\, W_{1313} + 2R_c^8 \left(R_c^2 - \alpha R_c + q^2\right) \sin^2\theta\, W_{2323}$$

$$W_{ijkl} = \begin{vmatrix} U^i & U^j \\ V^i & V^j \end{vmatrix} \begin{vmatrix} U^k & U^l \\ V^k & V^l \end{vmatrix}, \quad R_c = \left(|r - r_o|^n + \zeta^n\right)^{\frac{1}{n}}, \quad \zeta = \frac{\alpha + \sqrt{\alpha^2 - 4q^2}}{2},$$

$$q^2 < \frac{\alpha^2}{4}, \quad r, r_o \in \Re, \quad n \in \Re^+,$$

$$r = \sqrt{x_o^2 + y_o^2 + z_o^2} + \sqrt{(x - x_o)^2 + (y - y_o)^2 + (z - z_o)^2}$$

(9.76)

At $r = r_o$ this becomes,

$$K_S(r_o) = \frac{\alpha\zeta - 2q^2}{2\zeta^4} = \frac{4\left(\alpha^2 + \alpha\sqrt{\alpha^2 - 4q^2} - 4q^2\right)}{\left(\alpha + \sqrt{\alpha^2 - 4q^2}\right)^4} \tag{9.77}$$

which is independent of the direction vectors U^i and V^j. Therefore r_o is an isotropic point (the only isotropic point). This reaffirms that the Reissner-Nordström equivalence class cannot be extended to produce a black hole.

The Riemannian curvature for Reissner-Nordström equivalence class in isotropic coordinates is,

$$K_S = \frac{R_{0101}W_{0101} + R_{0202}\left(W_{0202} + W_{0303}\sin^2\theta\right) + R_{1212}\left(W_{1212} + W_{1313}\sin^2\theta\right) + R_{2323}W_{2323}}{G_{0101}W_{0101} + G_{0202}\left(W_{0202} + W_{0303}\sin^2\theta\right) + G_{1212}\left(W_{1212} + W_{1313}\sin^2\theta\right) + G_{2323}W_{2323}}$$

$$R_{0101} = \frac{8\left(16R_c^2 - \alpha^2 + 4q^2\right)\left(32q^2 - 32\alpha R_c - 8\alpha^2\right) + 4^4 R_c Y}{(4R_c + \alpha + 2q)^3 (4R_c - \alpha + 2q)^3} + \frac{32\left(16R_c^2 - \alpha^2 + 4q^2\right) Y}{(4R_c + \alpha + 2q)^3 (4R_c - \alpha + 2q)^4} +$$

$$+ \frac{32\left(16R_c^2 - \alpha^2 + 4q^2\right) Y}{(4R_c + \alpha + 2q)^4 (4R_c - \alpha + 2q)^3} + \frac{16\left(16R_c^2 - \alpha^2 + 4q^2\right)\left(4q^2 - 4\alpha R_c - \alpha^2\right) Y}{R_c (4R_c + \alpha + 2q)^4 (4R_c - \alpha + 2q)^4} -$$

$$- \frac{64 Y^2}{(4R_c + \alpha + 2q)^4 (4R_c - \alpha + 2q)^4}$$

253

9.6. THE MATHEMATICAL THEORY OF BLACK HOLES

$$Y = \left\{4R_c\left[(4R_c + \alpha)^2 - 4q^2\right] - (4R_c + \alpha)\left(16R_c^2 - \alpha^2 + 4q^2\right)\right\}$$

$$R_{0202} = -\frac{8R_c\left(16R_c^2 - \alpha^2 + 4q^2\right)^2 Y}{(4R_c + \alpha + 2q)^4(4R_c - \alpha + 2q)^4} \qquad R_{0303} = R_{0202}\sin^2\theta$$

$$R_{1212} = \frac{Y}{32R_c^3} \qquad R_{1313} = R_{1212}\sin^2\theta \qquad R_{2323} = -\frac{(Y + 16q^2 R_c)}{4^2 R_c}\sin^2\theta$$

$$G_{0101} = -\frac{(16R_c^2 - \alpha^2 + 4q^2)^2}{4^4 R_c^4} \qquad G_{0202} = -\frac{(16R_c^2 - \alpha^2 + 4q^2)^2}{4^4 R_c^2} \qquad G_{0303} = G_{0202}\sin^2\theta$$

$$G_{1212} = \frac{(4R_c + \alpha + 2q)^4(4R_c - \alpha + 2q)^4}{4^8 R_c^6} \qquad G_{1313} = G_{1212}\sin^2\theta$$

$$G_{2323} = \frac{(4R_c + \alpha + 2q)^4(4R_c - \alpha + 2q)^4 \sin^2\theta}{4^8 R_c^4}$$

$$W_{ijkl} = \begin{vmatrix} U^i & U^j \\ V^i & V^j \end{vmatrix}\begin{vmatrix} U^k & U^l \\ V^k & V^l \end{vmatrix}, \qquad R_c = (|r - r_o|^n + \zeta^n)^{\frac{1}{n}}, \qquad \zeta = \frac{\sqrt{\alpha^2 - 4q^2}}{4},$$

$$q^2 < \frac{\alpha^2}{4}, \qquad r, r_o \in \Re, \qquad n \in \Re^+, \qquad r = \sqrt{x_o^2 + y_o^2 + z_o^2} + \sqrt{(x - x_o)^2 + (y - y_o)^2 + (z - z_o)^2} \tag{9.78}$$

which depends upon θ.

Since $R_c(r_o) = \zeta \; \forall r_o \; \forall n$ it then reduces to,

$$K_S(r_o) = \frac{\frac{4\left(\alpha^2 - 4q^2 + \alpha\sqrt{\alpha^2 - 4q^2}\right)}{(\alpha^2 - 4q^2)}(W_{1212} + W_{1313}\sin^2\theta) - \frac{\left(\sqrt{\alpha^2 - 4q^2} + \alpha\right)^2}{4}W_{2323}\sin^2\theta}{\frac{\left[\left(\sqrt{\alpha^2 - 4q^2} + \alpha\right)^2 - 4q^2\right]^4}{4^2(\alpha^2 - 4q^2)^3}(W_{1212} + W_{1313}\sin^2\theta) + \frac{\left[\left(\sqrt{\alpha^2 - 4q^2} + \alpha\right)^2 - 4q^2\right]^4}{4^4(\alpha^2 - 4q^2)^2}W_{2323}\sin^2\theta} \tag{9.79}$$

If $q = 0$ this reduces to the Riemannian curvature invariant for the isotropic Schwarzschild equivalence class.

When $\theta = 0$ and $\theta = \pi$, the Riemannian curvature is,

$$K_S = \frac{R_{0101}W_{0101} + R_{0202}W_{0202} + R_{1212}W_{1212}}{G_{0101}W_{0101} + G_{0202}W_{0202} + G_{1212}W_{1212}} \tag{9.80}$$

and hence if also $r = r_o$ this reduces further to the Riemannian curvature invariant,

$$K_S = \frac{R_{1212}}{G_{1212}} = \frac{4\left(\alpha^2 - 4q^2\right)^2}{\left(\alpha^2 - 4q^2 + \alpha\sqrt{\alpha^2 - 4q^2}\right)^3} = \frac{4\left(\alpha^2 + \alpha\sqrt{\alpha^2 - 4q^2} - 4q^2\right)}{\left(\alpha + \sqrt{\alpha^2 - 4q^2}\right)^4} \tag{9.81}$$

which is the same as the isotropic Riemannian curvature invariant for the ordinary Reissner-Nordström equivalence class; and r_o is an isotropic point (the only isotropic point). This reaffirms that Reissner-Nordström spacetime cannot be extended to produce a black hole.

Then if $q = 0$ Eq.(9.81) reduces finally to the Riemannian curvature invariant for the isotropic Schwarzschild equivalence class,

$$K_S = \frac{1}{2\alpha^2} \tag{9.82}$$

CHAPTER 9. RELATIVISTIC COSMOLOGY AND EINSTEIN'S...

When $\theta = \pi/2$, the Riemannian curvature for the Reissner-Nordström equivalence class accordingly becomes,

$$K_S = \frac{R_{0101}W_{0101} + R_{0202}(W_{0202} + W_{0303}) + R_{1212}(W_{1212} + W_{1313}) + R_{2323}W_{2323}}{G_{0101}W_{0101} + G_{0202}(W_{0202} + W_{0303}) + G_{1212}(W_{1212} + W_{1313}) + G_{2323}W_{2323}} \tag{9.83}$$

The Kretschmann scalar f is also called the Riemann tensor scalar curvature invariant. It is defined by $f = R_{\mu\nu\rho\sigma}R^{\mu\nu\rho\sigma}$. Cosmologists incorrectly assert that their 'physical singularity' must occur where the Kretschmann scalar is 'infinite'. For the Schwarzschild equivalence class it is actually given by [43],

$$f = \frac{12\alpha^2}{R_c^6} = \frac{12\alpha^2}{(|r - r_o|^n + \alpha^n)^{\frac{6}{n}}} \tag{9.84}$$

$$R_c = (|r - r_o|^n + \alpha^n)^{\frac{1}{n}}$$

Hence, at $r = r_o$,

$$f(r_o) = \frac{12}{\alpha^4} \tag{9.85}$$

In the case of the Reissner-Nordström equivalence class it is given by [43],

$$f = \frac{8\left[6\left(\frac{\alpha R_c}{2} - q^2\right)^2 + q^4\right]}{R_c^8} \tag{9.86}$$

$$R_c = (|r - r_o|^n + \zeta^n)^{\frac{1}{n}}, \quad \zeta = \frac{\alpha + \sqrt{\alpha^2 - 4q^2}}{2}$$

Hence, at $r = r_o$,

$$f(r_o) = \frac{8\left[6\left(\frac{\alpha\zeta}{2} - q^2\right)^2 + q^4\right]}{\zeta^8} = \frac{8\left\{6\left[\frac{\alpha\left(\frac{\alpha+\sqrt{\alpha^2-4q^2}}{2}\right)}{2} - q^2\right]^2 + q^4\right\}}{\left(\frac{\alpha+\sqrt{\alpha^2-4q^2}}{2}\right)^8} \tag{9.87}$$

For the Kerr-Newman equivalence class the Kretschmann scalar is given by [43],

$$f = \frac{8}{(R_c^2 + a^2\cos^2\theta)^6} \left[\begin{array}{l} \frac{3\alpha^2}{2}\left(R_c^6 - 15a^2 R_c^4 \cos^2\theta + 15a^4 R_c^2 \cos^4\theta - a^6\cos^6\theta\right) - \\ -6\alpha q^2 R_c \left(R_c^4 - 10a^2 R_c^2 \cos^2\theta + 5a^4\cos^4\theta\right) + \\ +q^4\left(6R_c^4 - 34a^2 R_c^2 \cos^2\theta + 6a^4\cos^4\theta\right) \end{array} \right]$$

$$R_c = (|r - r_o|^n + \zeta^n)^{\frac{1}{n}}, \quad r, r_o \in \Re, \quad n \in \Re^+$$

$$\zeta = \frac{\alpha + \sqrt{\alpha^2 - 4(q^2 + a^2\cos^2\theta)}}{2}, \quad a^2 + q^2 < \frac{\alpha^2}{4},$$

$$r = \sqrt{x_o^2 + y_o^2 + z_o^2} + \sqrt{(x - x_o)^2 + (y - y_o)^2 + (z - z_o)^2}$$

$$\tag{9.88}$$

9.6. THE MATHEMATICAL THEORY OF BLACK HOLES

Note that here f depends upon θ.

Since $R_c(r_o) = \zeta \, \forall r_o \, \forall n$, when $r = r_o$ the Kretschmann scalar for the Kerr-Newman equivalence class becomes,

$$f(r_o) = \frac{8}{(\zeta^2 + a^2 \cos^2 \theta)^6} \begin{bmatrix} \frac{3\alpha^2}{2} \left(\zeta^6 - 15a^2 \zeta^4 \cos^2 \theta + 15a^4 \zeta^2 \cos^4 \theta - a^6 \cos^6 \theta \right) - \\ -6\alpha q^2 \zeta \left(\zeta^4 - 10a^2 \zeta^2 \cos^2 \theta + 5a^4 \cos^4 \theta \right) + \\ + q^4 \left(6\zeta^4 - 34 a^2 \zeta^2 \cos^2 \theta + 6a^4 \cos^4 \theta \right) \end{bmatrix} \quad (9.89)$$

The Kretschmann scalar is finite when $r = r_o$, irrespective of the values of r_o and n. When $\theta = 0$ and when $\theta = \pi$,

$$f(r_o) = \frac{8}{(\zeta^2 + a^2)^6} \begin{bmatrix} \frac{3\alpha^2}{2} \left(\zeta^6 - 15a^2 \zeta^4 + 15a^4 \zeta^2 - a^6 \right) - \\ -6\alpha q^2 \zeta \left(\zeta^4 - 10a^2 \zeta^2 + 5a^4 \right) + \\ + q^4 \left(6\zeta^4 - 34 a^2 \zeta^2 + 6a^4 \right) \end{bmatrix} \quad (9.90)$$

When $\theta = \frac{\pi}{2}$, $f(r_o)$ reduces to,

$$f(r_o) = \frac{8}{\zeta^8} \left[\frac{3\alpha^2 \zeta^2}{2} - 6\alpha q^2 \zeta + 6q^4 \right] \quad (9.91)$$

Note that this does not contain the 'angular momentum' term a and that the result is precisely that for the Reissner-Nordström equivalence class (where $a = 0$).

When $q = 0$ the Kretschmann scalar for the Kerr-Newman equivalence class reduces to that for the Kerr equivalence class,

$$f(r_o) = \frac{12\alpha^2}{(\zeta^2 + a^2 \cos^2 \theta)^6} \left(\zeta^6 - 15a^2 \zeta^4 \cos^2 \theta + 15a^4 \zeta^2 \cos^4 \theta - a^6 \cos^6 \theta \right) \quad (9.92)$$

This too depends upon the value of θ. When $\theta = 0$ and when $\theta = \pi$, $f(r_o)$ becomes,

$$f(r_o) = \frac{12\alpha^2}{(\zeta^2 + a^2)^6} \left(\zeta^6 - 15a^2 \zeta^4 + 15a^4 \zeta^2 - a^6 \right) \quad (9.93)$$

When $\theta = \frac{\pi}{2}$, $f(r_o)$ reduces to,

$$f(r_o) = \frac{12}{\alpha^4} \quad (9.94)$$

which is precisely the Kretschmann scalar for the Schwarzschild form, where both $q = 0$ and $a = 0$ in the Kretschmann scalar for Kerr-Newman spacetime.

In the case of the Schwarzschild equivalence class in isotropic coordinates, the Kretschmann scalar is given by [43],

$$f = 3 \cdot 4^{13} \frac{\alpha^2 R_c^6}{(4R_c + \alpha)^{12}} \quad (9.95)$$

$$R_c = \left[|r - r_o|^n + \left(\frac{\alpha}{4} \right)^n \right]^{\frac{1}{n}}, \quad r, r_o \in \Re, \quad n \in \Re^+$$

Since $R_c(r_o) = \frac{\alpha}{4} \, \forall r_o \, \forall n$,

$$f(r_o) = \frac{12}{\alpha^4} \qquad (9.96)$$

which is the very same finite value as that for the ordinary Schwarzschild equivalence class.

The Kretschmann scalar for the isotropic Reissner-Nordström equivalence class is [43],

$$f = \frac{4^{13} R_c^6 \left\{ [\alpha(4R_c+\alpha)^2 - 4q^2(8R_c+\alpha)]^2 + [\alpha(4R_c+\alpha)^2 - 4q^2(12R_c+\alpha)]^2 + [\alpha(4R_c+\alpha)^2 - 4q^2(4R_c+\alpha)]^2 \right\}}{(4R_c+\alpha+2q)^8 (4R_c+\alpha-2q)^8}$$

$$R_c = (|r - r_o|^n + \zeta^n)^{\frac{1}{n}}, \quad \zeta = \frac{\sqrt{\alpha^2 - 4q^2}}{4}, \quad 4q^2 < \alpha^2 \quad r, r_o \in \Re, \quad n \in \Re^+$$
(9.97)

Since $R_c(r_o) = \zeta \, \forall r_o \, \forall n$, the Kretschmann scalar reduces to the invariant,

$$f = \frac{4^{13} \zeta^6 \left\{ [\alpha(4\zeta+\alpha)^2 - 4q^2(8\zeta+\alpha)]^2 + [\alpha(4\zeta+\alpha)^2 - 4q^2(12\zeta+\alpha)]^2 + [\alpha(4\zeta+\alpha)^2 - 4q^2(4\zeta+\alpha)]^2 \right\}}{(4\zeta+\alpha+2q)^8 (4\zeta+\alpha-2q)^8}$$
(9.98)

The Kretschmann scalar is always finite. For it to be undefined by a division by zero, as contended by cosmologists, it requires that the positive real power of the absolute value of a real number must take on values less than zero, which is a violation of the rules of pure mathematics, as the case $r_o = 0, n = 2$ amplifies.

Geodesic incompleteness

A geodesic is a line in some space. In Euclidean space the geodesics are simply straight lines. This is because the Riemannian curvature of Euclidean space is zero. If the Riemannian curvature is not zero throughout the entire space, the space is not Euclidean and the geodesics are curved lines rather than straight lines. If a geodesic terminates at some point in the space it is said to be incomplete, and the manifold or space in which it lies is also said to be geodesically incomplete. If no geodesic in some manifold is incomplete then the manifold is said to be geodesically complete. More specifically,

> "A semi-Riemannian manifold M for which every maximal geodesic is defined on the entire real line is said to be geodesically complete - or merely complete. Note that if even a single point p is removed from a complete manifold M then $M - p$ is no longer complete, since geodesics that formerly went through p are now obliged to stop." O'Neill [89]

Consider now Hilbert's solution. In 1931, Hagihara [100] proved that all geodesics therein that do not run into the boundary at $r = 2Gm/c^2$ are complete. Hence this is also the case at $r = r_o$ for all the solutions generated by the Schwarzschild infinite equivalence class. This is also the case at $r = r_o$ for the isotropic forms. The geodesics terminate at the origin; the point from which

9.6. THE MATHEMATICAL THEORY OF BLACK HOLES

the radius emanates; $R_p = 0$. In other words, Hagihara effectively proved that all geodesics that do not run into the origin $R_p = 0$ are complete. This once again attests that Droste's solution cannot be 'extended' to produce a black hole.

The acceleration invariant

Doughty [101] obtained the following expression for the acceleration β of a point along a radial geodesic for the static spherically symmetric line-elements,

$$\beta = \frac{\sqrt{-g_{11}}\left(-g^{11}\right)\left|\frac{\partial g_{oo}}{\partial r}\right|}{2g_{oo}} \tag{9.99}$$

Since the Hilbert and Nordström metrics utilised by cosmologists are particular cases of their respective infinite equivalence classes, the foregoing expression becomes, in general,

$$\beta = \frac{\sqrt{-g_{11}}\left(-g^{11}\right)\left|\frac{\partial g_{oo}}{\partial R_c}\right|}{2g_{oo}}$$

$$R_c = (|r - r_o|^n + \zeta^n)^{\frac{1}{n}}, \quad r, r_o \in \Re, \quad n \in \Re^+ \tag{9.100}$$

$$\zeta = \frac{\alpha + \sqrt{\alpha^2 - 4q^2}}{2}, \quad q^2 < \frac{\alpha^2}{4}$$

The acceleration is therefore given by,

$$\beta = \frac{\alpha R_c - 2q^2}{2R_c^2\sqrt{R_c^2 - \alpha R_c + q^2}} \tag{9.101}$$

In all cases, whether or not $q = 0$, $r \to r_o \Rightarrow \beta \to \infty$, which constitutes an invariant condition, and therefore reaffirms that the Schwarzschild and Reissner-Nordström forms cannot be extended to produce black holes.

The expression for acceleration appears at first glance to be a first-order intrinsic differential invariant since it is superficially composed of only the components of the metric tensor and their first derivatives. This is however, not so, because the expression applies only to the radial direction, i.e. to the motion of a point along a radial geodesic. In other words, it involves a direction vector. Consequently, although the acceleration expression is a first-order differential invariant, it is not intrinsic. First-order differential invariants exist, but first-order intrinsic differential invariants do not exist [110, 111]. That the acceleration expression involves a direction vector is amplified by the Killing vector. Let X_a be a first-order tensor (i.e. a covariant vector). Then for it to be a Killing vector it must satisfy Killing's equations,

$$X_{a;b} + X_{b;a} = 0 \tag{9.102}$$

where $X_{a;b}$ denotes the covariant derivative of X_a. The condition for hypersurface orthogonality is [69, 71],

$$X_{[a}X_{b;c]} = 0 \tag{9.103}$$

The two foregoing conditions determine a unique time-like Killing vector that fixes the direction of time [69]. By means of this Killing vector the four-velocity v_i is,

$$v^a = \frac{X^a}{\sqrt{X_a X^a}} \qquad (9.104)$$

The absolute derivative of the four-velocity along its own direction gives the four-acceleration β^a,

$$\beta^a = \frac{Dv^a}{du} \qquad (9.105)$$

The norm of the four-acceleration is,

$$\beta = \sqrt{-\beta_a \beta^a} \qquad (9.106)$$

Applying this to the Reissner-Nordström equivalence class yields,

$$\beta = \frac{\alpha R_c - 2q^2}{2R_c^2 \sqrt{R_c^2 - \alpha R_c + q^2}}$$

$$R_c = (|r - r_o|^n + \zeta^n)^{\frac{1}{n}}, \quad r, r_o \in \Re, \quad n \in \Re^+ \quad \zeta = \frac{\alpha + \sqrt{\alpha^2 - 4q^2}}{2},$$

$$q^2 < \frac{\alpha^2}{4}$$

(9.107)

which is Eq.(9.101). Consequently, the acceleration expression is not intrinsic; it is a first-order differential invariant which is constructed with the metric tensor and an associated direction vector, as the motion of a point *along a radial geodesic* implies.

When $q = 0$ the acceleration expression reduces to,

$$\beta = \frac{\alpha}{2R_c^2 \sqrt{1 - \frac{\alpha}{R_c}}} \qquad (9.108)$$

$$R_c = (|r - r_o|^n + \alpha^n)^{\frac{1}{n}}, \quad r, r_o \in \Re, \quad n \in \Re^+$$

which can of course be calculated directly from the equations for the Schwarzschild equivalence class.

In the case of the isotropic Reissner-Nordström equivalence class the acceleration is given by [43],

$$\beta = \frac{8R_c^2 \left[64R_c(4R_c+\alpha+2q)(4R_c+\alpha-2q) - 16(16R_c^2 - \alpha^2 + 4q^2)(4R_c+\alpha) \right]}{(4R_c+\alpha+2q)^2 (4R_c+\alpha-2q)^2 (16R_c^2 - \alpha^2 + 4q^2)}$$

$$R_c = (|r - r_o|^n + \zeta^n)^{\frac{1}{n}}, \quad \zeta = \frac{\sqrt{\alpha^2 - 4q^2}}{4}, \quad 4q^2 < \alpha^2 \quad r, r_o \in \Re, \quad n \in \Re^+$$

(9.109)

If $q = 0$ this reduces to the acceleration for the isotropic Schwarzschild equivalence class [43].

In all cases $r \to r_o \Rightarrow \beta \to \infty$, which constitutes an invariant condition, and therefore reaffirms once again that the Schwarzschild and Reissner-Nordström forms cannot be extended to produce black holes.

The Newtonian 'black hole'

9.6. THE MATHEMATICAL THEORY OF BLACK HOLES

Cosmologists routinely assert, incorrectly, that the theoretical Michell-Laplace dark body, extracted from Newton's theory of gravity, is a black hole.

"Laplace essentially predicted the black hole ..." Hawking and Ellis [91]

"On this assumption a Cambridge don, John Michell, wrote a paper in 1683 in the Philosophical Transactions of the Royal Society of London. In it, he pointed out that a star that was sufficiently massive and compact would have such a strong gravitational field that light could not escape. Any light emitted from the surface of the star would be dragged back by the star's gravitational attraction before it could get very far. Michell suggested that there might be a large number of stars like this. Although we would not be able to see them because light from them would not reach us, we could still feel their gravitational attraction. Such objects are what we now call black holes, because that is what they are - black voids in space." Hawking [97]

"Eighteenth-century speculators had discussed the characteristics of stars so dense that light would be prevented from leaving them by the strength of their gravitational attraction; and according to Einstein's General Relativity, such bizarre objects (today's 'black holes') were theoretically possible as end-products of stellar evolution, provided the stars were massive enough for their inward gravitational attraction to overwhelm the repulsive forces at work." Cambridge Illustrated History of Astronomy [102]

"Two important arrivals on the scene: the neutron star (1933) and the black hole (1695, 1939). No proper account of either can forego general relativity." Minser, Thorne, and Wheeler [61]

"That such a contingency can arise was surmised already by Laplace in 1798. Laplace argued as follows. For a particle to escape from the surface of a spherical body of mass M and radius R, it must be projected with a velocity v such that $v^2/2 > GM/R$; and it cannot escape if $v^2 < 2GM/R$. On the basis of this last inequality, Laplace concluded that if $R < 2GM/c^2 = R_s$ (say) where c denotes the velocity of light, then light will not be able to escape from such a body and we will not be able to see it!

"By a curious coincidence, the limit R_s discovered by Laplace is exactly the same that general relativity gives for the occurrence of the trapped surface around a spherical mass" Chandrasekhar [45]

But it is not *"a curious coincidence"* that General Relativity gives the same R_s *"discovered by Laplace"* because the Newtonian expression for escape speed is deliberately inserted, *post hoc*, into Hilbert's solution by Einstein and his followers in order to make a mass appear in equations that contain no material source by mathematical construction.

The theoretical Michell-Laplace dark body is not a black hole. It possesses an escape speed at its surface, but the black hole has both an escape speed and no escape speed simultaneously at its 'surface' (i.e. event horizon); masses and light can leave the Michell-Laplace dark body, but nothing can leave the black hole; it does not require irresistible gravitational collapse to form, whereas the black hole does; it has no infinitely dense 'physical singularity', but the black hole does; it has no event horizon, but the black hole does; it has 'infinite gravity' nowhere, but the black hole has infinite gravity at its 'physical singularity'; there is always a class of observers that can see the Michell-Laplace dark body [116], but there is no class of observers that can see the black hole; the Michell-Laplace dark body persists in a space which by consistent theory contains other Michell-Laplace dark bodies and other matter and they can interact with themselves and other matter, but the spacetime of all types of black holes pertains to a universe that contains, supposedly, only one mass (but actually contains no mass by mathematical construction) and so cannot interact with any other masses (in other words, the Principle of Superposition holds for the theoretical Michell-Laplace dark body but does not hold for the black hole); the space of the Michell-Laplace dark body is 3-dimensional and Euclidean, but that of the black hole is a 4-dimensional non-Euclidean (pseudo-Riemannian) spacetime; the space of the Michell-Laplace dark body is not asymptotically anything whereas the spacetime of the black hole is asymptotically flat or asymptotically curved; the Michell-Laplace dark body does not 'curve' a spacetime, but the black hole does; the gravity of the theoretical Michell-Laplace dark body is a force whereas the 'gravity' of a black hole is not a force. Hence, the theoretical Michell-Laplace dark body does not possess any of the characteristics of the black hole, other than a finite mass, and so it is not a black hole.

9.7 The paradox of black hole mass

Although one violation of the rules of pure mathematics is sufficient to invalidate it, the black hole violates other rules of logic. Einstein maintains that although $R_{\mu\nu} = 0$ contains no terms for material sources (since $T_{\mu\nu} = 0$), a material source is nonetheless present, in order to cause a gravitational field. The material source is rendered present linguistically by the assertion that $R_{\mu\nu} = 0$ describes the gravitational field outside a body such as a star. Indeed, concerning Hilbert's solution, Einstein writes,

$$"ds^2 = \left(1 - \frac{A}{r}\right) dt^2 - \left[\frac{dr^2}{1 - \frac{A}{r}} + r^2 \left(\sin^2\theta \, d\varphi^2 + d\theta^2\right)\right] \quad (9.109a)$$

$$A = \frac{\kappa M}{4\pi}$$

M denotes the sun's mass centrally symmetrically placed about the origin of co-ordinates; the solution (9.109a) is valid only outside this mass, where all the $T_{\mu\nu}$ vanish." Einstein [85]

9.7. THE PARADOX OF BLACK HOLE MASS

Note that Einstein has incorrectly asserted, in the standard fashion of cosmologists, that his mass M in his Eq.(9.109a) is "*centrally symmetrically placed about the origin of co-ordinates*".

> "*In general relativity, the stress-energy or energy-momentum tensor T_{ab} acts as the source of the gravitational field. It is related to the Einstein tensor and hence to the curvature of spacetime via the Einstein equation.*" McMahon [42]

> "*Again, just as the electric field, for its part, depends upon the charges and is instrumental in producing mechanical interaction between the charges, so we must assume here that the metrical field (or, in mathematical language, the tensor with components g_{ik}) is related to the material filling the world.*" Weyl [103]

On the one hand, Einstein removes all material sources by setting $T_{\mu\nu} = 0$ and on the other hand immediately reinstates the presence of a massive source with words, by alluding to a mass "*outside*" of which equations $R_{\mu\nu} = 0$ apply; since his gravitational field must be caused by matter. This contradiction is reiterated by cosmologists.

> "*Einstein's equation, (6.26), should be exactly valid. Therefore it is interesting to search for exact solutions. The simplest and most important one is empty space surrounding a static star or planet. There, one has $T_{\mu\nu} = 0$.*" 't Hooft [63]

According to Einstein his equation (9.109a) contains a massive source, at "*the origin*", yet also according to Einstein, the universe modelled by $R_{\mu\nu} = 0$, from which (109a) is obtained, contains no material sources. The contradiction is readily amplified. That $R_{\mu\nu} = 0$ contains no material sources whatsoever is easily reaffirmed by the field equations,

$$R_{\mu\nu} = \lambda g_{\mu\nu} \qquad (9.110)$$

The constant λ is the so-called 'cosmological constant'. The solution for Eq.(9.110) is de Sitter's empty universe, which is empty precisely because the energy-momentum tensor for material sources is zero, i.e. $T_{\mu\nu} = 0$. De Sitter's universe contains no matter:

> "*This is not a model of relativistic cosmology because it is devoid of matter.*" d'Inverno [71]

> "*the de Sitter line element corresponds to a model which must strictly be taken as completely empty.*" Tolman [104]

> "*the solution for an entirely empty world.*" Eddington [54]

> "*there is no matter at all!*" Weinberg [105]

CHAPTER 9. RELATIVISTIC COSMOLOGY AND EINSTEIN'S...

Note that in $R_{\mu\nu} = 0$ and $R_{\mu\nu} = \lambda g_{\mu\nu}$, $T_{\mu\nu} = 0$. Thus, according to Einstein and the cosmologists, material sources are both present and absent by the very same mathematical constraint, which is a violation of the rules of logic. Since de Sitter's universe is devoid of material sources by virtue of $T_{\mu\nu} = 0$, the 'Schwarzschild' universe must also be devoid of material sources by the very same constraint. Thus, the universe modelled by $R_{\mu\nu} = 0$ contains no matter, whereby its solution is physically meaningless. But it is upon $R_{\mu\nu} = 0$ and its solution that the mathematical theory of black hole rests.

Not only does $R_{\mu\nu} = 0$ contain no matter, it also violates other 'physical principles' of General Relativity. According to Einstein his Principle of Equivalence and his Special Theory of Relativity must hold in his gravitational field.

"*Let now K be an inertial system. Masses which are sufficiently far from each other and from other bodies are then, with respect to K, free from acceleration. We shall also refer these masses to a system of co-ordinates K', uniformly accelerated with respect to K. Relatively to K' all the masses have equal and parallel accelerations; with respect to K' they behave just as if a gravitational field were present and K' were unaccelerated. Overlooking for the present the question as to the 'cause' of such a gravitational field, which will occupy us later, there is nothing to prevent our conceiving this gravitational field as real, that is, the conception that K' is 'at rest' and a gravitational field is present we may consider as equivalent to the conception that only K is an 'allowable' system of co-ordinates and no gravitational field is present. The assumption of the complete physical equivalence of the systems of coordinates, K and K', we call the 'principle of equivalence'; this principle is evidently intimately connected with the law of the equality between the inert and the gravitational mass, and signifies an extension of the principle of relativity to co-ordinate systems which are in non-uniform motion relatively to each other. In fact, through this conception we arrive at the unity of the nature of inertia and gravitation.*

"*Stated more exactly, there are finite regions, where, with respect to a suitably chosen space of reference, material particles move freely without acceleration, and in which the laws of the special theory of relativity, which have been developed above, hold with remarkable accuracy.*" Einstein [85]

"*We may incorporate these ideas into the principle of equivalence, which is this: In a freely falling (nonrotating) laboratory occupying a small region of spacetime, the laws of physics are the laws of special relativity.*" Foster and Nightingale [53]

"*We can think of the physical realization of the local coordinate system K_o in terms of a freely floating, sufficiently small, box which is not subjected to any external forces apart from gravity, and which*

> *is falling under the influence of the latter. ... It is evidently natural to assume that the special theory of relativity should remain valid in K_o.*" Pauli [106]

> "*General Relativity requires more than one free-float frame.*" Taylor and Wheeler [88]

>> "*Near every event in spacetime, in a sufficiently small neighborhood, in every freely falling reference frame all phenomena (including gravitational ones) are exactly as they are in the absence of external gravitational sources.*" Dictionary of Geophysics, Astrophysics and Astronomy [56]

Note that both the Principle of Equivalence and Special Relativity are defined in terms of the *a priori* presence of multiple arbitrarily large finite masses and photons. There can be no multiple arbitrarily large finite masses and photons in a spacetime that contains no matter by mathematical construction, and so neither the Principle of Equivalence nor Special Relativity can manifest therein. But $R_{\mu\nu} = 0$ is a spacetime that contains no matter by mathematical construction.

9.8 Localisation of gravitational energy and conservation laws

Without a theoretical framework by which the usual conservation laws for the energy and momentum of a closed system hold, as determined by a vast array of experiments, there is no means to produce gravitational waves by General Relativity. Einstein was aware of this and so devised a means for his theory to satisfy the usual conservation of energy and momentum for a closed system. However, Einstein's method of solving this problem is invalid because he violated the rules of pure mathematics. There is in fact no means by which the usual conservation laws for a closed system can be satisfied by General Relativity. Consequently the concept of gravitational waves has no valid theoretical basis in Einstein's theory.

It must first be noted that when Einstein talks of the conservation of energy and momentum he means that the sum of the energy and momentum of his gravitational field and its material sources is conserved *in toto*, in the usual way for a closed system, as experiment attests, for otherwise his theory is in conflict with experiments and therefore invalid.

> "*It must be remembered that besides the energy density of the matter there must also be given an energy density of the gravitational field, so that there can be no talk of principles of conservation of energy and momentum of matter alone.*" Einstein [85]

The meaning of Einstein's 'matter' needs to be clarified.

CHAPTER 9. RELATIVISTIC COSMOLOGY AND EINSTEIN'S...

> "*We make a distinction hereafter between 'gravitational field' and 'matter' in this way, that we denote everything but the gravitational field as 'matter'. Our use of the word therefore includes not only matter in the ordinary sense, but the electromagnetic field as well.*" Einstein [66]

> "*In the general theory of relativity the doctrine of space and time, or kinematics, no longer figures as a fundamental independent of the rest of physics. The geometrical behaviour of bodies and the motion of clocks rather depend on gravitational fields, which in their turn are produced by matter.*" Einstein [107]

Einstein himself is not free from contamination by his followers. He states clearly that in his theory his gravitational field is not matter and that only matter as he conceives of it can produce his gravitational field. Nevertheless, cosmologists alter Einstein's theory *ad arbitrium* in order to attempt justification of their own claims about his theory. For instance, according to 't Hooft Einstein's gravitational field can "*have a mass of its own*" [43, 108], in direct contradiction of Einstein's own account of his theory. Alteration of Einstein's theory to suit their purpose, pretending that their alterations are part of Einstein's theory and thereby have his seal of absolute authority, is another common method employed by his followers. Einstein's theory has the character of a chameleon, able to take on any colour required for any desired purpose. Einstein's enthusiastic followers have become "*more Einsteinich than he*" Heaviside [109]

The energy-momentum of Einstein's matter alone is contained in his energy-momentum tensor $T_{\mu\nu}$. To account for the energy-momentum of his gravitational field alone Einstein introduced his pseudotensor t_σ^α, defined by (Einstein [77]),

$$\kappa t_\sigma^\alpha = \frac{1}{2}\delta_\sigma^\alpha g^{\mu\nu}\Gamma_{\mu\beta}^\lambda \Gamma_{\mu\lambda}^\beta - g^{\mu\nu}\Gamma_{\mu\beta}^\alpha \Gamma_{\nu\sigma}^\beta \qquad (9.111)$$

where κ is a constant and δ_σ^α is the Kronecker-delta.

> "*The quantities t_σ^α we call the 'energy components' of the gravitational field.*" Einstein [66]

But t_σ^α is not a tensor. As such it is a coordinate dependent quantity, contrary to the basic coordinate independent tenet of General Relativity.

> "*It is to be noted that t as is not a tensor*" Einstein [66]

The justification is that t_σ^α acts 'like a tensor' under linear transformations of coordinates when subjected to certain strict conditions. Einstein then takes an ordinary divergence,

$$\frac{\partial t_\sigma^\alpha}{\partial x_\alpha} = 0 \qquad (9.112)$$

and claims a conservation law for the energy and momentum of his gravitational field alone.

9.8. LOCALISATION OF GRAVITATIONAL ENERGY AND...

> "*This equation expresses the law of conservation of momentum and of energy for the gravitational field.*" Einstein [66]

Einstein added his pseudotensor for his gravitational field alone to his energy-momentum tensor for matter alone to obtain the total energy-momentum equation for his gravitational field and its material sources.

$$E = (t_\sigma^\alpha + T_\sigma^\alpha) \tag{9.113}$$

Not being a tensor equation, Einstein cannot take a tensor divergence. He therefore takes an ordinary divergence, [66],

$$\frac{\partial (t_\sigma^\alpha + T_\sigma^\alpha)}{\partial x_\alpha} = 0 \tag{9.114}$$

and claims the usual conservation laws of energy and momentum for a closed system:

> "*Thus it results from our field equations of gravitation that the laws of conservation of momentum and energy are satisfied. ...here, instead of the energy components t_μ^σ of the gravitational field, we have to introduce the totality of the energy components of matter and gravitational field.*" Einstein [66]

The mathematical error is profound, but completely unknown to cosmologists. Contract Einstein's pseudotensor by setting $\sigma = \alpha$ to yield the invariant $t = t_\alpha^\alpha$, thus,

$$\kappa t_\alpha^\alpha = \kappa t = \frac{1}{2}\delta_\alpha^\alpha g^{\mu\nu}\Gamma_{\mu\beta}^\lambda \Gamma_{\mu\lambda}^\beta - g^{\mu\nu}\Gamma_{\mu\beta}^\alpha \Gamma_{\nu\alpha}^\beta \tag{9.115}$$

Since the $\Gamma_{\beta\sigma}^\alpha$ are functions only of the components of the metric tensor and their first derivatives, t is seen to be a first-order intrinsic differential invariant [41, 60, 61, 78], i.e. it is an invariant that depends solely upon the components of the metric tensor and their first derivatives. However, the pure mathematicians proved in 1900 that first-order intrinsic differential invariants do not exist [110]. Thus, by *reductio ad absurdum*, Einstein's pseudotensor is a meaningless collection of mathematical symbols. Contrary to Einstein and the cosmologists, it cannot therefore be used to represent anything in physics or to make any calculations, including those for the energy of Einstein's gravitational waves. Nevertheless, cosmology calculates:

> "*It is not possible to obtain an expression for the energy of the gravitational field satisfying both the conditions: (i) when added to other forms of energy the total energy is conserved, and (ii) the energy within a definite (three dimensional) region at a certain time is independent of the coordinate system. Thus, in general, gravitational energy cannot be localized. The best we can do is to use the pseudotensor, which satisfies condition (i) but not condition (ii). It gives us approximate information about gravitational energy, which in some special cases can be accurate.*" Dirac [90]

"*Let us consider the energy of these waves. Owing to the pseudo-tensor not being a real tensor, we do not get, in general, a clear result independent of the coordinate system. But there is one special case in which we do get a clear result; namely, when the waves are all moving in the same direction.*" Dirac [90]

Consider the following two equivalent forms of Einstein's field equations,

$$R^\mu_\nu = -\kappa \left(T^\mu_\nu - \frac{1}{2} T g^\mu_\nu \right) \quad (9.116)$$

$$T^\mu_\nu = -\frac{1}{\kappa} \left(R^\mu_\nu - \frac{1}{2} R g^\mu_\nu \right) \quad (9.117)$$

By Eq.(9.116), according to Einstein, if $T^\mu_\nu = 0$ then $R^\mu_\nu = 0$. But by Eq.(9.117), if $R^\mu_\nu = 0$ then $T^\mu_\nu = 0$. In other words, $R^\mu_\nu = 0$ and $T^\mu_\nu = 0$ must vanish identically - if either is zero then so is the other, and the field equations reduce to the identity $0 = 0$ [43, 73, 75, 111]. Hence, if there are no material sources (i.e. $T^\mu_\nu = 0$) then there is no gravitational field, and no universe. Bearing this in mind, with Eq.(9.113) and Eq.(9.114), consideration of the conservation of energy and momentum, and tensor relations, Einstein's field equations must take the following form [43, 73, 75, 111],

$$\frac{G^\mu_\nu}{\kappa} + T^\mu_\nu = 0 \quad (9.118)$$

where

$$G^\mu_\nu = R^\mu_\nu - \frac{1}{2} R g^\mu_\nu \quad (9.119)$$

Comparing Eq.(9.118) to Eq.(9.113) it is clear that the G^μ_ν/κ actually constitute the energy-momentum components of Einstein's gravitational field [43, 73, 75, 111], which is rather natural since the Einstein tensor G^μ_ν describes the geometry of Einstein's spacetime (i.e. his gravitational field). Eq.(9.118) also constitutes the total energy-momentum equation for Einstein's gravitational field and its material sources combined.

Spacetime and matter have no separate existence. Einstein's field equations,

"*...couple the gravitational field (contained in the curvature of spacetime) with its sources.*" Foster and Nightingale [53]

"*Since gravitation is determined by the matter present, the same must then be postulated for geometry, too. The geometry of space is not given a priori, but is only determined by matter.*" Pauli [106]

"*Mass acts on spacetime, telling it how to curve. Spacetime in turn acts on mass, telling it how to move.*" Carroll and Ostlie [60]

"*space as opposed to 'what fills space', which is dependent on the coordinates, has no separate existence*" Einstein [112]

9.8. LOCALISATION OF GRAVITATIONAL ENERGY AND...

> *"I wish to show that space-time is not necessarily something to which one can ascribe a separate existence, independently of the actual objects of physical reality."* Einstein [113]

Unlike Eq.(9.113), Eq.(9.118) is a tensor equation. The tensor (covariant derivative) divergence of the left side of Eq.(9.118) is zero and therefore constitutes the conservation law for Einstein's gravitational field and its material sources. However, the total energy-momentum by Eq.(9.118) is always zero, the G^μ_ν/κ and the T^μ_ν must vanish identically because spacetime and matter have no separate existence in General Relativity, and hence gravitational energy cannot be localised, i.e. there is no possibility of gravitational waves [43, 73, 75, 111]. Moreover, since the total energy-momentum is always zero the usual conservation laws for energy and momentum of a closed system cannot be satisfied [43, 73, 75, 111]. General Relativity is therefore in conflict with a vast array of experiments on a fundamental level.

The so-called 'cosmological constant' can be easily included as follows,

$$\frac{(G^\mu_\nu + \lambda g^\mu_\nu)}{\kappa} + T^\mu_\nu = 0 \qquad (9.120)$$

In this case the energy-momentum components of Einstein's gravitational field are the $(G^\mu_\nu + \lambda g^\mu_\nu)/\kappa$. When G^μ_ν or T^μ_ν is zero, all must vanish identically, and all the same consequences ensue just as for Eq.(9.118). Thus, once again, if there is no material source, not only is there no gravitational field, there is no universe, and Einstein's field equations violate the usual conservation of energy and momentum for a closed system.

The so-called 'dark energy' is attributed to λ by cosmologists. Dark energy is a mysterious aether *ad arbitrium*, because, according to Einstein [85, 114], λ is not a material source for a gravitational field, but is vaguely implicated by him in his gravitational field,

> *"... by introducing into Hamilton's principle, instead of the scalar of Riemann's tensor, this scalar increased by a universal constant"* Einstein [114]

The 'cosmological constant' however falls afoul of de Sitter's empty universe, which possesses spacetime curvature but contains no matter, and is therefore physically meaningless. By Eq.(9.120), if λg^μ_ν is to be permitted, for the sake of argument, it must be part of the energy-momentum of the gravitational field, which necessarily vanishes when $T^\mu_\nu = 0$. Recall that according to Einstein, everything except his gravitational field is matter and that matter is the cause of his gravitational field. The insinuation of λ can be more readily seen by writing Eq.(9.120) as,

$$\frac{[R^\mu_\nu - \frac{1}{2}(R - 2\lambda)g^\mu_\nu]}{\kappa} + T^\mu_\nu = 0 \qquad (9.121)$$

Einstein's *"scalar increased by a universal constant"* is clearly evident; it is the term $-(R - 2\lambda)/2$. Hence Einstein's field equations *"in the absence of matter"* [66], i.e. $R^\mu_\nu = 0$, once again, have no physical meaning, and so the

Schwarzschild solution too has no physical meaning, despite putative observational verifications. Consequently, the theories of black holes and gravitational waves are invalid.

9.9 Numerical relativity and perturbations on black holes

Numerical analysis of merging black holes and perturbation of black holes are ill-posed procedures for the simple fact that such mathematical means cannot validate a demonstrable fallacy. Numerical analysis and 'systematic perturbation expansions' on a fallacy produce fallacies still. Similarly, no amount of observation or experiment can legitimise entities that are the products of violations of the rules of pure mathematics and logic. However, cosmologists systematically perturb:

> "*In a systematic perturbation expansion one can compute the interactions, due to nonlinearity, between black holes.*" 't Hooft [115]

Since the premises are false and the conclusions drawn from them inconsistent with them, such numerical and perturbation procedures are consequently of no scientific merit. Nonetheless the LIGO-Virgo Collaboration has stated,

> "*The signal sweeps upwards in frequency from 35 to 250 Hz with a peak gravitational-wave strain of 1.0×10^{-21}. It matches the waveform predicted by general relativity for the inspiral and merger of a pair of black holes and the ringdown of the resulting single black hole.*" Abbott et al. [38]
>
> "*Using the fits to numerical simulations of binary black hole mergers in [129, 130], we provide estimates of the mass and spin of the final black hole, the total energy radiated in gravitational waves, and the peak gravitational-wave luminosity [76].*" Abbott et al. [38]
>
> "*Several analyses have been performed to determine whether or not GW150914 is consistent with a binary black hole system in general relativity [131]. A first consistency check involves the mass and spin of the final black hole. In general relativity, the end product of a black hole binary coalescence is a Kerr black hole, which is fully described by its mass and spin. For quasicircular inspirals, these are predicted uniquely by Einstein's equations as a function of the masses and spins of the two progenitor black holes. Using fitting formulas calibrated to numerical relativity simulations [129], we verified that the remnant mass and spin deduced from the early stage of the coalescence and those inferred independently from the late stage are consistent with each other, with no evidence for disagreement from general relativity.*" Abbott et al. [38]

Signal GW150914 was extracted from a database containing 250,000 numerically determined waveforms generated on the false assumptions of the existence of black holes and gravitational waves. A 'generic' signal cGW was initially reported by LIGO, after which powerful computers extracted GW150914 from the waveform database for a best fit element.

> "*The initial detection was made by low-latency searches for generic gravitational-wave transients [78] and was reported within three minutes of data acquisition [80]. Subsequently, matched-filter analyses that use relativistic models of compact binary waveforms [81] recovered GW150914 as the most significant event from each detector for the observations reported here.*" Abbott et al. [38]

With such powerful computing resources and so many degrees of freedom it is possible to best fit just about any LIGO instability with an element of its numerically determined waveform database. This is indeed the outcome for the LIGO-Virgo Collaborations, as they have managed to best fit a numerically determined waveform for and to entities that not only do not exist, but are not even consistent with General Relativity itself. This amplifies the futility of applying numerical and perturbation methods to ill-posed problems.

There are no known Einstein field equations for two or more masses and hence no known solutions thereto. There is no existence theorem by which it can even be asserted that Einstein's field equations contain latent capability for describing configurations of two or more masses [43, 73, 75, 116]. General Relativity cannot account for the simple experimental fact that two fixed suspended masses approach one another upon release. It is for precisely these reasons that all the big bang models treat the universe as a single mass, an ideal indivisible fluid of macroscopic density and pressure that permeates the entire universe. Upon this model the cosmologists simply superpose, where superposition does not hold.

> "*We may, however, introduce a more specific hypothesis by assuming that the material filling the model can be treated as a perfect fluid.*" Tolman [104]

> "*We can then treat the universe as filled with a continuous distribution of fluid of proper macroscopic density ρ_{oo} and pressure p_o, and shall feel justified in making this simplification since our interest lies in obtaining a general framework for the behaviour of the universe as a whole, on which the details of local occurrences could be later superposed.*" Tolman [104]

> "*...it must be remembered that these quantities apply to the idealized fluid in the model, which we have substituted in place of the matter and radiation actually present in the real universe.*" Tolman [104]

9.10 Big bang cosmology

The central dogma of big bang cosmology is that the Universe created itself out of nothing [117]. Often this nothingness is vaguely called a big bang

CHAPTER 9. RELATIVISTIC COSMOLOGY AND EINSTEIN'S...

'singularity'. Space, time, and matter, all came into existence with the big bang creation *ex nihilo*.

> "*General relativity plays an important role in cosmology. The simplest theory is that at a certain moment 't = 0', the universe started off from a singularity, after which it began to expand. ...All solutions start with a 'big bang' at t = 0.*" 't Hooft [49]

> "*At the big bang itself, the universe is thought to have had zero size, and to have been infinitely hot.*" Hawking [97]

That which has zero size has no volume and hence cannot possess mass or have a temperature. What is temperature? According to the physicists and the chemists it is the motion of atoms and molecules. Atoms and molecules have mass. The more energy imparted to the atoms and molecules the faster they move about and so the higher the temperature. In the case of a solid the atoms or molecules vibrate about their equilibrium positions in a lattice structure and this vibration increases with increased temperature.

> "*As the temperature rises, the molecules become more and more agitated; each one bounds back and forth more and more vigorously in the little space left for it by its neighbours, and each one strikes its neighbours more and more strongly as it rebounds from them.*" Pauling [118]

Increased energy causes atoms or molecules of a solid to break down the long range order of its lattice structure to form a liquid or gas. Liquids have short range order, or long range disorder. Gases have a great molecular or atomic disorder. In the case of an ideal gas its temperature is proportional to the mean kinetic energy of its molecules [119]- [121],

$$\frac{3}{2}kT = \frac{1}{2}m\langle v^2 \rangle \tag{9.122}$$

wherein $\langle v^2 \rangle$ is the mean squared molecular speed, m the molecular mass, and k is Boltzman's constant.

Now that which has zero size has no space for atoms and molecules to exist in or for them to move about in. And just how fast must atoms and molecules be moving about to be infinitely hot? An entity of zero size and infinite hotness has no scientific meaning whatsoever. Nonetheless, according to Misner, Thorne and Wheeler [61],

> "*One crucial assumption underlies the standard hot big-bang model: that the universe 'began' in a state of rapid expansion from a very nearly homogeneous, isotropic condition of infinite (or near infinite) density and pressure.*"

Just how close to infinite must one get to be "*near infinite*"? No object can possesses infinite or near 'infinite density' and pressure either, just as no object can possess infinite gravity or infinite temperature. Even Special Relativity forbids infinite density.

9.10. BIG BANG COSMOLOGY

Near infinities of various sorts are routinely and widely invoked by cosmologists and astronomers. Here is yet another example; this time from Professor Lawrence Krauss [122] of Arizona State University, on Australian national television:

> "*But is that, in fact, because of discovering that empty space has energy, it seems quite plausible that our universe may be just one universe in what could be almost an infinite number of universes and in every universe the laws of physics are different and they come into existence when the universe comes into existence.*"

Just how close to infinite is "*almost an infinite number*"? There is no such thing as "*almost an infinite number*".

> "*There's no real particles but it actually has properties but the point is that you can go much further and say there's no space, no time, no universe and not even any fundamental laws and it could all spontaneously arise and it seems to me if you have no laws, no space, no time, no particles, no radiation, it is a pretty good approximation of nothing.*" Krauss [122]

> "*There was nothing there. There was absolutely no space, no time, no matter, no radiation. Space and time themselves popped into existence which is one of the reasons why it is hard ...*" Krauss [126]

Thus, the Universe sprang into existence from absolutely nothing, by big bang creationism, "*at a certain moment 't = 0'*" [49] and nothing, apparently, is "*a good approximation of nothing*" [122]. And not only is nothing a good approximation of nothing, this rigmarole is pushed even further:

> "*But I would argue that nothing is a physical quantity. It's the absence of something.*" Krauss [122]

Professor Brian Schmidt is a Nobel Prize winning cosmologist [123]. The following question was put to him on Australian national television:

> "*How can something as infinitely large as the universe actually get bigger?*" Irvin [123]

Schmidt began his reply with the following:

> "*Ah, yes, this is always a problem: infinity getting bigger. So, if you think of the universe and when we measure the universe it, as near as we can tell, is very close to being infinite in size, that is we can only see 13.8 billion light years of it because that's how old the universe is, but we're pretty sure there's a lot more universe beyond the part we can see, which light just simply can't get to us. And our measurements are such that we actually think that very nearly that may go out, well, well, thousands of times beyond what we can see and perhaps an infinite distance.*" Schmidt [123, 124]

CHAPTER 9. RELATIVISTIC COSMOLOGY AND EINSTEIN'S...

However, an infinite universe cannot get bigger[7], bearing in mind that infinite simply means endless, and so is not even a real number. Professor Schmidt committed the very common cosmologist elementary error that *"very close to being infinite in size"* is a scientific quantity [125]. With this in mind, how likely is it that cosmologists actually measured the nearness to infinity that Professor Schmidt has claimed? Schmidt continued,

> *"So, ultimately, we're expanding into the future but think of it this way: in school you would have done this little experiment in math where you will put a ray starting at zero and it will go out one, two, three and off to infinity. You put a little arrow, it goes off forever. So I can multiply that by two. So zero stays at zero, one goes to two, two goes to four, four goes to eight and you can do that for any number you want all the way up to infinity. And that's sort of what the universe is doing. Infinity is just getting bigger and we're allowed to do that in mathematics. That's what's so cool about math."* [123, 124]

Consider the two infinite sequences of integers that Professor Schmidt utilised (where the dots mean, 'goes on in like manner without end'),

$$0, 1, 2, 3, 4, \ldots$$

$$0, 2, 4, 6, 8, \ldots$$

First, all Schmidt has done here is to place the non-negative even integers (the lower sequence) into a *one-to-one correspondence* with the non-negative integers (the upper sequence). This does not make infinity get bigger. Both sequences are infinite. For every number in the upper sequence there is one and only one corresponding number in the lower sequence, according to position. Second, since infinity is not a real number, contrary to Professor Schmidt's claim, it cannot even be multiplied by 2 because, ultimately, numbers on the real number line can only be multiplied by numbers. Infinity is often denoted by the symbol ∞. This is not a real number and so it cannot be used for the usual arithmetic or algebra. Substituting the symbol ∞ for the word 'endless' or the word 'infinity' or the word 'limitless' does not make ∞ a real number. Consequently, $2 \times \infty$ does not mean that infinity is doubled; it is a meaningless concatenation of symbols. In like fashion, multiply Professor Schmidt's first sequence by $\frac{1}{2}$. The resulting sequence is,

$$0, \frac{1}{2}, 1, \frac{3}{2}, 2, \ldots$$

Does this mean that infinity has been halved? Is not this sequence also infinite? Professor Schmidt's doubling of infinity by means of the real number line is just as nonsensical as halving infinity with the real number line.

Yet despite the zero size, the infinities and near infinities possessed by nothing, the absence of something, and big bang creation *ex nihilo*, Hawking admits that,

---

[7]I shall not consider the esoteric purely mathematical issues of Cantor's 'transfinite numbers', as they have no relevance here.

9.10. BIG BANG COSMOLOGY

"energy cannot be created out of nothing." Hawking [97]

Thus stands yet another cosmological contradiction.

The so-called 'Cosmic Microwave Background' (CMB) is inextricably intertwined with big bang cosmology. Without the 'CMB', big bang creationism and General Relativity are defunct. The reasons why the 'CMB' does not exist are simply stated:

1. Kirchhoff's Law of Thermal Emission is false [127].

2. Due to (1), Planck's equation for thermal spectra is not universal.

Consequently, when Penzias and Wilson [128] assigned a temperature to their residual signal and the theoreticians assigned that signal to the Cosmos, they violated the laws of thermal emission. It is a scientific fact that no monopole signal has ever been detected beyond $\approx$ 900 km of Earth. The signal is proximal (i.e. from the oceans on Earth [129]- [132]).

Nuclear Magnetic Resonance (NMR) and Magnetic Resonance Imaging (MRI) are thermal processes. That they exist is physical proof of the invalidity of Kirchhoff's Law of Thermal Emission and the non-universality of Planck's equation. If Kirchhoff's Law of Thermal Emission is true and Planck's equation is universal, then NMR and MRI would be impossible, because NMR and MRI utilise spin-lattice relaxation [133]. Hence, there is energy in the walls of an arbitrary cavity that is not available to thermal emission. Kirchhoff and Planck however, incorrectly permitted all energy in the walls of an arbitrary cavity to be available to the emission field. The very existence of clinical MRI, used in medicine everyday, proves that Kirchhoff's Law of Thermal Emission is false and that Planck's equation is not universal. This means that the 'CMB' does not exist because it requires the validity of Kirchhoff's Law of Thermal Emission and universality of Planck's equation. These facts alone invalidate big bang cosmology completely.

Put a glass of water inside a microwave oven then turn on the oven. The water gets hot because it absorbs microwaves [132]. That water absorbs microwaves is also well known for submarines, which is precisely why microwave radio communications cannot be used under water. It is well known from the laboratory that a good absorber is also a good emitter, and at the same frequencies. Approximately 70% of Earth's surface is covered by water. This water is not microwave silent. The reported 'CMB' is characterised by the monopole signal for the mean temperature of the microwave residue of the big bang. Its spectrum is a blackbody distribution at $\approx$ 2.725 K. Cosmologists claim the error bars of their CMB spectrum plot are some 400 times narrower than the thickness of the curve they have drawn for it. Yet it is a scientific fact that no monopole signal has ever been detected beyond $\approx$ 900 km of Earth. Without a monopole signal far from Earth, at say L2 (i.e. the second Lagrange point), all talk of a CMB and its alleged anisotropies is wishful thinking. All detections of the monople signal are of microwaves emitted by the oceans, scattered by the atmosphere.

The water molecule is bound by two bonds: (a) the hydroxyl bond, and (b) the hydrogen bond. The hydrogen bond weakly binds one water molecule to another. The hydroxyl bond strongly binds an oxygen atom to a hydrogen atom within the water molecule. Robitaille [134] has shown that the hydroxyl bond is $\approx$ 100 times stronger than the hydrogen bond. It is the hydrogen bond that is responsible for microwave emissions from water. If the oceans are at 300 K, then their microwave emission reports a temperature of $\approx$ 3 K. From this it is clear that a blackbody spectrum does not report the true temperature of the emission source, unless that source is a black material, such as soot; otherwise the temperature extracted from a blackbody spectrum is only apparent. Moreover, the Planckian (blackbody) distribution is continuous. Only condensed matter can emit a continuous spectrum. Gases can only emit in narrow bands, never a continuous spectrum, irrespective of pressure, and pressure requires the presence of a surface. Solar scientists and cosmologists believe the Sun and stars to be balls of hot gas, mostly hydrogen. Stars, they say, produce black holes by irresistible gravitational collapse. Liquids however are essentially incompressible. The photosphere of the Sun emits a Planckian spectrum. This alone is certain evidence that the Sun is condensed matter, not a ball of hot gas [135, 136]. The most likely candidate for the constitution of the Sun and stars is liquid metallic hydrogen [135, 136]. Furthermore, when a solar flare errupts it produces a radiating circular transverse wave emanating from its point of eruption in the Sun's surface, like that when a stone is flung into a pond. Gases cannot form or carry a transverse wave. The transverse wave produced by a solar flare too is certain evidence that the Sun is condensed matter.

9.11 Conclusions

LIGO did not detect Einstein gravitational waves or black holes. Black holes and Einstein's gravitational waves do not exist. The LIGO instability has been interpreted as gravitational waves produced by two merging black holes by a combination of theoretical fallacies, wishful thinking, and conformational bias. Black holes are products entire of violations of the rules of pure mathematics. Einstein's General Theory of Relativity is riddled with logical inconsistencies, invalid mathematics, and impossible 'physics'. The General Theory of Relativity violates the usual conservation of energy and momentum for a closed system and is thereby in conflict with a vast array of experiments, rendering it untenable at a fundamental level.

Arguments such as,

> "*What is more, astronomers have now identified numerous objects in the heavens that completely match the detailed descriptions theoreticians have derived. These objects cannot be interpreted as anything else but black holes.*" 't Hooft [63]

have no scientific merit [137].

LIGO is reported to have so far cost taxpayers $1,100,000,000.00 [138]. Just as with the Large Hadron collider at CERN, such large sums of public money demand justification by eventually finding what they said they would, despite the actual facts.

9.11. CONCLUSIONS

In the same fashion that people who believe in ghosts attribute the action of ghosts to that which they do not understand, cosmologists attribute the action of black holes and big bangs to that which they do not understand. No amount of experiment and observation can validate entities that have been extracted from theory by means of violations of the rules of pure mathematics, violations of basic logic, conflict with well established experimental findings, and just plain wishful thinking. With their litany of violations of scientific method it is perhaps not surprising that cosmologists, led by Professor Stephen Hawking, are now spending $100,000,000.00 of Milner's money, looking for aliens [139]; the very same aliens that fly saucers, man UFO's, and abduct human beings for experiments and vivisection, because they all come from outer space. Radio telescopes around the world will assist the cosmologists in their quest for alien contact [139]. One such telescope is the Parkes Radio Telescope in New South Wales, Australia. For 17 years, cosmologists at the Parkes facility mistook microwave signals from the microwave oven in their lunchroom for cosmic signals, and even called them 'perytons' [140]. Is there any doubt that the cosmologists will soon report alien contact? The aliens must be out there because the cosmologists even have a journal for them [141].

Kirchhoff's Law of Thermal Emission is false and Planck's equation for thermal spectra is not universal; physically proven by the clinical existence of MRI. It follows immediately from this that the Cosmic Microwave Background does not exist because it requires Kirchhoff's Law and universality of Planck's equation. The 'CMB' is due to microwave emission from the oceans on Earth [129]- [132]. For this reason no monopole signal has ever been detected beyond ≈ 900 km of Earth.

Planck's theoretical proof of Kirchhiff's Law of Thermal Emission is false, owing to violations of the physics of optics and thermal emission [127]. Einstein's derivation of Planck's equation is valid only for black materials because he invoked a Wien's field; a characteristic of black materials such as soot [127]. Only black materials such as soot emit a blackbody spectrum. Other materials emit an approximate black spectrum whilst yet others do not. The radiation within arbitrary cavities is not black and their radiation fields, at thermal equilibrium, depend upon the nature of the cavity walls.

The Sun and stars are not balls of hot gas; they are condensed matter [135, 136].

Modern physics is steeped in magic, mysticism and superstition. The proclivity of the Human Condition to magic and mysticism is well known to anthropologists:

> "*The reader may well be tempted to ask. How was it that intelligent men did not sooner detect the fallacy of magic? How could they continue to cherish expectations that were invariably doomed to disappointment? With what heart persist in playing venerable antics that led to nothing, and mumbling solemn balderdash that remained without effect? Why cling to beliefs which were so flatly contradicted by experience? How dare to repeat experiments that had failed so often? The answer seems to be that the fallacy was far from easy to detect, the failure by no means obvious, since in*

many, perhaps in most cases, the desired event did actually follow at a longer or shorter interval, the performance of the rite which was designed to bring it about; and a mind of more than common acuteness was needed to perceive that, even in these cases, the rite was not necessarily the cause of the event." Frazer [142]

9.11. CONCLUSIONS

Bibliography

Chapter 1 and General References

[1] M. W. Evans, "Generally Covariant Unified Field Theory: the Geometrization of Physics" (Abramis Academic, 2005 to present), vols. 1 to 4, vol. 5 in prep. (Papers 71 to 93 on www.aias.us).

[2] L. Felker, "The Evans Equations of Unified Field Theory" (Abramis Academic, 2007).

[3] K. Pendergast, "Crystal Spheres" (preprint on www.aias.us, Abramis to be published).

[4] Omnia Opera Section of www.aias.us.

[5] H. Eckardt, "ECE Engineering Model",
http://www.aias.us/documents/miscellaneous/ECE-Eng-Model.pdf.

[6] M. W. Evans, (ed.), "Modern Non-linear Optics", a special topical issue of I. Prigogine and S. A. Rice, "Advances in Chemical Physics" (Wiley Interscience, New York, 2001, second edition), vols. 119(1) to 119(3).

[7] M. W. Evans and S. Kielich (eds.), ibid., first edition (Wiley Interscience, New York, 1992, 1993, 1997), vols. 85(1) to 85(3).

[8] M. W. Evans and L. D. Crowell, "Classical and Quantum Electrodynamics and the B(3) Field" (World Scientific, 2001).

[9] M. W. Evans and J.-P. Vigier, "The Enigmatic Photon" (Kluwer, 1994 to 2002), in five volumes.

[10] M. W. Evans and A. A. Hasanein, "The Photomagneton in Quantum Field Theory" (World Scientific, 1994).

[11] M. W. Evans, "The Photon's Magnetic Field, Optical NMR" (World Scientific, 1992).

[12] M. W. Evans, Physica B, **182**, 227 and 237 (1992).

[13] S. P. Carroll, "Spacetime and Geometry, an Introduction to General Relativity" (Addison Wesley, New York, 2004), chapter three.

[14] J. B. Marion and S. T. Thornton, "Classical Dynamics of Particles and Systems" (Harcourt Brace College Publishers, 1988, third edition).

[15] S. Crothers, papers and references therein on the www.aias.us site and papers 93 of the ECE UFT series on www.aias.us.

[16] P. W. Atkins, "Molecular Quantum Mechanics" (Oxford University Press, 1983, 2nd edition and subsequent editions).

[17] J. R. Croca, "Towards a Non-linear Quantum Physics" (World Scientific, 2001).

[18] E. G. Milewski (Chief Editor), "The Vector Analysis Problem Solver" (Research and Education Association, New York, 1987, revised printing).

[19] J. D. Jackson, "Classical Electrodynamics" (Wiley, New York, 1999, third edition).

[20] M. W. Evans, Acta Phys. Polonica, **38**, 2211 (2007).

[21] M. W. Evans and H. Eckardt, Physica B, **400**, 175 (2007).

[22] M. W. Evans, Physica B, **403**, 517 (2008).

[23] Michael Krause (Director), "All About Tesla" (a film available on DVD that premiered in 2007).

Chapter 2

[24] L. H. Ryder, "Quantum Field Theory" (Cambridge University Press, 1996, 2nd ed.).

[25] M. W. Evans, Series Editor, "Contemporary Chemical Physics" (World Scientific).

Chapter 3

[26] Tadeusz Iwaniec, Gaven Martin: The Beltrami equation. In: Memoirs of the American Mathematical Society. Band 191, Nr. 893, 2008.

[27] D. Reed, "Beltrami-Trkalian Vector Fields in Electrodynamics" in M. W. Evans and S. Kielich, Eds., "Modem Nonlinear Optics" (Wiley, New York, 1992, 1993, 1997, 2001) in six volumes and two editions, vols. 85 and 119 of "Advances in Chemical Physics".

[28] G. E. Marsh, Force-Free Magnetic Fields, World Scientific, Singapore, 1994.

[29] S. Venkat et al., "Realistic Transverse Images of the Proton Charge and Magnetic Densities", NT@UW-10-15, 2010; http://arxiv.org/pdf/1010.3629v2.pdf

[30] G. Sardin, "Fundamentals of the Orbital Conception of Elementary Particles and of their Application to the Neutron and Nuclear Structure", Physics Essays Vol.12, no.2, 1999; http://uk.arxiv.org/ftp/hep-ph/papers/0102/0102268.pdf

[31] A. Proca, Sur la théorie ondulatoire des électrons positifs et négatifs, J. Phys. Radium 7, 347 (1936); Sur la théorie du positon, C. R. Acad. Sci. Paris 202, 1366 (1936).

[32] Dorin N. Poenaru, Proca Equations of a Massive Vector Boson Field, www.theory.nipne.ro/~poenaru/PROCA/proca_rila06.pdf

Chapter 4

[33] M. W. Evans and H. Eckardt, UFT paper 160, www.aias.us.

[34] P. L. Joliette and N. Rouze, Am. J. Phys., 62, 266 (1994).

[35] M. W. Evans, "Magnetization of an electron plasma by microwave pulses: Faraday induction", Foundations of Physics Letters, Volume 8, Issue 4, pp 359-364 (1995)

[36] H. Eckardt and D. W. Lindstrom, papers on www.aias.us, section Publications/Electromagnetic ECE Theory.

[37] A. S. Goldhaber and M. M. Nieto, Rev. Mod. Phys., 82, 939 (2010)

Chapter 9

[38] Abbott, B.P. *et al.*, Observation of Gravitational Waves from a Binary Black Hole Merger, PRL 116, 061102 (2016)

[39] Conover, E., Gravitational Waves Caught in the Act, APS News, February 11 2016, Accessed online 3rd March 2016

[40] Abbott, B.P. *et al.*, LOCALIZATION AND BROADBAND FOLLOW-UP OF THE GRAVITATIONAL-WAVE TRANSIENT GW150914, arXiv:1602.08492v1 [astro-ph.HE] 26 Feb 2016

[41] Landau, L. & Lifshitz, E., The Classical Theory of Fields, Addison-Wesley Publishing Co., Reading, MA, (1951)

[42] McMahon, D., Relativity Demystified, A Self teaching Guide, McGraw-Hill, New York, (2006)

BIBLIOGRAPHY

[43] Crothers, S.J., General Relativity: In Acknowledgement Of Professor Gerardus 't Hooft, Nobel Laureate, 4 August 2014, http://viXra.org/abs/1409.0072

[44] Crothers, S.J., A Critical Analysis of LIGO's Recent Detection of Gravitational Waves Caused by Merging Black Holes, *Hadronic Journal*, Vol. 39, (2016), http://vixra.org/pdf/1603.0127v4.pdf

[45] Chandrasekhar, S., The increasing role of general relativity in astronomy, *The Observatory*, 92, 168, 1972

[46] Hawking, S. W., The Theory of Everything, The Origin and Fate of the Universe, New Millennium Press, Beverly Hills, CA, (2002)

[47] Schutz, B., A First Course in General Relativity, 2nd Ed., Cambridge University Press, (2009)

[48] Townsend, P.K., Black Holes, lecture notes, DAMTP, University of Cambridge, Silver St., Cambridge, U.K., 4 July, 1977, http://arxiv.org/abs/gr-qc/9707012

[49] 't Hooft, G., Introduction to General Relativity, online lecture notes, 8/4/2002, http://www.phys.uu.nl/thooft/lectures/genrel.pdf, http://www.staff.science.uu.nl/~hooft101/lectures/genrel_2013.pdf

[50] Oppenheimer, J.R., Snyder, H., Phys. Rev. 56, 455 (1939)

[51] Hughes, S.A., Trust but verify: The case for astrophysical black holes, Department of Physics and MIT Kavli Institute, 77 Massachusetts Avenue, Cambridge, MA 02139, SLAC Summer Institute (2005)

[52] de Sitter, W., On Einstein's Theory of Gravitation, and its Astronomical Consequences, Monthly Notices of the Royal Astronomical Society, v. LXXIV, 9, pp. 699-728, (1916)

[53] Foster, J., Nightingale, J.D., A Short Course in General Relativity, 2nd Ed., Springer-Verlag, New York, (1995)

[54] Eddington, A.S., The mathematical theory of relativity, Cambridge University Press, Cambridge, (1963)

[55] Crothers, S.J., On Corda's 'Clarification' of Schwarzschild's Solution, http://vixra.org/abs/1602.0221

[56] Dictionary of Geophysics, Astrophysics, and Astronomy, Matzner, R. A., Ed., CRC Press LLC, Boca Raton, LA, (2001)

[57] Penrose, R., Gravitational Collapse: The role of General Relativity, *General Relativity and Gravitation*, Vol. 34, No. 7, July 2002

[58] Wald, R. M., General Relativity, The University of Chicago Press, (1984)

BIBLIOGRAPHY

[59] Dodson, C.T.J., Poston, T., Tensor Geometry - The Geometric Viewpoint and its Uses, 2nd Ed., Springer-Verlag, (1991)

[60] Carroll, B.W., Ostlie, D.A., An Introduction to Modern Astrophysics, Addison-Wesley Publishing Company Inc., (1996)

[61] Misner, C.W., Thorne, K.S., Wheeler, J.A., Gravitation, W.H. Freeman and Company, New York, (1973)

[62] Rees, M., Our Cosmic Habitat, Princeton University Press, (2001)

[63] 't Hooft, G., Introduction to The Theory of Black Holes, online lecture notes, 5 February 2009, http://www.phys.uu.nl/~thooft/

[64] Stern, D., NASA/JPL WISE Survey, www.nasa.gov/mission_pages/WISE/news/wise201220829.htm

[65] Abbott, B.P. et al., Tests of general relativity with GW150914, arXiv:1602.03841v1 [gr-qc] 11 Feb 2016

[66] Einstein, A., The Foundation of the General Theory of Relativity, *Annalen der Physik*, 49, (1916)

[67] Schwarzschild, K., On the Gravitational Field of a Point Mass According to Einstein's Theory, *Sitzungsber. Preuss. Akad. Wiss.*, Phys. Math. Kl: 189 (1916)

[68] Abrams, L.S., Black holes: the legacy of Hilbert's error, *Can. J. Phys.*, v. 67, 919, (1989)

[69] Antoci, S., David Hilbert and the origin of the "Schwarzschild solution", (2001), http://arxiv.org/abs/physics/0310104

[70] Loinger, A., On black holes and gravitational waves. La Goliardica Paves, Pavia, (2002)

[71] d'Inverno, R., Introducing Einstein's Relativity, Oxford University Press, (1992)

[72] Celloti, A., Miller, J.C., Sciama, D.W., Astrophysical evidence for the existence of black holes, (1999), http://arxiv.org/abs/astro-ph/9912186

[73] Crothers, S.J., The Schwarzschild solution and its implications for gravitational waves, for 'Mathematics, Physics and Philosophy in the Interpretations of Relativity Theory', Budapest, 4-6 September, 2009, www.sjcrothers.plasmaresources.com/Budapest09-b.pdf

[74] Crothers, S.J., The Schwarzschild solution and its implications for gravitational waves, for 'Conference of the German Physical Society', Munich, March 9-13, 2009, Verhandlungen der Deutsche Physikalische Gesellschaft Munich 2009: Fachverband Gravitation und Relativitötstheorie, http://vixra.org/pdf/1103.0051v1.pdf

BIBLIOGRAPHY

[75] Crothers, S.J., Flaws in Black Hole Theory and General Relativity, for the Proceedings of the XXIXth International Workshop on High Energy Physics, Protvino, Russia, 26-28 June 2013, http://www.worldscientific.com/worldscibooks/10.1142/9041, http://viXra.org/abs/1308.0073

[76] Crothers, S.J., On the General Solution to Einstein's Vacuum Field and Its Implications for Relativistic Degeneracy, *Progress in Physics*, v.1, pp. 68-73, 2005, http://vixra.org/abs/1012.0018

[77] Crothers, S.J., On the Ramifications of the Schwarzschild Space-Time Metric, *Progress in Physics*, v.1, pp.75-80, 2005, http://vixra.org/abs/1101.0002

[78] Crothers, S.J., Black Hole and big bang: A Simplified Refutation, 2013, http://viXra.org/abs/1306.0024

[79] Crothers, S.J., On the Geometry of the General Solution for the Vacuum Field of the Point-Mass, *Progress in Physics*, v.2, pp. 3-14, 2005, http://vixra.org/abs/1101.0003

[80] Crothers, S.J., A Short Discussion or Relativistic Geometry, *Bulletin of Pure and Applied Sciences*, Vol. 24E (No.2) 2005:P.267-273, http://vixra.org/abs/1208.0229

[81] Crothers, S.J., Gravitation on a Spherically Symmetric Metric Manifold, *Progress in Physics*, v.1, pp. 68-74, 2007, http://vixra.org/abs/1101.0005

[82] Crothers, S.J., On The 'Stupid' Paper by Fromholz, Poisson and Will, http://viXra.org/abs/1310.0202

[83] Crothers, S.J., On the Invalidity of the Hawking-Penrose Singularity 'Theorems' and Acceleration of the Universe with Negative Cosmological Constant, *Global Journal of Science Frontier Research Physics and Space Science*, Volume 13 Issue 4, Version 1.0, June 2013, http://viXra.org/abs/1304.0037

[84] Crothers, S.J., On the Generation of Equivalent 'Black Hole' Metrics: A Review, *American Journal of Space Science*, July 6, 2015, http://vixra.org/abs/1507.0098

[85] Einstein, A., The Meaning of Relativity, expanded Princeton Science Library Edition, (2005)

[86] Crothers, S.J., On Isotropic Coordinates and Einstein's Gravitational Field, *Progress in Physics*, v.3, pp.7-12, 2006, http://vixra.org/abs/1101.0008

[87] Collins Encyclopaedia of the Universe, Harper Collins Publishers, London, (2001)

BIBLIOGRAPHY

[88] Taylor, E.F., Wheeler, J.A., Exploring Black Holes - Introduction to General Relativity, Addison Wesley Longman, 2000 (in draft)

[89] O'Neill, B., Semi-Riemannian Geometry With Applications to Relativity, Academic Press, (1983)

[90] Dirac, P.A.M., General Theory of Relativit, Princeton Landmarks in Physics Series, Princeton University Press, Princeton, New Jersey, (1996)

[91] Hawking, S.W. and Ellis, G.F.R., The Large Scale Structure of Space-Time, Cambridge University Press, Cambridge, (1973)

[92] Bland-Hawthorn, J., ABC television interview with news reporter Jeremy Hernandez, 24 Sept. 2013, http://www.physics.usyd.edu.au/~jbh/share/Movies/Joss_ABC24_13.mp4

[93] Rennie, J., Why is a black hole black, http://physics.stackexchange.com/questions/28297/why-is-a-black-hole-black

[94] Rennie, J., Photons straight into black hole, http://physics.stackexchange.com/questions/101619/photons-straight-into-black-hole

[95] Vladmimirov, Yu., Mitskiévich, N., Horský, J., Space Time Gravitation, Mir Publishers, Moscow, (1984)

[96] Brillouin, M., The singular points of Einstein's Universe. *Journ Phys. Radium*, v. 23, 43, (1923)

[97] Hawking, S.W., A Brief History of Time from the Big Bang to Black Holes, Transworld Publishers Ltd., London, (1988)

[98] Einstein, A., Relativity: The Special and the General Theory, Methuen, London, (1954)

[99] Efimov, N.V., Higher Geometry, Mir Publishers, Moscow, (1980)

[100] Hagihara, Y., Jpn. J. Astron. Geophys. 8, 67, (1931)

[101] Doughty, N., Am. J. Phys., v.49, pp. 720, (1981)

[102] The Cambridge Illustrated History of Astronomy, Hoskin, M., Ed., Cambridge University Press, Cambridge, UK, (1997)

[103] Weyl, H., Space Time Matter, Dover Publications Inc., New York, (1952)

[104] Tolman, R.C., Relativity Thermodynamics and Cosmology, Dover Publications Inc., New York, (1987)

[105] Weinberg, S., Gravitation and Cosmology: Principles and Applications of the General theory of Relativity, John Wiley & Sons, Inc., (1972)

BIBLIOGRAPHY

[106] Pauli, W., The Theory of Relativity, Dover Publications, Inc., New York, (1981)

[107] Einstein, A., What is the Theory of Relativity, The London Times, November 28, 1919, reprinted in Ideas and Opinions, Crown Publishers, Inc., New York, (1985)

[108] 't Hooft, G., Strange Misconceptions of General Relativity, http://www.staff.science.uu.nl/~hooft101/gravitating_misconceptions.html

[109] Heaviside, O., letter to Professor Bjerknes, Monday 8th March 1920

[110] Ricci-Curbastro, G., Levi-Civita, T., Méthodes de calcul différentiel absolu ET leurs applications, *Matematische Annalen*, B. 54, p.162, (1900)

[111] Levi-Civita, T., Mechanics. - On the Analytical Expression that Must be Given to the Gravitational Tensor in Einstein's Theory, *Rendiconti della Reale* Accadmeia dei Lincei 26: 381, (1917)

[112] Einstein, A., 'Relativity and the problem of space', appendix 5 in the 15th edition of Relativity: The Special and the General Theory, Methuen, London, 1954, pp.135-157

[113] Einstein, A., Preface to the 15th edition of Relativity: The Special and the General Theory, Methuen, London, (1954)

[114] Einstein, A., Cosmological Considerations on the General Theory of Relativity, *Sitzungsberichte der Preussischen Akad. d. Wissenschaften*, (1917)

[115] Crothers, S.J., Gerardus 't Hooft, Nobel Laureate, On Black Hole Perturbations, http://vixra.org/pdf/1409.0141v2.pdf

[116] McVittie, G.C., Laplace's Alleged 'Black Hole', *Observatory* **98**: 272-274 (Dec 1978)

[117] Krauss, L., A Universe from Nothing: Why There is Something Rather than Nothing, Atria Books, 2012

[118] Pauling, L., General chemistry: an Introduction to Descriptive Chemistry and Modern Chemical Theory, 2nd Edition, W.H. Freeman and Company, San Francisco, (1953)

[119] Sears, F.W., An Introduction to Thermodynamics, The Kinetic Theory of Gases, and Statistical Mechanics, Addison-Wesley Publishing Company. Inc., Cambridge, Massachusetts, (1955)

[120] Beiser, A., Concepts of Modern Physics (International Edition), McGraw-Hill Book Company, (1987)

[121] Zemansky, M.W., Dittman, R.H., Heat and Thermodynamics (International Edition), McGraw-Hill Book Company, (1981)

BIBLIOGRAPHY

[122] Krauss, L., Q&A, television station ABC1, Australia, (Monday, 18 February, 2013), www.abc.net.au/tv/quanda/txt/s3687812.htm

[123] Crothers, S.J., A Nobel Laureate Talking Nonsense: Brian Schmidt, a Case Study, http://vixra.org/pdf/1507.0130v1.pdf

[124] Schmidt, B., Q&A, ABC television, 15 September 2014, (the section 'Expanding Universe'), www.abc.net.au/tv/qanda/txt/s4069393.htm

[125] Crothers, S.J., A Few Things You Need to Know to Tell if a Nobel Laureate is Talking Nonsense, 10 July 2015, http://vixra.org/pdf/1507.0067v2.pdf

[126] Krauss, L., Q&A, television station ABC1, Australia, (Monday, 27 May, 2013), www.abc.net.au/tv/qanda/txt/s3755423.htm

[127] Robitaille P.-M., Crothers S. J., "The Theory of Heat Radiation" Revisited: A Commentary on the Validity of Kirchhoff's Law of Thermal Emission and Max Planck's Claim of Universality, *Progress in Physics*, v.11, p.120-132, (2015), http://vixra.org/pdf/1502.0007v1.pdf

[128] Penzias, A.A. and Wilson R.W., A measurement of excess antenna temperature at 4080 Mc/s. Astrophys. J., 1965, v.1, 419-421.

[129] Robitaille, P.-M., WMAP: A Radiological Analysis, *Progress in Physics*, v.1, pp.3-18, (2007),
http://vixra.org/abs/1310.0121

[130] Robitaille, P.-M., COBE: A Radiological Analysis, *Progress in Physics*, v.4, pp.17-42, (2009),
http://vixra.org/abs/1310.0125

[131] Crothers, S.J., COBE and WMAP: Signal Analysis by Fact or Fiction?, *Electronics World*, March 2010, pp.26-31, http://vixra.org/pdf/1101.0009v1.pdf

[132] Crothers, S.J., The Temperature of the Universe before the Jury, *Principia Scientifica International*, http://www.principia-scientific.org/the-temperature-of-the-universe-before-the-jury.html

[133] Robitaille, P.-M., Kirchhoff's Law and Magnetic Resonance Imaging: Do Arbitrary Cavities Always Contain Black Radiation?, Abstract: D4.00002, *Bulletin of the American Physical Society*, 2016 Annual Spring Meeting of the APS Ohio-Region Section Friday-Saturday, April 8-9, 2016; Dayton, Ohio, http://meetings.aps.org/Meeting/OSS16/Session/D4

[134] Robitaille, P.-M., Water, Hydrogen Bonding, and the Microwave Background, *Progress in Physics*, Vol.2, April, 2009, http://vixra.org/abs/1310.0129

[135] Robitaille, P.-M., Forty Lines of Evidence for Condensed Matter – The Sun on Trial: Liquid Metallic Hydrogen as a Solar Building Block, *Progress in Physics*, v.4, pp.90-142, 2013, http://vixra.org/abs/1310.0110

BIBLIOGRAPHY

[136] Robitaille, P.-M., On the Corona & Chromosphere, https://www.youtube.com/watch?v=0Lg5eR7T61A&feature=share

[137] Crothers, S.J., General Relativity: A Case Study in Numerology, https://www.youtube.com/watch?v=QBorBKDnE3U

[138] Horgan, J., Is the Gravitational-Wave Claim True? And Was It Worth the Cost?, *Scientific American*, 12th February 2016, http://blogs.scientificamerican.com/cross-check/is-the-gravitational-wave-claim-true-and-was-it-worth-the-cost/

[139] Richards, V., Scientist warns the world to 'think twice before replying to alien signals from outer space', *Independent*, 23 July 2015, http://www.independent.co.uk/news/world/australasia/scientist-warns-world-to-think-twice-before-replying-to-alien-signals-from-outer-space-10408201.html

[140] Drake, N., Rogue Microwave Ovens Are the Culprits Behind Mysterious Radio Signals, *National Geographic*, 4th October 2015, http://phenomena.nationalgeographic.com/2015/04/10/rogue-microwave-ovens-are-the-culprits-behind-mysterious-radio-signals/

[141] The International Journal of Astrobiology, Cambridge University Press, http://journals.cambridge.org/action/displayJournal?jid=IJA

[142] Frazer, J.G., The Golden Bough, Vol.1, 2nd Ed., MacMillan & Co. Ltd., Norfolk, (1900)

CPSIA information can be obtained
at www.ICGtesting.com
Printed in the USA
BVHW071024180120
569853BV00003B/468